KU-281-924

Wildlife Population Growth Rates

Edited by
R. M. SIBLY
J. HONE
and
T. H. CLUTTON-BROCK

1642169

LIBRARY

ACC. No.	DEPT.
0113 0454	
CLASS No.	

UNIVERSITY COLLEGE CHESTER

THE ROYAL SOCIETY

In association with

CAMBRIDGE
UNIVERSITY PRESS

PUBLISHED BY THE PRESS SYNDICATE OF THE UNIVERSITY OF CAMBRIDGE
The Pitt Building, Trumpington Street, Cambridge, United Kingdom

CAMBRIDGE UNIVERSITY PRESS
The Edinburgh Building, Cambridge CB2 2RU, UK
40 West 20th Street, New York, NY 10011–4211, USA
477 Williamstown Road, Port Melbourne, VIC 3207, Australia
Ruiz de Alarcón 13, 28014 Madrid, Spain
Dock House, The Waterfront, Cape Town 8001, South Africa

http://www.cambridge.org

© The Royal Society 2003

This book is in copyright. Subject to statutory exception and to the provisions of relevant collective licensing agreements, no reproduction of any part may take place without the written permission of Cambridge University Press.

First published as *Population Growth Rate: Determining Factors and Role in Population Regulation*, Philosophical Transactions of The Royal Society, B, vol. 357, no. 1425, pp. 1147–1320.

This edition published 2003

Printed in the United Kingdom at the University Press, Cambridge

Typefaces Lexicon No.2 9/13 pt. and Lexicon No.1   *System* LATEX 2$_\varepsilon$ [TB]

A catalogue record for this book is available from the British Library

Library of Congress Cataloguing in Publication data
Wildlife population growth rates / edited by R. M. Sibly, J. Hone, and T. H. Clutton-Brock.
     p.   cm.
Includes bibliographical references (p.).
ISBN 0 521 82608 X – ISBN 0 521 53347 3 (pbk.)
1. Animal populations.   I. Sibly, R. M.   II. Hone, Jim.   III. Clutton-Brock, T. H.
QL752.W523   2003
591.7'88 – dc21   2003041953

ISBN 0 521 82608 x hardback
ISBN 0 521 53347 3 paperback

Accession no.
01130454

6
5.1.05

LIBRARY

Tel: 01244 375444   Ext: 3301

This book is to be returned on or before the
last date stamped below. Overdue charges
will be incurred by the late return of books.

UNIVERSITY COLLEGE
CHESTER

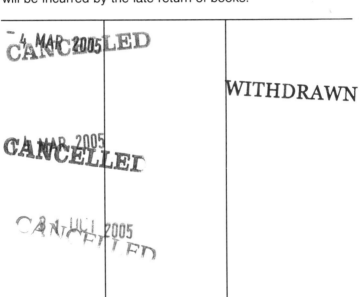

CANCELLED
4 MAR 2005

CANCELLED
MAR 2005

CANCELLED
2005

WITHDRAWN

Contents

Contributors

Peter Bayliss
Marsupial Cooperative Research Centre, PO Box 1927, Macquarie Centre NSW 2113, Australia.

Peter Calow
Department of Animal and Plant Sciences, Alfred Denny Building, Western Bank, Sheffield, S10 2TN, UK.

David Choquenot
Arthur Rylah Institute for Environmental Research, Natural Resources and Environment, PO Box 137 Heidelberg, Victoria 3084, Australia.

T. H. Clutton-Brock, FRS
Department of Zoology, Downing Street, Cambridge, CB2 3EJ, UK.

Stephen Davis
CSIRO Sustainable Ecosystems, GPO Box 284, Canberra, ACT 2601, Australia.

Valery E. Forbes
Department of Life Sciences and Chemistry, Roskilde University, Universitetsvej 1, 4000 Roskilde, Denmark.

H. C. J. Godfray, FRS
Centre for Population Biology, Department of Biology, Imperial College at Silwood Park, Ascot, Berks, SL5 7PY, UK.

Jim Hone
Applied Ecology Research Group, University of Canberra, Canberra, ACT 2601, Australia.

Peter Hudson
Institute of Biological Sciences, University of Stirling, Stirling, Scotland, FK9 4LA, UK.

Charles J. Krebs
Department of Zoology, University of British Columbia, 6270 University Blvd., Vancouver, B.C. V6T 1Z4, Canada.

Russell Lande
Division of Biology 0116, University of California, San Diego, 9500 Gilman Drive, La Jolla, CA 92093, USA.

Wolfgang Lutz
International Institute for Applied Systems Analysis (IIASA), A-2361 Laxenburg, Austria.

Ken Norris
School of Animal and Microbial Sciences, University of Reading, Reading, RG6 6AJ, UK.

Bernt-Erik Sæther
Department of Zoology, NTNU, Realfagsbygget, N-7491, Trondheim, Norway.

Richard Sibly
School of Animal and Microbial Sciences, University of Reading, Reading, RG6 6AJ, UK.

A. R. E. Sinclair
6270 University Boulevard, University of British Columbia, Vancouver, V6T 1Z4, Canada.

William J. Sutherland
School of Biological Sciences, University of East Anglia, Norwich, NR4 7TJ, UK.

1
———————

Introduction to wildlife population growth rates

This book takes a fresh approach to some of the classic questions in ecology. In particular, what determines where a species lives and what determines its abundance? Despite great progress in the twentieth century much more remains to be done before we can provide full answers to these questions so that reliable predictions can be made as to what will be found in unstudied areas or times. We believe that the methods described and deployed in this book point the way forward. The core message of the book is that key insights come from understanding what determines population growth rate (pgr). We believe that application of this approach will make ecology a more predictive science.

In this chapter we begin with an introduction to the major themes of the book, and then after briefly indicating how population growth rate is defined, we describe the contents of the book in more detail. Those who require more background material will find a brief guide to available texts at the end.

Major themes

In briefest outline, the approach taken in this book is as follows. Questions as to what determines where species live are questions about what values of environmental factors allow populations to persist. Populations persist where pgr ≥ 0, so the environmental conditions that produce pgr ≥ 0 are the conditions in which the species can live. This line of reasoning suggests that to understand what determines where species live, we should study the relationships between environmental factors and pgr. We elaborate on this in figure 1.1. Now consider questions about abundance. Introduced to a new area favourable to the species, a population

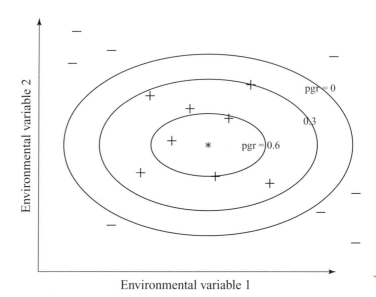

Figure 1.1. The fundamental ecological niche represented in a space whose axes are environmental variables. Laboratory experiments reveal the positions of contours (shown) on which population growth rate (pgr) at low density is constant. Occurrences in the field (+) are expected to lie within the pgr = r = o contour. Outside this contour populations go extinct and field occurrence is not expected (−). * represents the point of maximum pgr. Diagram after Maguire (1973) and Sibly & Hone (Chapter 2).

will initially grow (pgr > o), but eventually food or other resources become insufficient for the expanding population. Individual birth rates fall and death rates increase, and pgr, which is the integral of individual birth and death rates, declines. Eventually pgr becomes negative (pgr < o). The way in which pgr declines, as resources become limited, determines population abundance.

The central theme of the book is that the above relationships are best studied by identifying the determinants of pgr, and characterising their effects on pgr. The determinants of pgr can be thought of like factors in a regression analysis, in which some factors act on pgr in a positive manner, some in a negative manner, some in a linear manner and some in a nonlinear manner (figure 1.2). These mechanistic models give insight into the fundamental processes of population ecology and should not only improve the quality of the underlying science, but also increase the realism and accuracy of prediction in key applied areas. For conservation biologists, human demographers, wildlife managers and

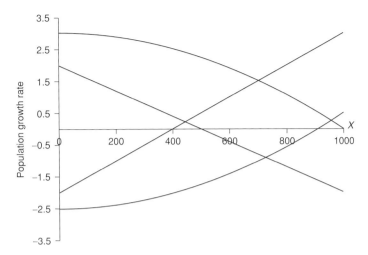

Figure 1.2. Examples of hypothetical linear, nonlinear, positive and negative relationships between demographic, mechanistic and density-dependent determinants (X) of population growth rate.

ecotoxicologists, it is of practical importance to know whether populations are increasing or declining, and to identify the factors that determine growth rates.

What is population growth rate (pgr)?

Population growth rate, pgr, measures the *per capita* rate of growth of a population. It tells us whether population size is increasing, stable or decreasing, and indicates how fast it is changing. The simplest definition is that pgr is the multiplication factor by which population size increases per year, conventionally given the symbol λ ($= N_{t+1}/N_t$, where N_t refers to the number of individuals in the population at time t). Alternatively, pgr can be defined as $r = \log_e \lambda$. Generally in this book pgr refers to pgr $= r$, although there are some exceptions. The definition and estimation of pgr are considered in more detail in Chapter 2.

It is important to note that measured values of pgr are specific to the environment in which they are measured. This may seem obvious but it is of great importance and constitutes one of the major themes running through this book. Change the environment and the value of pgr will change, and there may also be changes in the relationship between pgr and the factors that determine it. We shall return to this point.

Description of book contents

Let us now consider in more detail the general questions outlined above. What determines where a species lives? This can be considered together with the closely related questions: what limits a species' geographic distribution, and why is one species found in one type of habitat and others elsewhere? Classically, these questions have been addressed through the concept of the ecological niche, descriptions of which feature in every ecological textbook. Briefly, species differ in their tolerance of environmental variables such as pH, moisture and temperature regimes. Therefore, we can plot the regions in which each species can persist in a space with the axes being the environmental variables. These regions show the range of pH, moisture and temperature regimes that each species can tolerate.

In Chapter 2, Sibly & Hone argue that a more detailed picture of the ecological niche is obtained by defining it in terms of pgr, as shown in figure 1.1. Populations persist where pgr $\geq$ 0, but since species differ in their tolerance of environmental variables, there are differences among species in how environmental variables affect their population growth rates. This leads to different species occupying different regions of niche space. There is an attractive predictive aspect to this approach, because measurements of the fundamental niche made in the laboratory provide a basis for prediction of the habitats in which the organism will be found in the field (figure 1.1). Predictions of this kind are used more in botany than zoology, the reason being that plants are less mobile than animals, so that it is relatively straightforward to grow them in a range of environments and to evaluate their success in relation to environmental variables. From this a picture is built up of the range of tolerance of each species. On this basis field guides can give an indication of the range of pH, and of winter frost and summer drought that each species can survive.

The identification of factors that determine pgr, and the measurement of their effects, has generally been approached in three main ways, each incorporating observations, experiments and theoretical analyses. These are contrasted as alternative paradigms, with case studies, in the chapters by Sibly & Hone, and Hone & Sibly (Chapters 2 and 3). Those seeking a reductionist understanding of the effects of external factors on pgr are working within a *mechanistic paradigm*. The net effect of the external factors acting on individuals may however be a simple relationship between population density and pgr, and the study of such relationships constitutes a *density paradigm*. Others have concentrated on elucidation of the link between

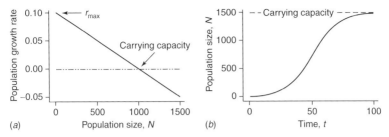

Figure 1.3. (*a*) Straight line relationship between *population size* and pgr giving rise to (*b*) logistic population growth. The intercepts of the relationship between *population size* and pgr have special importance. The intercept with the *x*-axis (when pgr = r = 0) gives carrying capacity and the intercept with the *y*-axis gives the maximum rate at which the population can grow, usually denoted r_{max}. These intercepts retain their meaning even if the relationship between *population size* and pgr is a curve, provided it is monotonic decreasing.

pgr and individual life histories, and this can be seen as a *demographic paradigm*.

The simplest line of attack on the determination of pgr is to look for and characterize the effects of population density (the 'density paradigm') in a constant environment, and the book begins with analyses of this type. Populations for which there is a negative relationship between density and pgr are regulated by negative feedback. The operation of this negative feedback mechanism will now be described. When population size is low there are plenty of resources for everyone, so birth rates are high and no-one dies of starvation. High birth rate and low death rate mean that pgr is high, because pgr is approximately equal to *birth rate* minus *death rate*. Since pgr is high, population size increases. As population size increases, more individuals are competing for the available resources so there is a reduction in the resources available per individual, and as a result birth rates decline and death rates increase. Thus, pgr declines as population size increases. The simplest relationship between population size and pgr is a linear relationship, and this gives the well known logistic population growth curve, according to which small populations grow exponentially initially, but thereafter more and more slowly, eventually reaching a plateau at carrying capacity (figure 1.3). Population growth stops completely at carrying capacity, so at carrying capacity pgr = 0.

If there is a negative relationship between density and pgr, then the population is controlled by negative feedback in the same way that in the mechanical world the motions of pendulums and springs are governed by gravity and elasticity. A profitable line of theoretical attack has exploited and developed this analogy (May *et al.* 1974). One of the salient results of these analyses is the insight that stability, over-compensation, under-compensation and so on all depend critically on the slope of pgr vs. $\log_e$ density, measured at carrying capacity. Mathematically, the simplest case is that generations are discrete and are censused once each generation, so that pgr can be measured as $\text{pgr} = r = \log_e (N_{t+1}/N_t)$, with time measured in generations. May *et al.* (1974) show that the stability properties of the system depend on the negative of the slope of pgr vs. $\log_e$ density, which they designate the 'characteristic return rate'. In our notation, equation 22 of May *et al.* (1974) becomes:

$$\text{characteristic return rate} \equiv -[d\text{pgr}/d\log_e N]_K \equiv -[1/N \, d\text{pgr}/dN]_K .$$

May *et al.* (1974) show that the properties of the regulated system, in the neighbourhood of carrying capacity, all depend on the value of the characteristic return rate. The equilibrium point – which is carrying capacity – is stable provided that the characteristic return rate is less than two but greater than zero. If it is less than two but greater than one then the system overshoots and oscillates before settling at carrying capacity. If the characteristic return rate is positive but less than one, there is no oscillation and the system approaches carrying capacity smoothly as in figure 1.3*b*. These properties depend on the discrete generation property, and if this does not hold, the formulae have to be modified appropriately. This approach is developed in the chapters by Sæther & Engen (Chapter 5) and Lande, Engen & Sæther (Chapter 4), the characteristic return rate being referred to as the 'strength of density-dependence'. Lande *et al.* show how the 'total' density dependence in a life history can be estimated from the sum of the autoregression coefficients in census data.

Sæther & Engen (Chapter 5) consider the important question of whether population dynamics vary between organisms with different types of life history. They show that much of the variation in ability of birds to grow from low density can be explained by position in the 'slow–fast continuum'. Populations of 'fast' species – those with large clutch sizes and mortality rates – grow faster at low density and are more vulnerable to environmental stochasticity effects than 'slow' species. In a related paper Sæther *et al.* (2002) consider the related question of whether the strength of density dependence varies between organisms with different types of life history. Fowler (1981) set the scene for this type of analysis by suggesting that organisms with fast life histories – his examples were insects and fish – would have relatively weak density dependence, whereas those with slow life histories – his example was mammals – would have relatively strong density dependence. Sæther *et al.* (2002) analyse the case of solitary birds to show that, as Fowler predicted, those with faster life histories have weaker density dependence. This is an important result, since it potentially gives the ability to predict volatility or stability of population dynamics from knowledge of life history.

To what extent might density dependence operate in man? This is clearly a question of importance to all, since the mechanisms by which it operates could determine the quality of life of future generations. Surprisingly, the question has not been considered in human demography in recent years, even though Pearl & Reed's (1920) important paper on density dependence of pgr used the population of the United States 1870–1920 as their worked example. In an intriguing analysis of pgr, birth rate and density, Lutz & Qiang (Chapter 6) show that density dependence does appear to operate in man. How it operates is, of course, an extremely interesting question. It may be, as Lutz & Qiang suggest, that there is an important psychological and/or perceptual component, so that with the availability of birth control people plan their family sizes and life styles with regard to local norms of behaviour.

At this point the book begins to widen the scope of the analyses to include mechanistic, reductionist approaches. In the simplest case, density achieves its effects simply by resource competition, generally competition for food. Methods of analysis of the resultant relationships were developed in particular by G. Caughley (e.g. Caughley & Krebs, 1983), and prior to and subsequent to his death in 1994 his work has much influenced the development of mammalian ecology, especially in Australia and New Zealand (see the chapters by Bayliss & Choquenot (Chapter 9), and Davis, Pech & Catchpole (Chapter 10)). Krebs here (Chapter 7) argues strongly in favour of such mechanistic reductionist analyses, backed with experiment wherever possible, and some of his points are also developed by Sutherland & Norris in Chapter 12. Krebs' chapter will make ecologists using the density paradigm think very carefully about the assumptions underlying their analyses. Since measured values of pgr are specific to the environment in which they are measured, it follows that analyses of long-term data sets are of little value unless the environment has been constant or nearly so. He provides an example in which the relationship between pgr and population density appears to have changed with time, over decades, and other examples in which density-dependent relationships vary from one place to another.

Starting with the premise that food supply is the primary factor determining population growth rate in animal populations, Sinclair & Krebs (Chapter 8) consider the complications introduced by varying predation, social interactions and stochastic disturbances. In a series of diagrams that will fascinate many ecologists they compare density-dependent relationships in contrasted environments. Typically, the contrast is in the presence or absence of predators. In this way we begin to get a feel for how pgr depends jointly on the densities of the study animals and their predators.

Bayliss & Choquenot look in depth at how two important herbivorous marsupial populations are regulated: kangaroos and possums. These species are high profile and of conservation and commercial importance, but despite intensive work for many years, scientific knowledge has made only a limited contribution to their management until now. A key feature of the studies described is that food availability is explicitly estimated for wildlife populations. Therefore, the mechanistic determinants of population growth rate are estimated directly thus avoiding the need to use surrogates of resource availability, such as density or climate. The authors suggest that future progress will be obtained by a marriage between mechanistic consumer–resource and density-dependent models. The combined effects of food and density on pgr are also analysed in the case studies of *Tisbe* by Sibly & Hone (Chapter 2), and barn owls by Hone & Sibly (Chapter 3). Davis, Pech & Catchpole (Chapter 10) extend the argument by considering the effects of temporal variability in the food supply on herbivores' distribution and abundance.

Hudson *et al.* (Chapter 11) provide a comprehensive account of one of the great studies of cycling populations, the red grouse in England and Scotland. Major themes are why the populations cycle, why the cycle lengths vary between populations, and why some populations are more stable than others. A feature of this work has been the inclusion of experiments that identify the mechanisms that affect productivity and pgr. Food quality appears to account for variations in the maximum pgr between areas. Year to year variations within areas are generated by the effects of parasites on the fecundity of the grouse, though interactions with other natural enemies and spacing behaviour are also important. Man, who maintains the population as a game bird, has a more limited effect.

Sutherland & Norris (Chapter 12) propose the application of a new principle to mechanistic analyses, obtained by adding relevant features of an animal's behavioural ecology. Behavioural ecology takes into account that evolution is an optimising process, selecting behavioural features adaptive to the environments in which organisms find themselves (Krebs & Davies 1993). Special attention is therefore paid to models of optimal behaviour, and models of this kind have been particularly successful in predicting the ways that animals, especially birds, distribute themselves in patchy environments. The approach appears to solve one of the core problems, understanding the dependence of pgr on the environment in which the population lives. This might in turn open up the chance to make more secure generalizations between environments and, perhaps, between species. In this approach the relationship between population growth rate and density can be extracted from the analysis, but it does not have to be – the relationship is an epiphenomenon of the mechanistic understanding.

Clutton-Brock & Coulson (Chapter 13) also obtain a fairly complete reductionistic understanding of the factors governing population change without invoking pgr as an intervening variable. Studying sheep and deer on two small islands from which and to which there is no migration, they have tracked numbers in each age class, and shown how the vital rates of each age class are determined, by extrinsic factors such as weather, parasites and so on. Knowing population structure and the state of the extrinsic factors, successful prediction of the next year's population structure is possible. This is achieved directly using a population matrix approach, without explicit calculation of pgr.

Many of the contributors have interests in conservation, and Forbes & Calow (Chapter 14) are interested in the ways in which we protect the environment against potentially harmful chemicals. In practice, the ecological risks of toxic chemicals are most often assessed on the basis of their effects on survival, reproduction and somatic growth, but which vital rate is the most sensitive varies with species and chemical, and pgr analysis offers a more robust approach. They conclude that, given sufficient time and resources, pgr analysis should form the basis of ecological risk assessment.

Godfray & Rees (Chapter 15) remind us that there is an important analogy between pgr and Darwinian fitness, provide perspective on the whole subject area, and judiciously consider the ways ahead.

Conclusion

Finally, we would like to emphasize some points made by many contributors. Although pgr is an important summary parameter, it is specific to the time and place in which it is measured, and the same will be true of knowledge of the determinants of pgr. To generalize we have to have further mechanistic understanding, and this can only come through knowledge of the factors that affect the births and deaths of individuals. When environmental effects are understood at this level, it should become possible to use demography to build a reliable overall picture of how extrinsic and intrinsic factors contribute to pgr. As in all areas of science, model-based understanding needs to be checked by experimentation, even though this is peculiarly difficult in population ecology because of the scale, time and effort involved.

This book focusses on population growth rate as a key variable in ecology. The parameter pgr links together seemingly disparate areas of ecology – habitat ecology, population ecology, conservation ecology, ecotoxicology, and this allows us to see ecology as a unified whole. We hope our book will inspire field ecologists to collect and analyse more data using the approaches presented here.

Further reading

For those who require more background material we now provide a brief guide to some of the available texts. Campbell & Reece (2002) provide an excellent one chapter introduction at the first-year undergraduate level in a general biology text. At the advanced undergraduate or postgraduate level we particularly like the treatment by Ricklefs & Miller (1999). Caughley *et al.* (1987) describe the results of a detailed study of herbivore dynamics, exemplifying the mechanistic paradigm, and the combined use of observational, experimental and modelling approaches. Berryman (1999) provides a short introduction to population dynamics in a book rich in data examples and exercises, and takes a similar approach to that adopted here, as does the paper by Turchin (1999). Caswell (2001) has written a comprehensive account of the population matrix approach to demography. For those with an interest in the historical background we recommend Cole (1958) and Hutchinson (1978). Finally, many of the issues discussed here are prefigured to a greater or lesser degree in Andrewartha & Birch's (1954) classic text with the resonant title *The Distribution and Abundance of Animals*. Building on the work of earlier ecologists, maybe the subtitle of the present book should be 'the distribution, abundance and population growth rate of animals'.

2
_____

Population growth rate and its determinants: an overview

2.1. Introduction

With the persistent increase of the human population – now exceeding six billion – all species face increased pressure on resources. Understanding the factors responsible for limiting populations or causing species' extinctions therefore has increased urgency. Recent developments in population analysis, described below, have refined our understanding of the determinants of population growth rate and linked the theory to field data, and there is increasing interest in applying methods of this kind in conservation, wildlife management and ecotoxicology. This paper emphasizes the central role of population growth rate and reviews the use of data to test relevant theory and models primarily for wildlife populations. In this section we consider the definition and importance of population growth rate and briefly examine its historical background.

(a) Definitions and estimation of population growth rate

Population growth rate is the summary parameter of trends in population density or abundance. It tells us whether density and abundance are increasing, stable or decreasing, and how fast they are changing. Population growth rate describes the per capita rate of growth of a population, either as the factor by which population size increases per year, conventionally given the symbol λ ($= N_{t+1}/N_t$), or as $r = \log_e \lambda$. Generally here, population growth rate will refer to r. λ is referred to variously as 'finite growth rate', 'finite rate of increase', 'net reproductive rate' or 'population multiplication rate'. r is known as 'rate of natural increase', 'instantaneous growth rate', 'exponential rate of increase' or 'fitness'. In the simplest population model all individuals in the population are assumed equivalent, with the same death rates and birth rates, and there is no migration in or out of the population, so exponential growth occurs; in this model, population growth rate $= r =$ instantaneous birth rate $-$ instantaneous death rate.

Population growth rate is typically estimated using either census data over time or from demographic (fecundity and survival) data. Census data are analysed by the linear regression of the natural logarithms of abundance over time, and demographic data using the Euler–Lotka equation (Caughley 1977) and population projection matrices (Caswell 2001). The two methods of estimation can give similar values, as shown in studies of the northern spotted owl, *Strix occidentalis caurina* (Lande 1988; Burnham *et al.* 1996), greater flamingo, *Phoenicopterus ruber roseus* (Johnson *et al.* 1991), and pea aphids, *Acyrthosiphon pisum* (Walthall & Stark 1997). The census method has greater statistical power to detect a population decline than the demographic method when applied to high-density populations and the reverse occurs with low-density populations (Taylor & Gerrodette 1993). The estimation of the finite population growth rate (λ) may be biased by several factors: the manner of analysing census data with zeros, by the existence of spatial variance and spatial–temporal covariance, and by the scale of the area studied (Steen & Haydon 2000).

r varies from a minimum value of $-\infty$ to a value of 0.0 for a stable population, to a maximum value (r_{max}) when the population increases at the maximum possible rate, when food is abundant and there are no predators, pathogens and competitors. The corresponding values of λ are 0.0, 1.0 and λ_{max}. The frequency distribution of λ of a population persisting through time and space is positively skewed, with the maximum value further to the right of the mode than the minimum value. The frequency distribution of r is generally closer to a normal distribution (Hone 1999).

(b) Importance in projection of future population sizes

The importance of population growth rate is that it allows qualified projection of future population sizes. If there were no density dependence then population growth would be exponential, at a rate calculable from the Euler–Lotka equation using demographic data from the existing population. However, in a finite world the resources needed to support exponential growth must eventually become inadequate, and population growth rate then declines, giving density dependence of population growth rate as discussed in §2.3. If the form of density dependence were constant and known, then the future population dynamics could to some degree be predicted. To what extent this can be realized in practice is discussed below; here we wish only to emphasize the central role of population growth rate in the projection of future population sizes.

Negative feedback between population growth rate and population density is a necessary condition for population regulation (Turchin 1999), and Sinclair (1996) has suggested that we recognize population regulation as occurring when population growth rate is negatively density dependent. According to this view, density dependence is seen as providing an explicit negative feedback mechanism which regulates the population. When density is below carrying capacity, population growth rate is positive and the population increases towards carrying capacity; conversely when density is above carrying capacity, population growth rate is negative and the population declines. In this way population density is controlled by the mechanism of the negative density dependence of population growth rate, and in the absence of environmental changes or time-lags, population density remains in the vicinity of carrying capacity.

(c)   Historical background

The pivotal importance of population growth rate has been recognized for a long time. The historical background has been described by Cole (1958) and Hutchinson (1978), on which some of the following outline is based. The idea of geometric population growth restrained at higher densities by the carrying capacity of the environment was put forward in a book by Botero (1588), and was famously elaborated and brought to general attention by Malthus (1798). Detailed interest in tables of mortality began in the late seventeenth century, with mathematical analyses by Huygens and later Buffon among others. Interestingly, Cole suggests that Newton notably 'failed to grasp the basic concept that life expectancy is a function of age'. Euler (1760), in deriving the equation that bears his name, established the mathematical dependence of population growth rate on age-specific birth rates and death rates, and commented that 'it always comes back to these two principles, that of mortality and that of fertility, which, once they have been established for a certain place, make it easy to resolve all the questions which one could propose . . .'. The proposal that population growth rate declines linearly with population density, known as the logistic equation, was put forward by Verhulst (1838). Population growth rate is central to the work of the modern founding fathers of ecology (Lotka 1925; Fisher 1930; Nicholson 1933; Andrewartha & Birch 1954), but the complexities of its dependence on age-specific birth and death rates have until recently prevented thorough examination of the role of population growth rate. With the advent of modern computing and the development of matrix methods for the analysis of life tables, the importance of population growth rate in the study of population ecology is becoming more widely appreciated (Caswell 2001).

(d)   Scope and layout of paper

In attempting to establish the central role of population growth rate in population ecology, we first consider how it should be used in the definition of basic concepts such as environmental stress and ecological niche, and then consider its role in population dynamics, examining in particular the form of the relationship between population growth rate and population density. Using these as foundations we then attempt to characterize recent studies in terms of three approaches to the identification of the determinants of population growth rate.

In practice, this means studying the causes of variation in population growth rate, attributing observed variation to known sources of variation, while bearing in mind that the specific sources of variation may change over time and in space. We make no attempt to review the extensive related literature on theoretical population dynamics and food webs; for an introduction to these see Berryman *et al.* (1995) and Rees *et al.* (1996). We conclude with an example illustrating our ideas in practice, and some consideration of applications to conservation biology and other fields of wildlife management, and ecotoxicology.

2.2.   Environmental stress and the ecological niche

Environmental stressors can be defined as factors that, when first applied to a population, reduce population growth rate (Sibly & Calow 1989; Hoffmann & Parsons 1991). Examples of stressors may be climatic conditions, toxicants, food quality, pathogens and so on. The advantages of explicit operational definitions of the terms 'stress' and 'stressor' are obvious, and it is encouraging that identical definitions are now used from the population level down to that of molecular responses (e.g. animal welfare (Broom & Johnson 1993) and stress proteins (Hengge-Aronis 1999)). Some examples of the effects of stressors on population growth rate are shown in figure 2.1.

At the individual and population level the effects of environmental stressors are commonly measured using life table response experiments (Levin *et al.* 1987; Caswell 1989, 2001). The first true example of this

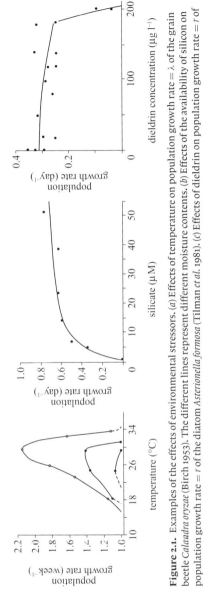

Figure 2.1. Examples of the effects of environmental stressors. (*a*) Effects of temperature on population growth rate = λ of the grain beetle *Calandra oryzae* (Birch 1953). The different lines represent different moisture contents. (*b*) Effects of the availability of silicon on population growth rate = *r* of the diatom *Asterionella formosa* (Tilman *et al.* 1981). (*c*) Effects of dieldrin on population growth rate = *r* of the cladoceran *Daphnia pulex*. (After Daniels & Allan (1981).)

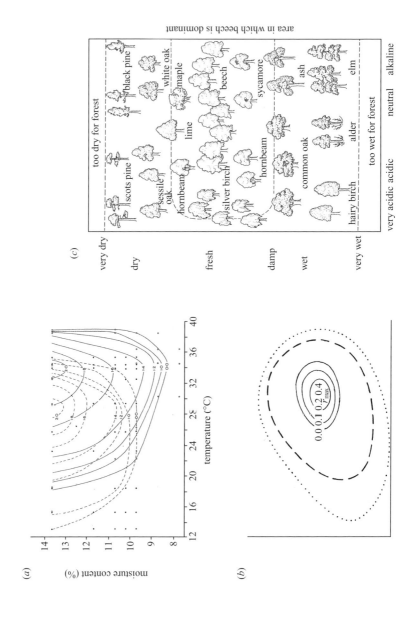

(a)

moisture content (%)

temperature (°C)

(b)

F_{max}
0.0 0.1 0.2 0.4

(c)

very dry

dry

fresh

damp

wet

very wet

too dry for forest

black pine

white oak
maple

beech

sycamore

ash

elm

scots pine

sessile
oak

lime

hornbeam

silver birch

hornbeam

common oak

alder

hairy birch

too wet for forest

very acidic   acidic   neutral   alkaline

area in which beech is dominant

approach was Birch's (1953) classic study of the effects of temperature and wheat moisture content on two species of grain beetles (figures 2.1*a* and 2.2*a*). The approach has been much developed by the elaboration of matrix population models (Caswell 2001), and in recent years, increasing concern about the environment has led to numerous studies of the population growth rate effects of pollutants (Forbes & Calow 1999).

The combined action of environmental stressors can be thought of as defining an organism's ecological niche. Although textbooks often treat the concept of an organism's ecological niche historically, in terms of Hutchinson's (1957) rectilinear definition of the niche, a more straightforward approach is to define an organism's niche as the set of points in 'niche space' where the population growth rate is greater than zero (Maguire 1973; Hutchinson 1978; figure 2.2*b*). The axes of niche space are physical or chemical variables such as temperature, food size or pH. When the niche is characterized at low population density and in the absence of predators, parasites and interspecific competitors, it is referred to as the 'fundamental niche'. In the presence of predators, parasites and interspecific competitors the set of points for which the population growth rate is greater than zero is reduced, and this set of points defines the 'realized niche' (Maguire 1973).

Birch's (1953) summary of his experimental results depicting the effects of environmental stressors on population growth rate (figure 2.2*a*) looks very similar to figure 2.2*b*, and it is curious that while Birch's work is generally referred to in discussions of the ecological niche, the explicit link is rarely made. Birch (1953) concluded: 'the significance of this information [i.e. figure 2.2*a*] is that it provides background experimental data for studies on distribution and abundance. The limits of distribution of the beetles are determined, so far as temperature and moisture are concerned, by the combination of temperature and moisture beyond which the finite rate of increase (λ) is less than 1'. He also showed that the known geographical distribution of the two beetle species conformed to the predictions of the

Figure 2.2. The ecological niche defined by population growth rate contours. (*a*) The study by Birch (1953) of the effects of temperature and wheat moisture content on population growth rate = λ of two species of grain beetle: dotted lines, *Calandra oryzae*; solid lines, *Rhizopertha dominica*. The points on the graph show conditions at which measurements were made. (*b*) Maguire's (1973) definition of the ecological niche in terms of population growth rate = r contours (after Hutchinson 1978). Axes are environmental variables as described in the text. (*c*) Niches of the major northern European trees. (From Polunin & Walters (1985), copyright owned by Ellenberg (1988).)

laboratory work. A very similar interpretation of species distribution in terms of population growth rate is described by Caughley *et al.* (1988), in which the edge of distribution is identified as occurring where the maximum value of population growth rate is zero. Populations can however persist with a population growth rate greater than zero if they are maintained by immigration ('sink' populations; Pulliam 1988).

An example of how the niche concept can be used in practice is given in figure 2.2c. The fundamental niches of the trees shown overlap in the central region of figure 2.2c (Ellenberg 1988) but in nature, competition between tree species restricts the occurrence of individual species as illustrated.

2.3. The form of the relationship between population growth rate and population density

As we have seen, environmental stressors have negative effects on population growth rate, and the same is true of population density. Indeed the nature of the negative relationship between population density and population growth rate is at the heart of population ecology. If the relationship is linear, the describing equation is the logistic, and many theoretical analyses are built on this assumption. Fowler (1981, 1987) has argued on theoretical and some empirical grounds that relationships between population growth rate and density are expected to be concave, viewed from above, for species with life histories like those of insects, convex for large mammals and similar species, and linear for intermediate species (see figure 2.3 for examples of concave and convex). Concave and convex correspond to values less or greater than one respectively for the shape parameter θ in the generalized logistic equation (Eberhardt 1987).

Given its fundamental importance in population ecology, it is surprising that data on the form of the dependence of population growth rate on density have only rarely been plotted out and analysed, though Berryman (1999) and Turchin (1999) give examples. No doubt this is partly because the relationship cannot be seen without adequate variation in density in a fairly constant environment; thus the best evidence may come from population recoveries after experimental perturbations (Sinclair 1996). Of the 25 cases we found (examples in figure 2.3), 12 relationships are linear, and in all but two of the remainder the relationship was concave viewed from above (table 2.1). Note, however, that there is considerable scatter about each relationship (e.g. figure 2.3), indicating that other factors besides

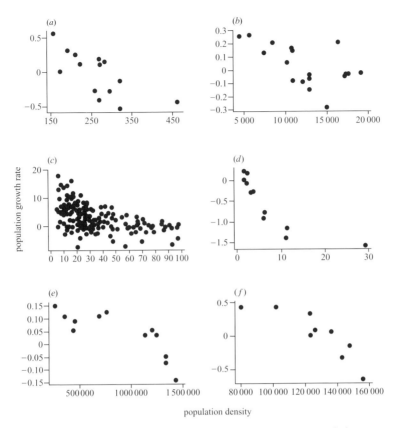

Figure 2.3. Examples of the form of the relationship between population growth rate (*r*) and population density. Linear relationships in (*a*) magpie goose and (*b*) elk; concave viewed from above in (*c*) meadow vole and (*d*) arctic ground squirrel; convex viewed from above in (*e*) wildebeest and (*f*) sandhill crane. Sources in table 2.1.

density affect population growth rate. Interestingly, there does seem to be some support for Fowler's hypothesis, but the data are not conclusive, and there are counter-examples (e.g. elk and sparrowhawk are linear not convex). Further support has recently been provided by studies of birds (Sæther *et al.* 2002). There are also studies of the relationship between λ and density, in for example the dipper *Cinclus cinclus* (Sæther *et al.* 2000b), but to relate these to the logistic equation, λ has to be $\log_e$ transformed.

Although a priori density dependence can be expected in real populations that persist long term (Sinclair 1996), recognizing the operation

Table 2.1. *Examples of linear and nonlinear density dependence in the relationship between population growth rate and population density. (Our classification is in some cases provisional and original references should be consulted. Some more complicated cases involving time-lags can be found in Berryman (1999) and Turchin (1999).)*

Species	Reference
Linear	
cladoceran *Daphnia pulex*	Frank *et al.* (1957)
treehole mosquito *Aedes triseriatus*	Livdahl (1982) and Edgerley & Livdahl (1992)
annual plant *Salicornia brachystachys*	Crawley (1990)
guppy *Poecilia reticulata*	Barlow (1992)
magpie goose *Anseranas semipalmata*	Bayliss (1989)
Yellowstone elk *Cervus elaphus*	Coughenour & Singer (1996)
European rabbit *Oryctolagus cuniculus*	Barlow & Kean (1998)
lake whitefish *Coregonus clupeaformis*	Berryman (1999)
red pine cone beetle *Conophthorus resinosae*	Berryman (1999)
mountain pine beetle *Dendroctonus ponderosae*	Berryman (1999)
sparrowhawk *Accipiter nisus*	Sibly *et al.* (2000c)
field vole *Microtus agrestis*	Klemola *et al.* (2002)
Concave viewed from above	
cladoceran *Daphnia magna*	Smith (1963)
wood mouse *Apodemus sylvaticus*	Montgomery (1989)
fruitfly *Drosophila melanogaster*	Turchin (1991)
winter moth *Operophtera brumata*	Roland (1994)
grey-sided vole *Chlethrionomys rufocanus*	Saitoh *et al.* (1997)
meadow vole *Microtus pennsylvanicus*	Turchin & Ostfeld (1997)
leaf-eared mouse *Phyllotis darwini*	Lima & Jaksic (1999)
cowpea weevil *Callosobruchus maculatus*	Berryman (1999)
marine copepod *Tisbe battagliai*	Sibly *et al.* (2000b); cf. figure 2.7a
brushtail possum *Trichosurus vulpecula*	Efford (2000)
Arctic ground squirrel *Spermophilus parryii plesius*	Karels & Boonstra (2000)
Convex viewed from above	
wildebeest *Connochaetes taurinus*	Sinclair (1996)
sandhill crane *Grus canadensis*	Berryman (1999)

of density dependence in real datasets has been the subject of much controversy. A number of meta-analyses have been performed applying these methods to data from different taxa, including birds and insects (Hassell *et al.* 1989; Greenwood & Baillie 1991; Woiwod & Hanski 1992). The largest such study is that of Woiwod & Hanski (1992), which analysed 5715 time-series of 447 species of moths and aphids in the UK, and found good agreement between three analytical methods which differ in their assumptions about the form of density dependence, namely those of

Bulmer (1975), Pollard *et al.* (1987) and simple regression of $\log_e(N_{t+1}/N_t)$ against N_t.

The general conclusion of these meta-analyses is that the existence of density dependence can be established from census data provided sufficient data are available. The objective in analysing density dependence should therefore be the discovery of the form of density dependence, and the effect, if any, of time-lags. The existence of time-lags can be investigated by adding terms such as N_{t-1} into the regression of $\log_e(N_{t+1}/N_t)$ against N_t (Turchin 1990; Woiwod & Hanski 1992; Berryman 1999; Erb *et al.* 2001).

Despite the successes of the meta-analyses, worries remain about the effects of measurement errors and of fluctuations in environmental variables. None of the existing methods allows for the effects of measurement errors, which may exaggerate the effects of density dependence, and could produce the illusion of density dependence in invariant populations (Woiwod & Hanski 1992). Some authors recommend Bulmer's test but all methods have strengths and weaknesses (Lebreton & Clobert 1991; Fox & Ridsdill-Smith 1996). An additional serious concern is that fluctuations in environmental variables may obscure density effects and make it difficult to locate the positions of equilibrium densities ('density vagueness'; Strong 1986; Murdoch 1994; Chapter 7).

2.4. Contrasting approaches to identifying the determinants of population growth rate

Because so many factors affect population growth rate, it is never going to be easy to separate their effects when they operate simultaneously. Ecologists have used a variety of methods to study the combined and relative effects on population growth rate of determining factors. Here, we compare and contrast the three main approaches that have been used to identify the determinants of population growth rate in wildlife populations with overlapping generations and with, usually, no explicit spatial elements. For convenience we label these three approaches the 'density paradigm', the 'demographic paradigm' and the 'mechanistic paradigm'. These are best understood as different simplifications of a rather complicated reality. Intuitively it is clear that population growth rate is fully determined by the complete record of age-specific birth and death rates, and mathematically the link is made by the Euler–Lotka equation. Exploration of the link between population growth rate and the age-specific life table gives the demographic paradigm. Age-specific birth and death rates depend causally on such factors as food supply per individual, parasite burdens, predation, environmental stressors and interference competition, and some of these depend directly or indirectly on population density. Looking directly at the link between these causal factors and population growth rate gives the mechanistic paradigm, and focusing on the link between population growth rate and population density gives the density paradigm; these two paradigms were identified and contrasted by Krebs (1995). Any or all of these approaches can be undertaken using observational, experimental and modelling studies. In this section we first consider use of these approaches on their own, and then how they have been used in combination.

The density paradigm, discussed in §2.3 and illustrated in figure 2.3, aims primarily to describe density effects on population growth rate, and is used to make predictions taking account of population density, where detailed mechanistic understanding may be unnecessary or impossible because of lack of appropriate data. It assumes, however, constancy of important features of the environment such as population food supply; where such assumptions are untenable, as occurs in most populations of wild animals, the mechanistic approach is preferable (see §2.4b).

(a)   The demographic paradigm

The demographic paradigm focuses on the relationship between population growth rate and age-specific fecundity and survival. Population growth rate is increased by an increase in fecundity or survival, or by breeding earlier. Detailed examination of the link between population growth rate and demographic parameters is often made using population projection matrices as described by Caswell (2001), although other methods are also available employing the classical Euler–Lotka equation (e.g. Lande 1988; Burnham et al. 1996; Calow et al. 1997). Because stable age structure is necessary to the derivation of this equation it is often supposed that population growth rate cannot be estimated without stable age structure. However, Sibly & Smith (1998) have argued that even when the age structure is not stable and the various age classes are growing at different rates, population growth rate defined as the solution of the Euler–Lotka equation still provides a useful index of population growth, because it represents an appropriate weighted average of the growth rates of the different age classes.

Many examples of work within the demographic paradigm are described by Caswell (2001), and this approach has also been important in studies of the northern spotted owl (*Strix occidentalis caurina*), in which λ is related to estimates of fecundity and survival obtained in mark–recapture studies (Lande 1988; Lahaye *et al.* 1994; Burnham *et al.* 1996; Franklin *et al.* 2000; Seamans *et al.* 2001).

The relative effects on population growth rate (λ) of proportional changes in fecundity and survival rates have been examined in elasticity analysis, which was reviewed in a series of recent studies (e.g. Caswell 2000; De Kroon *et al.* 2000; Easterling *et al.* 2000; Grant & Benton 2000; Heppell *et al.* 2000; Sæther & Bakke 2000; Van Tienderen 2000; Wisdom *et al.* 2000). These studies compared the contributions of fecundity and survival to population growth in species with differing life histories; in short-lived species fecundity can make a greater proportional contribution than survival, and the reverse in longer-lived species. Elasticity analysis is used alongside observation of actual levels of variation in demographic parameters (Gaillard *et al.* 1998) to determine which demographic factors have most effect on population growth rate. In a similar vein Sibly & Smith (1998) have argued that traditional 'key factor' analysis should be redesigned to identify the life-history trait whose variation has most effect on population growth rate.

Spatial elements have also been incorporated into efforts to explain variation in population growth rate. The contributions to population growth rate of *in situ* recruitment, survival and immigration can be estimated from mark–recapture studies, as described for meadow voles (*Microtus pennsylvanicus*) by Nichols *et al.* (2000). Populations for which emigration exceeds immigration are referred to as source populations; where the reverse obtains, we have sinks (Pulliam 1988). Thomas & Kunin (1999) have demonstrated a graphical method of representing the concepts.

(b)   The mechanistic paradigm

The mechanistic paradigm focuses on the relationship between population growth rate and variables such as climate, food availability, predator abundance, pathogens and parasites, and competitors ('extrinsic factors'). Turchin (1999) provides an excellent discussion of what we can treat as variables: ideally one might keep track of the fates of all relevant individuals, but apart from the impossibility of this, there is often more heuristic value in using summary variables such as 'numbers in each age class' and so on. Historically, the mechanistic paradigm was followed in early models of the effects of competition by Lotka and Volterra in the 1920s, of predation (Lotka 1925; Volterra 1926), and of pathogens (Kermack & McKendrick 1927). Monod (1950) considered the relationship between

Table 2.2. *Examples of wildlife population dynamics, analysed to study determinants of population growth rate, using one or more of the density, demographic and mechanistic analyses, and their combinations. (The yes/no classification refers to whether an analysis was or was not used.)*

Demographic analysis	Mechanistic analysis	Density analysis	
		No	Yes
no	no	exponential; large mammals (Eberhardt 1987)	logistic; magpie goose (Bayliss 1989), sparrowhawk (Sibly et al. 2000c)
no	yes	numerical response; red kangaroo (Bayliss 1987), barn owl (Taylor 1994)	ratio; wolf (Eberhardt & Peterson 1999), modified numerical response; house mouse (Pech et al. 1999)
yes	no	modified Lotka; northern spotted owl (Lande 1988)	modified logistic; white-tailed deer (Hobbs et al. 2000)
yes	yes	modified Lotka; red fox (Pech et al. 1997)	modified logistic; kit fox (White & Garrott 1999)

population growth rate and resource availability in populations of bacteria, and in an influential experiment Tilman et al. (1981) related population growth rate of the diatom *Asterionella formosa* to the availability of silicon, which diatoms need to secrete their silicate 'shells' (figure 2.1b). Later, in a theoretical synthesis, Tilman (1982) considered more generally the relationship between population growth rate and the availability of resources. These ideas are very similar to the formulations of ecological niche theory and stressors discussed earlier.

In the simplest case, the effect of population density on population growth rate is replaced, in the mechanistic paradigm, by resource availability. A collation of the results of many studies of mammals (table 2.2; Sinclair 1996) showed that food is very often a cause of density dependence. In these cases density may have no effects other than those on food availability. We shall refer to the relationship between population growth rate and food availability as the 'numerical response' (table 2.2), although it is important to note that not all authors use the term this way, as there has been an evolution in its meaning since Solomon (1949) first used it to describe the increase in numbers of animals as their resources increased. Application to the relationship between population growth rate and food availability came later, via an intermediate step of May (1974, p. 83) who used the term to describe the effect of food availability on fecundity, and Caughley (1976, p. 207) who described it as the effect of food availability on the rate of amelioration of population decline. Later, the term was commonly used to describe the relationship between population growth rate and food availability (May 1981a). The divergence is well illustrated by Caughley & Sinclair (1994): in ch. 10, p. 172 they use the Solomon definition and in ch. 6, p. 75 they use the population growth rate definition. The relationship between these concepts is considered by Sinclair & Krebs (Chapter 8). Methods of analysis of the numerical response were developed in particular by Caughley (e.g. Caughley 1976, 1987b; Caughley & Lawton 1981; Caughley & Krebs 1983), and prior to and subsequent to his death in 1994 his work has much influenced the development of mammalian ecology, especially in Australia and New Zealand.

Examples of the numerical response, plotting population growth rate against food availability, are shown in figure 2.4. Note that as food availability increases, population growth rate generally increases to a maximum. Availability is availability to individuals, and is not always best measured by the population food supply (Abrams & Ginzburg 2000). If the population is food limited, for instance, it may be more appropriate to divide the population food supply by population density (see §2.4d and §2.5). In addition to the examples in figure 2.4, positive generally non-linear relationships between population growth rate and food availability have been reported for European rabbit (*Oryctolagus cuniculus*; Pech & Hood 1998), red fox (*Vulpes vulpes*; Pech & Hood 1998) and house mouse (*Mus domesticus*; Pech et al. 1999). A positive linear relationship between the annual percentage population growth of a barn owl (*Tyto alba*) population and abundance of voles (*Microtus agrestis*), their main food, was reported by Taylor (1994) and this is converted in figure 2.4e to a relationship between population growth rate and vole abundance. In these examples of the mechanistic paradigm, density, previously seen as a surrogate of resource availability, has been replaced by a direct measure of the availability of food.

In other studies of the numerical response, food availability is replaced by a surrogate, such as rainfall. Examples have been reported for red kangaroo (Caughley et al. 1984; Bayliss 1985b; Cairns & Grigg 1993; McCarthy 1996), Pacific black duck (*Anas superciliosa*) and maned duck (*Chenonetta jubata*; Briggs & Holmes 1988), magpie goose (Bayliss 1989), feral buffalo (*Bubalus bubalis*) (Freeland & Boulton 1990; Skeat 1990) and feral cattle

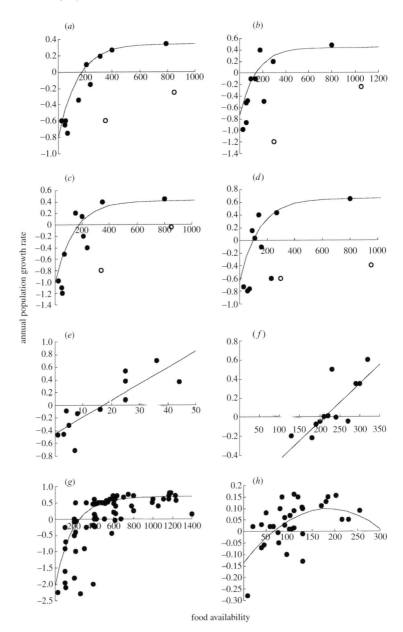

(*Bos taurus*) (Skeat 1990), feral pig (Caley 1993) and house mouse (Brown & Singleton 1999). Lebreton & Clobert (1991) urge caution in simple regression analysis of the effects of environmental variables, because although estimates of slope are not biased, their variances are, and this leads to problems in significance testing.

The above examples of the mechanistic paradigm have been observational or correlative studies. Stronger inferences of cause and effect can be obtained using experiments. The effects on population dynamics of extrinsic factors have been studied in field-based experiments, such as adding food or removing predators (the snowshoe hare (*Lepus americanus*) studies of Krebs *et al.* (1995)) or controlling pathogens (the red grouse (*Lagopus lagopus*) and nematode (*Trichostrongylus tenuis*) studies of Hudson *et al.* (1998)). There is considerable scope for further testing of the cause and effect basis of relationships in the mechanistic paradigm with such experimental studies, especially the relationship between population growth rate and food availability (Eberhardt 1988).

Controversy between the supporters of the density and mechanistic paradigms dominated population ecology for decades during the middle 1900s, as described by Sinclair (1989, 1996), Krebs (1995) and Kingsland (1996). The debate was partly about approaches and partly about differences in terminology and understanding, especially limitation versus regulation (Sinclair 1989, 1996). The more recent emphasis on biological mechanisms has shed new light on population dynamics (Bjornstad & Grenfell 2001).

(c)   Analyses of demography and density

Density and demography are linked in studies of density dependence operating on mortality, survival or fecundity; k-value analyses fall into this category as k-values are measures of mortality rates (see Sinclair (1989, 1996) for reviews of k-value analyses). Density dependence of fecundity is

---

Figure 2.4. Examples of the numerical response of populations to food availability, plotting annual population growth rate (r) against pasture biomass (kg ha^{-1}), except in (*e*) the x-axis is vole abundance and in (*h*) the x-axis is per capita food availability. (*a*) Red kangaroo in Kinchega National Park (Bayliss 1987); (*b*) red kangaroo next to Kinchega National Park (Bayliss 1987); (*c*) western grey kangaroo in Kinchega National Park (Bayliss 1987); (*d*) western grey kangaroo next to Kinchega National Park (Bayliss 1987); (*e*) barn owl (*Tyto alba*) (modified from Taylor (1994); $r^2 = 0.72$, $p < 0.01$); (*f*) feral goat (Maas 1997); (*g*) feral pig (*Sus scrofa*) (Choquenot 1998); (*h*) wildebeest (*Connochaetes taurinus*) (Krebs *et al.* 1999). (*a–d*) Open circles represent data not used in the estimation.

studied within k-value analysis by positing a maximum possible fecundity, and treating observed fecundities as falling short in consequence of 'mortality'. This manoeuvre, while awkward, is of course realistic in the case of abortions. There would seem no reason nowadays not to examine the relationships between life-history traits, density and population growth rate directly (Sibly & Smith 1998). For instance if we write r for population growth rate, x for density and b and m for two independent life-history traits, then

$$\frac{\mathrm{d}r}{\mathrm{d}x} = \frac{\partial r}{\partial b}\frac{\mathrm{d}b}{\mathrm{d}x} + \frac{\partial r}{\partial m}\frac{\mathrm{d}m}{\mathrm{d}x}. \tag{2.4.1}$$

This shows how changes in density ($\mathrm{d}x$) which change life-history traits (e.g. $\mathrm{d}m$) result in changes in population growth rate ($\mathrm{d}r$). In practice, $\mathrm{d}m/\mathrm{d}x$ would be measured as the regression coefficient in, for example, a k-value analysis. The contribution this makes to changing population growth rate is given by the sensitivity of population growth rate to the life-history trait ($\partial r/\partial m$). Formulae for sensitivities are available for many life histories (e.g. Caswell 2001; Sibly $et\ al.$ 2000a), calculated by implicit differentiation of appropriate forms of the Euler–Lotka equation. Alternatively, the effect of density on population growth rate acting through demographic parameters can be described in density-dependent Leslie matrix models (Lebreton & Clobert 1991; Caswell 2001).

The life-history stages at which density-dependent effects occur in mammals were reviewed by Sinclair (1996: table 2.1). Density was found to affect fecundity in over half the large terrestrial herbivores and large marine mammals, and in some species had effects in the early juvenile phase. By contrast, in small mammals and carnivores the effects of density were felt most in the late juvenile phase.

Demography and density have been linked for white-tailed deer (*Odocoileus virginianus*) in the modified logistic equations (table 2.2) of Hobbs $et\ al.$ (2000) which assume positive effects of survival and fecundity on finite population growth rate (λ) and a linear negative effect of density on recruitment (the net result of fecundity and survival to first reproduction). Detailed field studies have demonstrated that density affects fecundity and mortality in red deer (*Cervus elaphus*; Clutton-Brock & Albon 1989; Clutton-Brock $et\ al.$ 1997a) and Soay sheep (*Ovis aries*; Clutton-Brock $et\ al.$ 1997a; Milner $et\ al.$ 1999). Lebreton & Clobert (1991) have argued that it is more efficient to analyse the effect of density on population growth rate via its effects on demographic parameters.

Demography and density are used in population viability analyses to model the probable dynamics of small populations. In one implementation, in the software program V O R T E X (Lacy 1993), there is the option to make the demographic parameters density dependent, and outputs include population growth rate and estimated mean time to extinction. V O R T E X has been used to assess the feasibility of reintroducing wild boar (*Sus scrofa*) to Scotland (Howells & Edwards-Jones 1997).

(d) Analyses of mechanisms and density

The 'numerical response' discussed above in §2.4b calculates the effects on population growth rate of food availability, and as noted there, availability may sometimes be best measured by dividing the population food supply by population density. Thus, McCarthy (1996) reported for red kangaroo a positive relationship between population growth rate and the resources/density ratio (a per capita numerical response), and Barlow & Norbury (2001) found in ferrets (*Mustela furo*) a negative relationship between population growth rate and the ferrets/rabbit ratio. These analyses derive ultimately from Leslie's (1948) ratio model of the predator–prey relationship, developed subsequently by May (1974), Caughley & Lawton (1981) and Caughley & Krebs (1983). In similar vein, Pech *et al.* (1999) showed for the house mouse (*M. domesticus*) a positive effect of food availability and a negative effect of density on population growth rate (table 2.2). The population growth rate of Yellowstone elk was positively related to autumn precipitation and negatively related to density (Coughenour & Singer 1996), and similarly population growth rate of the San Joaquin kit fox (*Vulpes macrotis mutica*) was positively related to annual growing season rainfall and negatively related to density (Dennis & Otten 2000). The analyses of McCarthy (1996) and of Barlow & Norbury (2001) assumed a multiplicative effect of food and the reciprocal of density, and the analyses of Coughenour & Singer (1996) and of Pech *et al.* (1999) assumed an additive effect of food and density. The finite rate of population growth (λ) of wolf (*Canis lupus*) populations has been negatively related to the ratio of wolves per deer (Eberhardt 1998; Eberhardt & Peterson 1999). All these studies corrected for density to produce a measure of food availability relative to abundance. If more animals were competing for the food, less was available to each, so there was competition for the resource. Density may, however, have negative effects on population growth rate additional to those of resource competition, as a result of interference competition. This is discussed further in §2.5.

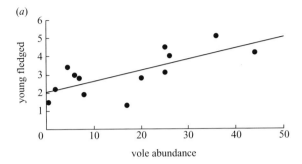

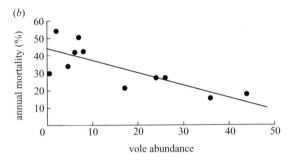

Figure 2.5. An example of use of the combined demographic and mechanistic paradigm. (*a*) First-brood young barn owls fledged per year per breeding pair plotted against abundance of voles. (*b*) Annual mortality of breeding adult barn owls and vole abundance. (After Taylor (1994).)

(e)   Analyses of demography and mechanisms

The demographic and mechanistic approaches are linked by showing that demographic parameters are affected by variables from different trophic levels, e.g. food availability. The population growth rate of red fox has been related to separate effects of rainfall on fecundity and on survival (table 2.2; Pech *et al.* 1997). In a barn owl population in southern Scotland, Taylor (1994) showed that an increase in vole abundance increased fecundity (figure 2.5*a*), decreased adult mortality rate (figure 2.5*b*) and decreased juvenile mortality. The net effect of increasing food abundance on population growth rate of the owls was positive (figure 2.4*e*). In a study of the northern spotted owl the finite population growth rate (λ) was estimated from survival and recruitment rates with these demographic rates influenced by climate and/or habitat features (Franklin *et al.* 2000). Food availability was not estimated.

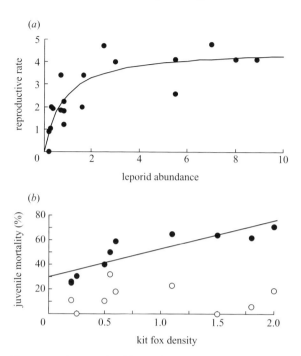

Figure 2.6. An example of the combined density, demographic and mechanistic paradigm. (*a*) Adult reproductive rate of kit fox and the abundance of leporids, their main food. (*b*) Juvenile mortality (%) of kit fox and kit fox density. The solid line and solid points demonstrate density-dependent mortality caused by predation by coyotes. The open circles show density independence associated with other causes of juvenile mortality. (After White & Garrott (1999).)

(f)   Combined analyses of demography, mechanisms and density

A full analysis shows the effects of trophic factors and density on demography and hence on population growth rate (table 2.2). For instance the rat, *Mastomys natalensis*, has dynamics determined by the effects of rainfall on fecundity and the effects of rainfall and density on survival (Leirs *et al.* 1997). The population dynamics of the kit fox are determined by demographic rates which are age-structured. White & Garrott (1999) showed that reproductive rates were positively related to food (leporid) availability (figure 2.6*a*), which in turn depended on rainfall. Juvenile mortality (figure 2.6*b*) was positively density dependent (because of predation by coyotes, *Canis latrans*). Soay sheep (*O. aries*) have population dynamics influenced by demography which in turn depends on weather conditions and density (Coulson *et al.* 2001). Harsh winters kill young and old sheep but only in high-density years (Bjornstad & Grenfell 2001).

The many mathematical models that have been used to describe relationships between population growth rate and demographic parameters, mechanistic and density effects suggest a need for a more detailed comparative evaluation of the fit of the models. Steps have been made in this direction, such as the comparative analysis by assessing goodness of fit using coefficients of determination (r^2) (water buffalo and feral cattle (Skeat 1990), feral goats (*Capra hircus*; Maas 1997), red kangaroo populations (McCarthy 1996), *Mastomys* rat (Leirs *et al.* 1997) and the house mouse (Pech *et al.* 1999)). An alternative is to use an information theoretic model-selection procedure (kit fox; Dennis & Otten 2000).

2.5.   Case study

The following example illustrates some of the ideas discussed in the paper so far, including density dependence, the numerical response and the action of environmental factors. The case study is based on the results of an experiment into the dynamics of a marine copepod, *Tisbe battagliai*. When the data were presented by Sibly *et al.* (2000b) they were analysed solely in terms of density dependence and environmental factors. Here, we analyse the numerical response, and show that some of the negative effects of population density are due to partitioning of food between competitors. It turns out, however, that interference competition further restricts access to resources and produces additional negative effects on population growth rate. The study provides an example of analysis in terms of mechanism and density.

(a)   Methods

A factorial experimental design was used, comprising two factors (prey species richness and food concentration), each applied for 11 weeks to 10 replicate laboratory populations of the marine copepod *T. battagliai*. Food was either the alga *Isochrysis galbana*, which alone will sustain cultures indefinitely, or a mix of two algal species, *I. galbana* and *Rhodomonas reticulata*. Each food was given at two concentrations (1300 and 3250 µg Cl^{-1}), replaced daily. Each *Tisbe* population was founded at low density (usually two pairs), and surviving copepods were transferred daily to a duplicate culture plate containing 10 ml of freshly prepared test solution. Population biomass was calculated as the sum of the dry weights of all age classes of copepods. Population growth rate (r) was estimated as the natural logarithm of (biomass in week $t + 1$)/(biomass in week t). Further details are given in Williams (1997) and Sibly *et al.* (2000*b*).

In analysing the numerical response a realistic measure of food availability is needed (§2.4b and §2.4d). Here, we use food per copepod because carrying capacity increased linearly with population food supply in this experiment (Sibly *et al.* 2000*b*), suggesting that the copepods ate all the food supplied to them, at least at higher densities. The amount of food available to each copepod ('food per copepod') was therefore calculated as the population food supply divided by the copepod population biomass. This approach to measuring food availability when predators eat all the food supplied was introduced by Leslie (1948).

(b)   Results and discussion

Population growth rate decreased as density increased, the expected density-dependent response (figure 2.7*a*). Population density alone did not account for all the observed variation, however, because population growth rate was higher at the higher food concentration (crosses above circles in figure 2.7*a*). This is readily understood because individuals living at higher food availability are expected to grow and reproduce faster, resulting in higher population growth rate.

The numerical responses of population growth rate to food availability (figure 2.7*b*) show that populations declined (population growth rate < 0) when food per copepod was low, but increased (population growth rate > 0) when food availability was higher. The increase in population growth rate with food supply was not linear, because as food increased, the copepods showed diminishing ability to transform extra food into further population growth. When the relationship is linearized using a log transformation (figure 2.7*c*), it appears that while food availability (food per copepod) accounts for a great deal of the variation in population growth rate, there is still some unexplained variation: note the crosses below the circles in figure 2.7*c*. The combined effects on population growth rate of food availability, prey species richness and population density, the data of figure 2.7*c*, were analysed by multiple regression, the regression equation being

$$\text{population growth rate} = -2.89 + 1.46 \log_{10} (\text{food per copepod})$$
$$+ 0.43 \text{ prey species richness}$$
$$- 0.24 \text{ density} \qquad (2.5.1)$$

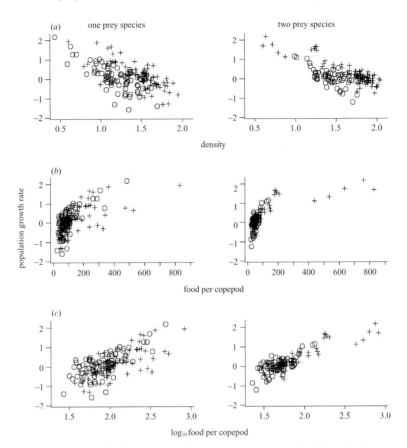

Figure 2.7. The determinants of population growth rate (r) in *Tisbe battagliai*. Population growth rate per week is plotted in relation to (a) density, measured as $\log_{10}$ copepod population biomass, μg; (b) food availability, measured as food supply per unit biomass of copepod, μg C l^{-1} μg^{-1}; (c) as (b) with food per copepod transformed to $\log_{10}$ on the x-axis. Food was either the alga *Isochrysis galbana* (left-hand panels) or a mix of two algal species (*I. galbana* and *Rhodomonas reticulata*, right-hand panels). Food concentrations were: crosses, 3250 μg C l^{-1}; circles, 1300 μg C l^{-1}.

The multiple regression was highly significant ($F_{3,303} = 128.1$; $p < 0.0001$; $r^2_{adj} = 56\%$). In the regression, prey species richness was 1 or 2, density was $\log_{10}$ (copepod biomass) and other predictors were defined as in figure 2.7. Statistical significance of regression coefficients were food per copepod: $t_{300} = 11.05$, $p < 0.0001$; prey species richness: $t_{300} = 8.46$, $p < 0.0001$; density: $t_{300} = -2.02$, $p < 0.05$. Plots of residuals against predictor variables show no indication of departure from model assumptions and there was no evidence of serial correlation (Durbin–Watson statistic 1.81).

The population's food supply is shared between its members through a process of resource competition. This is why we assess food availability by 'food per copepod'. The additional negative effects of density on population growth rate of *T. battagliai* could occur because of interference competition influencing reproductive performance, survival or somatic growth or foraging efficiency. Crowding is known to increase swimming activity in *T. battagliai*, with an associated increase in energy costs (Gaudy & Guerin 1982) possibly as a result of antagonism between larvae (Brand *et al.* 1985); crowding is also known to reduce reproductive output and depress larval viability (Fava & Crotti 1979; Zhang & Uhlig 1993) and to influence the sex ratio from a female bias at low density to a male bias at high density in *T. holothuriae* (Hoppenheit 1976; Heath 1994). Negative effects of population density on population growth rate could occur in nature because of density-dependent effects of predators, parasites, pathogens, interspecific competitors and mutualists, but they were excluded in this experiment. We therefore conclude that the negative effects of density, additional to effects through food availability, were caused in this experiment by interference competition.

A graphical model of how food availability, density and prey species richness jointly affect population growth rate in this example is given in figure 2.8. Note that population growth rate is related to *per capita* food availability in contrast to the numerical responses of figure 2.4, which were not per capita relationships except in figure 2.4*h*. The results suggest that the ratio model of Leslie (1948) should be modified to incorporate other trophic effects, such as prey species richness and interference competition. For a given level of per capita food availability in the present study, increasing interference competition reduces population growth rate (compare solid and dashed lines) and increases the level of food availability needed to support a stable (population growth rate = 0) population. Increasing prey species richness increases population growth rate and lowers the level of per capita food availability needed to support a stable population (figure 2.8*a,b*). This graphical model illustrates our results, and shows that the density-dependent and the mechanistic approaches of figures 2.3 and 2.4 are not incompatible. Instead, figure 2.8 shows that the two approaches are complementary and can be combined by joint analysis of the effects of mechanism and density, thus clarifying the mechanistic links that connect density, food availability and population growth rate.

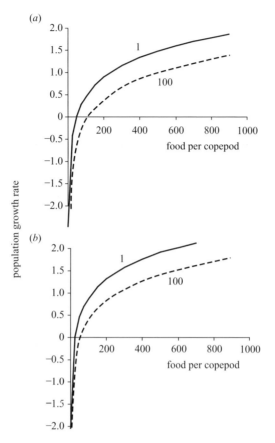

Figure 2.8. Graphical model showing how food availability (food per copepod), density and prey richness together determine population growth rate. The curves are derived directly from equation (2.5.1). Density has an indirect effect via food availability, and a direct negative effect through interference competition, illustrated here by the contrasting solid and dashed lines (solid line, copepod biomass = 1; dashed line, biomass = 100). (*a*) refers to one and (*b*) to two prey species.

2.6.   Applications in conservation biology, wildlife management, ecotoxicology and human demography

A better understanding of how to estimate population growth rate, and what determines population growth rate, is fundamental to the application of population growth rate to wildlife management and other topics in population ecology. The three classical applications of conservation, harvesting and pest control aim to increase, maintain and decrease population growth rate, respectively (Caughley 1976, 1977). Each of these applications also affects the frequency distribution of population growth rate (Hone 1999).

Of general concern in conservation biology is population performance at low density, which is the danger zone for endangered species. Population growth rate may be submaximal when density is low because of the difficulty of finding mates (the Allee effect; Courchamp *et al.* 1999; Stephens & Sutherland 1999) or because of an effect of predation on a prey population, wherein the predator has a type III functional response (Sinclair *et al.* 1998). An example of a decline in population growth rate at low density was described for pronghorn antelope (*Antilocapra americana*; Sinclair 1996). It had been thought that an Allee effect might have kept the North Atlantic right whale at low density, but this possibility has recently been eliminated by demographic analysis (Fujiwara & Caswell 2001). Another potentially important issue in conservation biology is the effect of inbreeding on population growth. Although there is debate about the practical significance of inbreeding in wild animal populations (Caughley 1994; Gundersen *et al.* 2001), it is easy to see that it could be important in small isolated populations, because we know from the Euler–Lotka equation that any reductions in fecundity and survival caused by inbreeding would necessarily depress population growth rate. Inbreeding is included as a determinant of population growth rate in population viability analyses such as VORTEX (Howells & Edwards-Jones 1997). These topics in conservation biology focus on the small population size being a cause of conservation problems (Caughley 1994). Also of interest are the questions as to why the population is small, why it declined, and what management can do to reverse the decline.

Another common theme in conservation biology is the decline of populations because of high mortality caused by people. The rate of population growth of populations subject to intensive illegal harvest, such as the African elephant (*Loxodonta africana*) and black rhinoceros (*Diceros bicornis*) has been related to the effort expended in reducing or preventing poaching (Leader-Williams & Albon 1988). The dynamics of seabirds, such as the wandering albatross (*Diomedea exulans*), are influenced by fishing activities. For instance, Weimerskirch *et al.* (1997) showed that the rate of population decline depended on the cumulative number of longline fishing hooks set in albatross foraging areas. These studies use the mechanistic approach to identify the causes of population declines.

Also of general interest are the effects of habitat loss, and research here aims to identify the main features of wildlife habitats, and to establish the way they affect distribution and abundance, i.e. to characterize each species' ecological niche. Traditionally such research has involved measuring a large number of habitat features though typically not including food, predators, pathogens and competitors. In an encouraging effort to advance wildlife habitat studies and hence wildlife management in general, Morrison (2001) recommended a change to a focus on resources such as food, and their effects on survival and reproduction. Such a focus would be a shift to demographic–mechanistic analysis as described above, and if implemented would lead to greater understanding of mechanisms determining population growth rate.

Wildlife management has two other fields of study, harvesting and pest control. These are conceptually similar (Caughley 1976, 1977), being based on ideas of reducing abundance and expecting compensatory responses in one or both of fecundity and survival. The sustainable harvest of a wildlife population growing according to logistic growth is a function of abundance, carrying capacity and the maximum rate of population growth (r_{max}; Caughley 1977). Species at the same abundance having the same carrying capacity have higher sustainable annual harvests if their r_{max} values are higher. When r_{max} is very low, as in long-lived species like whales and elephants, a sustained harvest may not be economic (Clark 1973; Caughley & Sinclair 1994). Variation in the effects of environmental factors lowers the level of sustained harvest (Bayliss 1989) and of course increases the variation of the harvest (Beddington & May 1977). Negative effects of harvesting on population growth rate have been described in white-tailed deer in Ontario (Fryxell et al. 1991), mallard (Anas platyrhynchos) in North America (Reynolds & Sauer 1991), and moose (Alces alces) in Norway (Solberg et al. 1999).

Pest control is another field where population growth rate is central (Hone 1994); indeed, several early population ecologists (e.g. Howard, Fiske, Nicholson, Andrewartha and Birch) studied pests. A pest control programme may become a sustained harvest if pest abundance is reduced and kept at low levels. Pest populations may, however, adapt to control, for example, by developing resistance to pesticides. The rate of development of resistance depends on population growth rate and generation interval (Dobson & May 1986).

Population growth rate can also be seen as the key unifying concept in ecotoxicology (Walker et al. 2001). Thus, pollutants can be defined as environmental chemicals that exceed normal background levels and have the potential to adversely affect birth, growth or mortality rates, with consequent reduction in population growth rate. Defined in this way, pollutants appear as a particular case of environmental stressors (see §2.2). Although the impact of pollutants on organisms is studied within eco- toxicology at different organizational levels, from biochemistry through to communities, each with its own measure of pollutant effect, it can be argued that population growth rate provides the best summary statistic. After reviewing the experimental literature, Forbes & Calow (1999) con- clude that population growth rate is a better measure of responses to toxicants than are individual-level effects, because it integrates poten- tially complex interactions among life-history traits and provides a more relevant measure of ecological impact. By contrast, Sutherland & Norris (Chapter 12) suggest that there are advantages in building up from the level of individual behaviour to a population level perspective on popu- lation growth rate.

Where complex models of the determinants of population growth rate are used for management purposes, model validation is clearly of crucial importance: how do we know we can believe the model's predictions? For model validation, long-term field datasets are needed, ideally from inde- pendent sites/populations. In this respect it is interesting to revisit the predictions of the classic early paper of Pearl & Reed (1920), in which they rediscovered the logistic equation and applied it to data on the size of the human population of the continental United States. They implied that the logistic equation results from human competition for the means of sub- sistence, namely food, clothing material and fuel. They calculated the car- rying capacity of the continental United States as 197 million, but noted that at the levels of production then obtaining it would be necessary to import about half the food supply. Their projections look a little low now, but this can be attributed to technological advances. There would seem to be a need for more studies of the type initiated by Pearl & Reed (1920) that consider the match between global levels of sustainable resources and individual levels of human consumption (cf. Lutz & Qiang, Chapter 6). A recent projection for the global human population based on existing downward trends in fertility indicates that population growth will slow over the coming decades, and peak in *ca.* 2070 at a population size around nine billion (Lutz *et al.* 2001).

Further examples of model validation can be found in the modelling of a *Mastomys* rat (Leirs *et al.* 1997) and of house mouse population dy- namics (Pech *et al.* 1999). In each study, observed rodent abundances were compared with predictions from a model of dynamics constructed using data from the same site but at an earlier time period. Model testing and validation are clearly essential wherever possible to increase the reliability of model projections.

2.7.   Conclusion

Our central thesis is that population growth rate is the unifying variable linking the various facets of population ecology; thus analyses of population regulation, density dependence, resource and interference competition and the effects of environmental stress are all best undertaken with population growth rate as the response variable. Throughout, we have emphasized the key role of statistical analyses of observational or experimental data, and we hope in the future to see more studies analysed by multiple regression and allied nonlinear techniques to estimate and distinguish the effects on population growth rate of food availability, environmental stressors, density and so on. Effects of time-lags can also, in principle, be treated within this framework. Such analyses are needed not only to improve the quality of the underlying science, but also to increase the realism and accuracy of prediction in key applied areas such as conservation, wildlife management and ecotoxicology. Much practical concern in controlling, managing or conserving populations has to do with limiting, managing or encouraging population growth; one key indicator of effect in such enterprises is population growth rate. There are, however, still many unresolved issues in identifying the causes of variation in population growth rate, stemming largely from the need to distinguish the effects of many causal factors with distinctly limited datasets. Much challenging research remains.

We thank T. D. Williams and M. B. Jones for generously allowing us to use the *Tisbe* data, M. Rees for pointing us towards useful data sources, M. Begon, P. Doncaster, P. Caley, C. J. Krebs, A. R. E. Sinclair and M. Walter for comments on an earlier draft manuscript, and J. C. Whittaker and H. Grubb of the Department of Applied Statistics, University of Reading, for statistical advice. We also thank The Royal Society and the University of Canberra for financial assistance.

3

Demographic, mechanistic and density-dependent determinants of population growth rate: a case study in an avian predator

3.1. Introduction

The empirical estimation of the determinants of how quickly or slowly population density increases or decreases, has used a variety of approaches. Krebs (1995, Chapter 7) suggested that, historically, two approaches (paradigms) have been used: a density paradigm, which focuses on the effects of density on population growth rate, and a mechanistic paradigm, which focuses on the effects of trophic factors such as food, predators and parasites on population growth rate. These paradigms, and a related approach, the demographic paradigm (Chapter 2) that focus on the effects of demographic rates (fecundity and survival) on population growth rate, have been widely used, though rarely compared and contrasted.

This paper demonstrates the application of the various approaches to identify the determinants of population growth rate in a closed population. A range of alternative hypotheses, as expressed in mathematical models, are described and evaluated empirically. Data from Taylor's exemplary (1994) study of the barn owl (*Tyto alba*) are used to illustrate the approaches.

3.2. Models

The patterns described in five ecological models, and the field data, are compared. The first four models are specific examples of the general discrete-time population model of Dennis & Otten (2000; eqn (3.2.1)). The general model is

$$N_{t+1} = N_t e^{a+bNt+cv},$$

(3.2.1)

where abundance at time $t + 1$ (N_{t+1}) is related to abundance at time t (N_t) and food availability (v) at time $t + 1$. Coefficients a, b, c are parameters to be estimated. The model of Dennis & Otten (2000) has been modified here to be a deterministic model, and by substituting food availability (v) for their weather term (W).

The first *a priori* model is exponential growth, which occurs in density-independent growth. Abundance at time $t + 1$ (N_{t+1}) is related to abundance at time t (N_t) by

$$N_{t+1} = N_t \lambda^t = N_t e^{rt}, \tag{3.2.2}$$

where the finite population growth rate is $\lambda (= N_{t+1} / N_t)$ and the instantaneous population growth rate is $r(= \ln \lambda)$. Exponential growth occurs in equation (3.2.1) when $b = c = 0$, and $a = r$ and there is a time-step of 1 yr. This and other models assume the population is closed, with no emigration or immigration. The barn owl population studied by Taylor (1994) was sedentary with very little movement into and out of the population.

The second *a priori* model is logistic growth, as a form of density-dependent population growth. Logistic growth is described by equation (3.2.1) when $c = 0$. Evidence of logistic growth is an example of the density paradigm of Krebs (1995, Chapter 7).

The third *a priori* model is a numerical response. Two types of numerical response were estimated: (i) a Solomon-type, after Solomon (1949), relating predator (owl) abundance to prey (vole) abundance (v); and (ii) a Caughley-type relating the predator (owl) population growth rate $(\lambda$ or $r)$ to food availability (vole abundance). The types of numerical responses are reviewed briefly by Sibly & Hone (Chapter 2), and Bayliss & Choquenot (Chapter 9). The Caughley-type numerical response using r is described by equation (3.2.1) when $b = 0$. Use of the numerical response is an example of the mechanistic paradigm of Krebs (1995, Chapter 7). Alternative hypotheses were, that there was a positive linear relationship between λ and vole abundance, and a positive linear relationship between r and vole abundance. The latter relationship is the linear assumption in the original Lotka–Volterra model (May 1981a; table 5.1). The linear regression of λ and voles is the finite difference analogue of the linear Lotka–Volterra model. The Caughley-type numerical responses were examined with, and without, a time-lag of 1 year in the vole data, to examine whether there was any evidence of lagged effects of food on owl dynamics.

The fourth *a priori* model is the full Dennis–Otten model (2000) as described in equation (3.2.1). In that model, there are additive effects of food and of owl abundance.

The fifth *a priori* model is the two-stage Euler–Lotka equation,

$$\lambda^{\alpha}(1 - s/\lambda) = lb, \qquad\qquad (3.2.3)$$

where α is the age at first reproduction, s is annual adult survival, l is survival from birth to age at first reproduction, and b is annual fecundity (mean female young per female) (Lande 1988; Sibly *et al.* 2000a). When the age at first reproduction equals 1.0 then the equation can be rearranged to give

$$\lambda = s + lb, \qquad\qquad (3.2.4)$$

which states that the finite population growth rate (λ) equals the sum of annual adult survival (s) and recruitment (lb).

The implementation of equations (3.2.3) or (3.2.4) is an example of the demographic paradigm. It can be combined with the ideas in the mechanistic paradigm by assuming or demonstrating that one or more of the demographic rates in equations (3.2.3) or (3.2.4) are determined by mechanistic factors, such as food. In the study of barn owls in southern Scotland (Taylor 1994), the demographic rates, s, l and b, were shown to be correlated with the abundance of the main food of the owls, field voles (*Microtus agrestis*). If it is assumed that the demographic parameters in equations (3.2.3) or (3.2.4) are related to food availability, then it follows that the finite population growth rate (λ) must be determined by food availability.

In the analyses described here, model fit is assessed firstly by statistical significance and then by the coefficient of determination (R^2). An information-theoretic analysis, such as Akaike's information criterion, was not used as the x- and y-terms in analyses were not consistent. Burnham & Anderson (1998) advised against use of Akaike's information criterion with such analyses.

(a)   Parameter estimation

(i)   Population growth rates (λ, r)

The finite population growth rate (λ) was estimated as the ratio of successive annual estimates of abundance of pairs of barn owls. Data on abundance of pairs of barn owls (Taylor 1994; fig. 15.1; Newton 1998; fig. 7.4) were used. Total abundance of pairs was the sum of the number of breeding pairs and the number of non-breeding pairs. The mean annual population growth rate (r) was estimated by the regression of the natural logarithm of abundances over time (Caughley 1980).

(ii)   Intrinsic population growth rate (r_m) and carrying capacity (K)

Evidence of logistic growth was assessed by the regression of observed instantaneous growth rate $(r = \ln(N_{t+1}/N_t))$ on total owl abundance (N_t) (Caughley 1980). The fitted regression has an intercept on the y-axis of the intrinsic rate of population growth (r_m), and on the x-axis of carrying capacity (K).

(iii)   Numerical responses

The Solomon-type numerical response was estimated as the linear regression of abundance of pairs of owls in year t on vole abundance in year t, following Solomon (1949). The Caughley-type numerical response relationships between the finite population growth rate (λ) and vole abundance, and the instantaneous growth rate (r) and vole abundance were estimated by linear regression. The abundance of pairs of barn owls was estimated by the minimum number known to be alive in the breeding season (spring and summer) each year during intensive observation and mark–recapture of owls. An index of abundance of field voles (v) was estimated annually by trapping in spring (Taylor 1994).

(iv)   Age at first reproduction (α)

Taylor (1994) reported that owls fledged in spring and summer first bred the following spring or summer. Hence, it was assumed that $\alpha = 1$ year.

(v)   Annual adult survival (s)

Annual survival is the complement of annual mortality. Taylor (1994) reported that annual mortality of adult owls ranged from 15 to 55%, and was strongly negatively correlated (correlation coefficient $= -0.78$, d.f. $= 9$, $p < 0.01$) with vole abundance (Taylor 1994; fig. 14.5). The linear regression was estimated to be,

$$\text{annual adult mortality} = 1 - s = 0.44 - 0.0071v, \qquad (3.2.5)$$

where vole abundance is v.

(vi)   Juvenile survival (l)

Survival from birth to age at first reproduction (l) was significantly negatively correlated (correlation coefficient $= -0.78$, $p < 0.01$) with vole abundance (Taylor 1994; p. 210). It was assumed, for analysis, that in the absence of voles many, but not all, owls died so survival was low (0.1). When vole abundance was high (30) it was assumed that survival was 0.3. The linear regression was then estimated to be

$$l = 0.1 + 0.007v. \tag{3.2.6}$$

Equation (3.2.6) estimates that when mean vole abundance is 20, then survival is 0.24, and therefore juvenile mortality equals 0.76. This is similar to the higher juvenile survival estimates reported (Taylor 1994; p. 207).

(vii)   Annual fecundity (b)
The clutch size (correlation coefficient = 0.87, d.f. = 11, $p < 0.001$) (Taylor 1994; fig. 11.8) and the number of young fledging (correlation coefficient = 0.72, $p < 0.01$) (Taylor 1994; fig. 12.9), varied positively with vole abundance. The estimated linear regression relating the number of female young fledged per female per year to vole abundance was

$$b = 1.02 + 0.03v. \tag{3.2.7}$$

(viii)   Sensitivity analysis
The effects of a small change in each demographic parameter (α, s, l or b) on the finite population growth rate (λ) were estimated in a sensitivity analysis using the equations of Lande (1988). The assumed parameter values, based on data in Taylor (1994) were as follows: age at first reproduction (α) was 1 year, mean annual adult survival (s) was 0.65, mean juvenile survival (l) was 0.3 and mean annual fecundity (b) was 1.5 females per female. These values provide an estimated annual finite population growth rate (λ) of 1.1 and a generation interval of 2.4 years.

3.3.   Results

(a)   Exponential growth
The regression of the natural logarithms of owl abundance over time, was not quite statistically significant ($F = 4.69$, d.f. $= 1,11$, $p = 0.053$, $R^2 = 0.30$) (table 3.1). The slope of the regression, an estimate of mean r, was -0.06 yr^{-1} with 95% CI of -0.115 to 0.001. Hence, there was no strong evidence against the proposal that $r = 0$; however, with additional data that conclusion would be reinforced. The observed trends in abundance of barn owls and field voles are illustrated in figure 3.1, showing apparently linked oscillations over time.

LIBRARY, UNIVERSITY COLLEGE CHESTER

Table 3.1. *The coefficients of determination* (R^2) *and statistical significance* (p) *of each model of barn owl dynamics. Also shown is the dependent variable* $(\ln N_t, \lambda, r \text{ or } N_t)$ *in each analysis.*

Model	λ, r or N_t	R^2	p
exponential	$\ln N_t$	0.30	0.053
logistic	r	0.21	0.137 (NS)
Solomon-type numerical response (no time-lag)	N_t	0.44	0.014
Caughley-type numerical response (no time-lag)	λ	0.72	0.0005
Caughley-type numerical response (1 year time-lag)	λ	0.05	0.476 (NS)
Caughley-type numerical response (no time-lag)	r	0.72	0.0005
Caughley-type numerical response (1 year time-lag)	r	0.01	0.720 (NS)
Caughley-type numerical response (vole and owl effects with no time-lag: full model)	r	0.82	0.0004

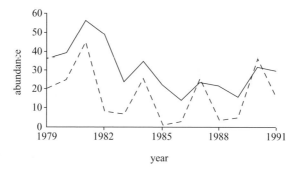

Figure 3.1. Observed trends in abundance of pairs of barn owls (solid line) and field voles (dashed line) in southern Scotland. Modified from Taylor (1994).

(b)   Logistic growth

The analysis for evidence of logistic growth showed a non-significant result ($F = 2.62$, d.f. $= 1,10$, $p = 0.137$, $R^2 = 0.21$) (table 3.1). Hence, there was no strong evidence supporting the proposal of logistic growth. Therefore, there were no estimates of r_m or K estimated by the regression analysis. Given the observed trend in owl abundance (figure 3.1) it was not surprising that there was no empirical support for logistical growth.

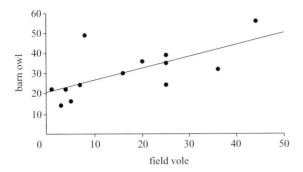

Figure 3.2. The Solomon-type numerical response of pairs of barn owls in year *t* to variation in field vole abundance in year *t*. Data are estimated from Taylor (1994) and Newton (1998) and are based on censuses.

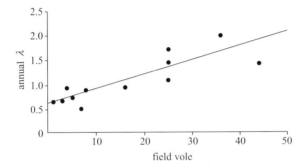

Figure 3.3. The Caughley-type numerical response of a barn owl population to variation in field vole abundance. The response variable is the annual finite population growth rate (λ) of the owl population. Data are estimated from Taylor (1994) and Newton (1998) and are based on censuses.

(c) Numerical responses

The Solomon-type numerical response of the relationship between the number of pairs of owls and vole abundance was significant ($F = 8.52$, d.f. $= 1,11$, $p = 0.014$, $R^2 = 0.44$) (table 3.1). The relationship and the fitted regression are shown in figure 3.2.

The Caughley-type numerical response of the annual finite population growth rate (λ) and vole abundance (*v*) was highly significant ($F = 25.16$, d.f. $= 1,10$, $p = 0.0005$, $R^2 = 0.72$) (table 3.1). The fitted regression was

$$\lambda = 0.607 + 0.028v, \tag{3.3.1}$$

which is illustrated in figure 3.3. The regression estimates that owl abundance declines when vole abundance drops below 14, and when voles are

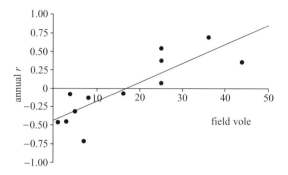

Figure 3.4. The Caughley-type numerical response of a barn owl population to variation in field vole abundance. The response variable is the annual instantaneous population growth rate (r) of the owl population. Data are estimated from Taylor (1994) and Newton (1998) and are based on censuses.

absent the annual growth rate (λ) of the owl population equals 0.607. The numerical response using vole abundance in the previous year as the independent variable was not significant ($F = 0.55$, d.f. $= 1,10$, $p = 0.476$, $R^2 = 0.05$) (table 3.1).

The Caughley-type numerical response of the annual instantaneous population growth rate (r) and vole abundance (v) was highly significant ($F = 25.12$, d.f. $= 1,10$, $p = 0.0005$, $R^2 = 0.72$) (table 3.1). The fitted regression was

$$r = -0.446 + 0.026v, \tag{3.3.2}$$

which is illustrated in figure 3.4. The regression estimates that owl abundance declines when vole abundance drops below 17, and when voles are absent, r of the owl population equals -0.446 yr^{-1}. The numerical response using vole abundance in the previous year as the independent variable was not significant ($F = 0.14$, d.f. $= 1,10$, $p = 0.720$, $R^2 = 0.01$) (table 3.1).

The Caughley-type numerical response of the annual instantaneous population growth rate (r) and vole abundance (v) and owl abundance in year t (equation (3.2.1)) was highly significant ($F = 20.82$, d.f. $= 2,9$, $p = 0.0004$, $R^2 = 0.82$) (table 3.1).

The fitted regression was

$$r = -0.069 - 0.011\,\text{owls} + 0.024v, \tag{3.3.3}$$

which is illustrated in figure 3.5. The population growth rate increased when vole abundance increased and decreased as owl abundance

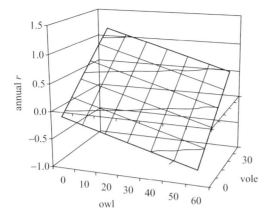

Figure 3.5. The Caughley-type numerical response of a barn owl population to variation in field vole abundance and owl abundance (equation (3.3.3)). The response variable is the annual instantaneous population growth rate (*r*) of the owl population.

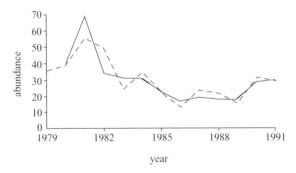

Figure 3.6. The observed (dashed line) and reconstructed (solid line) abundance of pairs of barn owls over years. There is no estimate of owl abundance in 1979 because of the absence of data on owl abundance in the prior year.

increased. The standard errors, and associated *t*-values, on each parameter (a, b, c in equation (3.3.1)) were 0.1871 ($t = 0.37$, NS), 0.0049 ($t = -2.33$, $p = 0.045$) and 0.0044 ($t = 5.49$, $p = 0.0004$), respectively. These indicate the intercept (*a*) was not different to $r = 0$, but the coefficient for the effect of owl abundance (*b*) was significant and for the effect of vole abundance (*c*) was highly significant. The observed (dashed line) and reconstructed (using equation (3.3.3), solid line) abundance of barn owls are shown in figure 3.6, showing the close agreement.

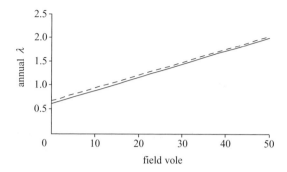

Figure 3.7. A comparison of the Caughley-type numerical responses, estimated by two methods, of a barn owl population to variation in field vole abundance. The numerical response estimated from census data (equation (3.3.1) and shown in figure 3.3) is shown by the solid line and the numerical response estimated from demographic data (equation (3.3.4)) is illustrated by the dotted line. The response variable is the annual finite population growth rate (λ) of the owl population.

(d)   Demographic and mechanistic determinants of population growth rate

The relationship between the finite population growth rate (λ) and vole abundance was estimated by substituting equations (3.2.5) to (3.2.7) into equation (3.2.4). The resultant equation showed a positive relationship between λ and vole abundance (figure 3.7). The estimated regression was

$$\lambda = 0.662 + 0.0172v + 0.00021v^2 . \tag{3.3.4}$$

The estimated relationship is slightly curved being upwardly concave, though this is not obvious in figure 3.7. Because of the substitution of equations, a quadratic equation with a positive intercept and positive regression coefficients for the linear and quadratic terms is produced.

The demographic numerical response relationship shown in figure 3.7 is compared with that estimated from the annual census data (figure 3.3). The two numerical response relationships are very similar (figure 3.7) with that derived from demographic and mechanistic data giving slightly higher estimates of λ than that estimated from the census data.

(e)   Sensitivity analysis

The sensitivity analysis showed that the largest effect on λ was produced by a small change in annual adult survival (1.7), followed by survival to age at first reproduction (1.5), annual fecundity (0.3) and age at first reproduction (−0.2).

3.4.   Discussion

The population growth rate of a closed population is determined by demographic rates and the influence of extrinsic mechanistic factors, such as food, and intrinsic factors, such as spacing behaviour. The results reported here show that for the avian predator, the barn owl, changes in abundance are closely related to the variation in food supply (especially of field voles). The Caughley-type numerical response, as estimated by the finite population growth rate (λ), to food supply can be estimated directly from a field census of owls (figure 3.3). The Caughley-type numerical response can also be estimated indirectly (figure 3.7) from empirically estimated relationships between demographic rates and food supply. Hence, use of the demographic and mechanistic approaches to estimating the determinants of population growth rate can yield similar and complementary results.

The ecological model (equation (3.3.3)) with the best fit, as assessed by the coefficient of determination R^2, was the Caughley-type numerical response of population growth rate, as r, on vole abundance and owl abundance (table 3.1). That relationship accounted for 82% of the variation in population growth rate. The model combined the mechanistic and density paradigms of Krebs (1995, Chapter 7). The additive and negative effect of owl abundance on population growth rate (r) in the barn owl study may have been associated with some form of density-dependent competition for food and possibly breeding sites. The barn owls have overlapping home ranges but do defend nesting sites (Taylor 1994).

The numerical response relationships showed that the barn owls responded to variation in food availability within a year. The analyses using lagged vole data were non-significant. This result occurred through within-year changes in survival and fecundity of owls. With higher vole abundance, owls laid their first egg earlier in the breeding season (Taylor 1994; fig. 9.3b), mean clutch size was higher (Taylor 1994; fig. 11.8) and the mean number of young fledged per pair increased (Taylor 1994; fig. 12.9). Such a quick response, with no time-lag, has also been recorded in Tengmalm's owl (*Aegolius funereus*) (Korpimäki & Norrdahl 1989) and red kangaroo (*Macropus rufus*) and western grey kangaroo (*M. fuliginosus*) populations (Bayliss 1987) in response to variation in food availability.

Variation in prey availability has been shown to be related to demographic rates of predatory wildlife in several studies. For example, microtine rodent abundance was related to fecundity in Tengmalm's owls (Korpimäki & Norrdahl 1989), variation in lifetime reproductive success in male Tengmalm's owl (Korpimäki 1992), survival of male Tengmalm's owl (Hakkarainen *et al.* 2002) and survival of breeding females and age at first reproduction in Ural owl (*Strix uralensis*) (Brommer *et al.* 1998). Fecundity in kit fox (*Vulpes macrotis*) was positively related to leporid abundance (White & Garrott 1999).

The Solomon-type numerical response of predators (barn owl) and prey (voles) was positive (figure 3.2). Such a positive relationship has also been reported for Tengmalm's owl and *Microtus* spp. (Korpimäki & Norrdahl 1989).

The results reported here suggest that approaches used in other studies of the determinants of population growth rate may be usefully modified. For example, the demographic studies of population growth rate in northern spotted owl (*Strix occidentalis caurina*), such as those of Lande (1988) through to Seamans *et al.* (2001), focused on using demographic rates to estimate population growth rate, but did not estimate food availability or a numerical response relationship. In contrast, studies of kangaroos, such as those of Bayliss (1987), Caughley (1987*b*), Cairns & Grigg (1993) and McCarthy (1996), focused on estimating a Caughley-type numerical response without any detailed data on demographic rates. The study here shows that such different approaches can be combined. Such a combination of demographic and mechanistic approaches occurred in a graphical, not quantitative, form in the study of Himalayan thar (*Hemitragus jemlahicus*) of Caughley (1970b; fig. 5).

The models and analyses described in the present study have strengths and limitations. The estimated numerical responses reported here (figures 3.3, 3.4, 3.5 and 3.7) are linear. Over a broader range of food availability, such numerical responses should be curved with the population growth rate approaching a maximum, the intrinsic rate of growth (r_m). The estimated relationships between the demographic rates and food availability should also be curved relationships, as survival clearly has a maximum value of 1.0 and fecundity has a maximum value determined by genotype. There was no clear evidence of curved relationships in the data; however, perhaps such curves would be evident if food (vole) availability occurred at higher levels than reported in the study by Taylor (1994).

The numerical response may also be influenced by other mechanistic factors. For example, fecundity may be partly determined by the availability of nest sites, as well as food. If that occurred then a modified version of equation (3.2.7) would need to be estimated showing the effects of food and nest sites. Expressing the effects of food and nest sites on population growth rate would be similar to the description of ecological niche given by Sibly & Hone (Chapter 2) and the modelling of the effects of two obligate resources on population growth rate by Tilman (1982). Similarly, effects of pesticides on owl abundance, or wildlife generally, could be incorporated into the modelling by evaluating whether adult and/or juvenile survival are affected by food and pesticide exposure. In many studies of birds, including avian predators, the effects of nest sites and pesticides have been reported (Newton 1998).

Most models of the numerical response had no explicit effects of density, though the full model (equations (3.2.1) and (3.3.3)) did. Future research could evaluate whether one or more demographic rates (such as adult survival and fecundity) are density dependent, as has been reported in wildebeest (*Connochaetes taurinus*) in which annual adult survival was negatively related to per capita food availability (Mduma *et al.* 1999).

The numerical response relationship of r versus food, could be used in a model of trophic interactions, such as described by a modified Lotka–Volterra model (Caughley 1987*b*). That has not been done here as the data in the original study do not allow description of vole dynamics in the absence of owls, and the functional response of owls to a variation in food supply was not described in the original study by Taylor (1994). Such use of the numerical response in a modified Lotka–Volterra model would include a density-dependent response of the owls and their food supply. That could occur in one of two ways: an increase in vole abundance would cause an increase in owl abundance that would cause, in the following year, an increase in predation of voles by owls that may lower vole abundance; hence, there would be negative feedback on vole abundance. Alternatively, use of the full model in equation (3.3.3) describes an instantaneous effect of owl abundance on r within a year.

The sensitivity analysis indicated that a change in annual adult survival had the greatest effect on finite population growth rate (λ). That is similar to the results reported for the northern spotted owl (Lande 1988). The sensitivity analyses for both species ranked the demographic parameters in the same descending order of effects on finite population growth rate: s, l, b and α. The sensitivity of population growth rate to juvenile survival was higher for barn owls, and was presumably associated with the shorter life expectancy for that species. The results of the sensitivity analysis and of the demographic and numerical response analysis demonstrate an important result from this study. The demographic and numerical response analyses show the causes of changes in demographic rates, while the sensitivity analysis shows the effects of changes in demographic rates. Those distinctions show why it is useful to combine aspects of the mechanistic and demographic paradigms in identifying the determinants of population growth rates.

The authors thank The Royal Society, the Novartis Foundation and the Universities of Canberra and Reading for their support. J. Olsen, P. Caley, V. Forbes and P. Bayliss provided useful comments on draft manuscripts.

4

Estimating density dependence in time-series of age-structured populations

4.1. Introduction

Detection and estimation of density dependence is complicated because it usually operates with a time-lag due to intrinsic factors in individual development and life history (May 1973, 1981b; MacDonald 1978; Renshaw 1991; Nisbet 1997; Jensen 1999; Claessen *et al.* 2000) and extrinsic factors in an autocorrelated environment (Williams & Liebhold 1995; Berryman & Turchin 1997), including interspecific ecological interactions (Turchin 1990, 1995; Royama 1992; Turchin & Taylor 1992; Kaitala *et al.* 1997; Ripa *et al.* 1998; Hansen *et al.* 1999). The life history of a species may largely determine the relative importance of intrinsic and extrinsic factors in contributing to time-lags in population dynamics. For short-lived species with high population growth rates, such as some insects, ecological interactions may best explain time-lags longer than a generation (Turchin 1990, 1995; Royama 1992). For long-lived species with low population growth rates, such as many vertebrates, most time-lags may occur within a generation because of life history (Jensen 1999; Coulson *et al.* 2001; Thompson & Ollason 2001). Understanding density dependence has been impeded by the lack of a general quantitative definition that would allow comparisons among species with different life histories and forms of density dependence (Murdoch 1994).

Time-lags in population dynamics caused by life history have not, to our knowledge, previously been incorporated into methods for detecting and estimating density dependence from population time-series (Bulmer 1975; Pollard *et al.* 1987; Turchin 1990, 1995; Royama 1992; Turchin & Taylor 1992; Hanski *et al.* 1993; Dennis & Taper 1994; Zeng *et al.* 1998). However, Jensen (1999) demonstrated by simulation that stochastic fluctuations in the life history of Walleye fish (*Stizostedion vitreum*) could produce the pattern of delayed density dependence detected by autoregression analysis. Coulson *et al.* (2001) and Thompson & Ollason (2001) showed that time-lags in life history are important in explaining temporal patterns of population fluctuations in Soay sheep (*Ovis aries*) and in Northern fulmars (*Fulmarus glacialis*).

Here, we analyse a density-dependent stage-structured life history to derive linearized autoregressive dynamics of small or moderate population fluctuations around a stable equilibrium. We estimate density dependence in observed time-series of avian populations reproducing at discrete annual intervals. Vertebrate species with mean adult body mass greater than 1 kg usually have $r_{max} \leq 0.1$ per year (Charnov 1993) and, even for highly fecund species, such as many fishes, insects and plants, maximum population growth rates are limited by high density-independent mortality (Myers *et al.* 1999). Such species tend to show damped fluctuations around a stable equilibrium (May 1981*b*) and often have a small or moderate coefficient of variation in population size (Pimm & Redfearn 1988; Pimm 1991).

4.2. Definition of density dependence

Consider a simple deterministic population model with no age structure, where individuals that reach the age of one year reproduce and then die, as for univoltine insects or annual plants with no seed bank. With population size in year t denoted as $N(t)$, the dynamics are given by $N(t) = \lambda[N(t-1)]N(t-1)$, where $\lambda[N(t-1)]$ is the density-dependent finite rate of population increase, the product of the probability of survival to maturity and the mean fecundity. We assume that fluctuations in the population size are sufficiently small for a linearized model to have good accuracy. For populations without age structure, a linearized model gives results that are accurate to within 10% if the coefficient of variation is as high as 30% (Lande *et al.* 1999). Denoting the equilibrium population size or carrying capacity as K and the deviation from equilibrium as $x(t) = N(t) - K$, Taylor expansion of λ produces the linearized dynamics

$$x(t) = (1 - \gamma) x(t-1), \tag{4.2.1a}$$

where $\gamma = -(\partial \ln \lambda / \partial \ln N)_K$ is the rate of return toward the equilibrium. In the simple model with no age structure (and a generation time of one year), γ can be used to define the strength of density dependence as the negative elasticity (De Kroon *et al.* 1986; Caswell 2001) of population growth rate with respect to change in population density at equilibrium.

A new measure of total density dependence of the life history of an age-structured population can be defined by interpreting λ in equation (4.2.1a) as the asymptotic multiplicative growth rate of the population per year. Analysis of a general age-structured life-history model with density dependence in age-specific fecundity and first-year survival (Lande *et al.* 2002) indicates that the total density dependence in the life history, D, should be defined as the negative elasticity of population growth rate per generation, λ^T, with respect to change in population density of adults, evaluated at equilibrium. The generation time, T, is the mean age of mothers of newborn individuals when the population is in a stable age distribution (Caswell 2001). Because $\ln \lambda^T = T \ln \lambda$ and at equilibrium $\lambda = 1$ or $\ln \lambda = 0$, we find

$$D = -\left(\frac{\partial \ln \lambda^T}{\partial \ln N}\right)_K = -\left(T\frac{\partial \ln \lambda}{\partial \ln N} + \ln \lambda \frac{\partial T}{\partial \ln N}\right)_K$$
$$= -\left(T\frac{\partial \ln \lambda}{\partial \ln N}\right)_K. \tag{4.2.1b}$$

Equations (4.2.1a,b) show that, with age structure, the asymptotic rate of return to equilibrium is the total density dependence in the life history divided by the generation time at equilibrium, $\gamma = D/T$. This definition of total density dependence in the life history also applies in the stage-structured life history analysed below (§4.3), with density dependence in juvenile and adult survival as well as recruitment.

4.3.   Stage-structured life history

The limited duration of most population time-series requires analysis of a stage-structured life history to reduce the number of parameters to be estimated in comparison with a general age-structured life history (Lande *et al.* 2002). Avian populations often have life histories in which the annual survival and reproductive rates of adults are roughly constant and independent of age (Deevey 1947; Nichols *et al.* 1997). In such populations, most individuals die before reaching the age of senescence, which is therefore of little demographic consequence. We assume, as appears roughly accurate for some populations, that all density dependence is exerted by the adult population density (Sæther 1997). This would apply, at least approximately, if juveniles do not compete with adults or if adults are long-lived and juveniles compose a small fraction of the population.

Defining α as the age at first breeding and $N(t)$ as the number of adults (of age $\geq \alpha$) in year t, the stochastic density-dependent dynamics are described by the nonlinear recursion

$$N(t) = s(N, t-1)N(t-1) + \phi(N, t-\alpha, \ldots, t-1)N(t-\alpha). \quad (4.3.1a)$$

The time dependence of population sizes on the right-hand side of (4.3.1a) is specified in the functional definitions (4.3.1b) and (4.3.1c), below. The probability of adult annual survival is s. The adult annual recruitment rate ϕ is the product of annual fecundity (female offspring per adult female per year) multiplied by first-year survival, f, and the probabilities of annual survival from age i to $i+1$ during the juvenile stages, s_i,

$$\phi(N, t-\alpha, \ldots, t-1) = f(N, t-\alpha)\prod_{i=1}^{\alpha-1} s_{\alpha-i}(N, t-i). \quad (4.3.1b)$$

Environmental and demographic stochasticity affect these vital rates through additive perturbations $\zeta(t)$, $\varepsilon(t)$ and $\delta_\tau(t)$ with $\bar{\zeta} = \bar{\varepsilon} = \bar{\delta}_\tau = 0$,

$$
\begin{aligned}
f(N, t) &= \bar{f}[N(t)] + \varepsilon(t), \\
s_\tau(N, t) &= \bar{s}_\tau[N(t)] + \delta_\tau(t) \quad \text{for } 1 \leq \tau \leq \alpha - 1, \\
s(N, t) &= \bar{s}[N(t)] + \zeta(t).
\end{aligned}
\quad (4.3.1c)
$$

Similar models with or without stochasticity and density dependence have been applied to a variety of species (Caswell 2001, p. 192).

The coefficient of total density dependence in equation (4.2.1b) can be derived by implicit differentiation of the deterministic Euler equation for this life history (Lande 1988), $\bar{\phi}(N)\lambda^{-\alpha} = 1 - \bar{s}(N)/\lambda$, where

$$\bar{\phi}(N) = \bar{f}(N)\prod_{i=1}^{\alpha-1} \bar{s}_i(N)$$

is the adult recruitment rate in the average environment. Finding $\partial\lambda/\partial N$, evaluating the result at equilibrium ($N = K$ and $\lambda = 1$) and using the generation time for this life history at equilibrium in the average environment,

$T = \alpha + \bar{s}/(1-\bar{s})$ (Lande 1988), produces

$$D = -\left(\frac{\partial \ln \bar{f}}{\partial \ln N} + \sum_{\tau=1}^{\alpha-1} \frac{\partial \ln \bar{s}_\tau}{\partial \ln N} + \frac{\bar{s}}{1-\bar{s}}\frac{\partial \ln \bar{s}}{\partial \ln N}\right)_K$$

$$= -\left(\frac{\partial \ln \bar{\phi}}{\partial \ln N} - \frac{\partial \ln \bar{\mu}}{\partial \ln N}\right)_K = -\left(\frac{\partial \ln(\bar{\phi}/\bar{\mu})}{\partial \ln N}\right)_K, \qquad (4.3.2a)$$

where $\bar{\mu} = 1 - \bar{s}$ is the adult mortality rate. Thus, density dependence in the stage-structured life history can be measured by the negative elasticity of the ratio of adult recruitment rate to mortality rate with respect to change in adult population density at equilibrium.

From the Euler equation at the deterministic equilibrium with $\lambda = 1$, it can be seen that the recruitment rate of adults equals their mortality rate. Denoting equilibrium values as $\hat{s} = \bar{s}(K)$ and $\hat{\phi} = \bar{\phi}(K)$,

$$\hat{\phi} = 1 - \hat{s} = \hat{\mu}. \qquad (4.3.2b)$$

Expanding the vital rates in equation (4.3.1a) in Taylor series around K, with the deviation from equilibrium denoted as $x(t) = N(t) - K$, gives the linearized autoregression with up to α time-lags

$$x(t) = \sum_{\tau=1}^{\alpha} b_\tau x(t-\tau) + \xi(t), \qquad (4.3.3)$$

with constant coefficients

$$b_1 = \left[1 + \left(\frac{\partial \ln \bar{s}}{\partial \ln N}\right)_K\right]\hat{s} + \left(\frac{\partial \ln \bar{s}_{\alpha-1}}{\partial \ln N}\right)_K \hat{\phi},$$

$$b_\tau = \left(\frac{\partial \ln \bar{s}_{\alpha-\tau}}{\partial \ln N}\right)_K \hat{\phi} \text{ for } \tau = 2,\ldots,\alpha-1,$$

$$b_\alpha = \left[1 + \left(\frac{\partial \ln \bar{f}}{\partial \ln N}\right)_K\right]\hat{\phi}.$$

The noise term has time-lags of between one year and α years,

$$\xi(t) = \left[\zeta(t-1) + \hat{\phi}\sum_{i=1}^{\alpha-1}\frac{\delta_{\alpha-1}(t-i)}{\hat{s}_{\alpha-i}} + \hat{\phi}\frac{\varepsilon(t-\alpha)}{\hat{f}}\right]K.$$

Hence, even with no autocorrelation in the vital rates, the noise in the autoregression – equation (4.3.3) – will be autocorrelated if the vital rates operating at different time-lags are cross-correlated at a given time.

For species with age at maturity of one year ($\alpha = 1$), the form of the autoregression and the interpretation of the regression coefficient is different, with the new b_1 being the sum of the old b_1 through b_α. Using equation (4.3.2*b*), the autoregressive equation is $x(t) = b_1 x(t-1) + \xi(t)$, where $b_1 = 1 - \hat{\mu} D$ and the noise $\xi(t) = [\zeta(t-1) + \varepsilon(t-1)]K$ has only a single time-lag and no autocorrelation.

Statistical analysis of population dynamics frequently is performed using ln N rather than N (Royama 1992; Turchin 1995). The linearized autoregression for the dynamics of ln N is identical to that for N. This can be seen by dividing both sides of equation (4.3.3) by K, noting that, for small fluctuations, $x/K \approx \ln(1 + x/K) = \ln(N/K) = \ln N - \ln K$.

4.4.   Estimating density dependence

The maximum-likelihood estimator of the autoregression coefficients is identical to the least-squares estimator for a standard regression, if we ignore end effects as proposed by Kendall *et al.* (1983). The autoregression coefficients can be expressed in terms of population autocorrelations for time-lag τ, denoted as ρ_τ. For $\alpha = 1$, the autoregression coefficient is estimated by $b_1 = \rho_1$. More generally, the autoregression coefficients can be estimated as the solution of the Yule–Walker equations (Box *et al.* 1994), $\mathbf{b} = P^{-1}\rho$, where ρ and $\mathbf{b}$ are column vectors with elements $\rho_1, \ldots, \rho_\alpha$ and $b_1, \ldots, b_\alpha$, and P is the population autocorrelation matrix with elements $P_{ij} = \rho_{i-j}$ for $i, j = \{1, \ldots, \alpha\}$ and $\rho_0 = 1$. However, these estimators of the autoregression coefficients are biased because population sizes at a given time enter the autoregression as both dependent and independent variables. This time-series bias can be removed and standard errors and significance tests on the autoregression coefficients can be obtained by computer simulation (Lande *et al.* 2002).

The autoregression coefficients in equation (4.3.3) do not directly reveal the strength of density dependence in population dynamics because the coefficients of density dependence in the vital rates also depend on the life-history parameters $\hat{\phi}$ and $\hat{s}$. For example, even in the absence of density dependence in adult fecundity, $\partial \ln f / \partial \ln N = 0$, the autoregression coefficient for lag α years is not zero but equals the adult recruitment rate at equilibrium, $b_\alpha = \hat{\phi}$.

The time-series analysis described above can be used with life-history estimates of adult annual survival and recruitment rates to determine where in the life cycle density dependence has acted. However, the limited

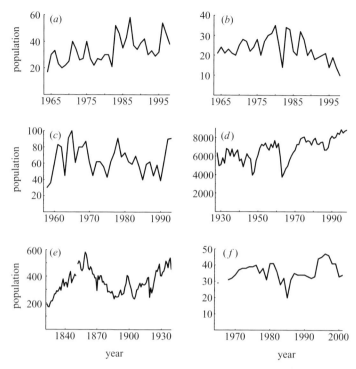

Figure 4.1. Time-series for annual census of adult population in six avian species. (*a*) Great tit (*Parus major*); (*b*) blue tit (*Parus caeruleus*); (*c*) tufted duck (*Aythya fuligula*); (*d*) grey heron (*Ardea cinerea*); (*e*) mute swan (*Cygnus olor*); (*f*) South Polar skua (*Catharacta maccormicki*).

duration of most ecological time-series reduces the statistical accuracy of such assessments (Lande *et al.* 2002).

There are α autoregression coefficients and, from equations (4.3.2) and (4.3.3), the product of the adult mortality rate and the total density dependence in the life history can be estimated, with reasonable accuracy, as one minus the sum of the autoregression coefficients,

$$\hat{\mu}D = 1 - \sum_{\tau=1}^{\alpha} b_{\tau}. \qquad (4.4.1)$$

Time-series were analysed for six avian populations with three or more decades of accurate annual census data and few missing observations (figure 4.1). Counts of the great tit (*Parus major*) and blue tit (*Parus caeruleus*) at Ghent, Belgium, and the tufted duck (*Aythya fuligula*) at Engure Marsh,

Table 4.1. *Bias-corrected estimates of total density dependence, D, with 95% confidence intervals in parentheses.*

(Population time-series fitted to the stage-structured life-history model – equation (4.3.3) – with age of first reproduction α and annual adult survival rate $\hat{s}(= 1 - \hat{\mu})$) obtained from the literature (Clobert et al. 1988; Dhondt et al. 1990; Blums et al. 1993, 1996; Owen 1960; North & Morgan 1979; Bacon & Andersen-Harild 1989; Bacon & Perrins 1991; Jouventin & Guillotin 1979; H. Weimerskirch, unpublished data). ** *p < 0.01 for hypotheses that $\hat{\mu} D > 0$, by one-tailed test. CV, coefficient of variation; R^2, proportion of total variance explained.)*

Species	Years	CV	α	$\hat{\mu} D$	R^2	$\hat{s}$	D
great tit	35	0.30	1	0.535** (0.184, 0.879)	0.16	0.46	0.99
blue tit	35	0.26	1	0.507** (0.157, 0.849)	0.15	0.49	0.99
tufted duck	36	0.27	1	0.564** (0.221, 0.907)	0.14	0.65	1.61
grey heron	71	0.18	2	0.081 (0.000, 0.246)	0.73	0.70	0.27
mute swan	116	0.26	4	0.062 (0.000, 0.184)	0.80	0.85	0.41
South Polar skua	35	0.16	5	0.397** (0.103, 1.114)	0.29	0.85	2.65

Latvia, are for the total adult population (≥ 1 year old). Counts of the tits are almost exact since nearly all pairs breed in nest boxes, but there is considerable exchange of individuals with other populations. The grey heron (*Ardea cinerea*) counts are for the breeding adult population (≥ 2 years old) in southern Britain, which therefore constitutes a relatively closed population. Counts of the mute swan (*Cygnus olor*) on the Thames, England, are for the total population minus fledglings. The time-series for the mute swan was truncated following a period of no data during World War II, after which large population increases occurred in both adults and yearlings (Cramp 1972). Some of the mute swan annual counts may be biased (Birkhead & Perrins 1986) and fledglings were wing-clipped during the counts to reduce emigration (Cramp 1972); this series is included mainly for illustrative purposes because of its length. The South Polar skua (*Catharacta maccormicki*) population at Pointe Géologie archipelago, Terre Adélie, Antarctica, has significant recruits from outside of the archipelago. The strong territoriality of adults helps to ensure that all birds in the archipelago are ringed and the counts of breeding adults are exact. Years without a complete census were excluded from the analysis of the mute swan and South Polar skua.

Employing basic statistical methods for stationary autoregressive time-series analysis, we found evidence of density dependence in four out of the six species (table 4.1). Although the theory indicates that the noise in the population process may be autocorrelated – see equation (4.3.3) – so that, using life-history information, the population time-series could in principle be analysed as an autoregressive moving average process (ARMA, cf. Box *et al*. 1994), residuals from the simple autoregression showed no significant autocorrelations in the noise, justifying the approximation of independent errors in estimation and significance testing.

4.5. Discussion

Turchin (1990, 1995), Royama (1992), Turchin & Taylor (1992), Zeng *et al*. (1998) and others fitted nonlinear autoregressive models with time-lags of one, two or three years to population time-series. They interpreted a significant autoregression coefficient for a time-lag greater than one year as evidence for density dependence with a time-lag. However, their models are phenomenological and not based on explicit demographic mechanisms. Our results – see equation (4.3.3) – demonstrate that the interpretation of autoregression coefficients is clarified by deriving the form of the linearized autoregressive equation from a nonlinear stochastic life-history model. In the stage-structured model, density dependence operates with time-lags up to α years. Contrary to previous interpretations, the autoregression coefficients do not directly measure density dependence operating at particular lags. The autoregressive coefficients depend on parameters of the life history as well as density dependence of a particular stage class.

It is instructive to consider a species with $\alpha = 1$ and autoregression coefficient $b_1 = 0$, which implies that all autocorrelations are zero (except $\rho_0 = 1$), corresponding to a flat power spectrum or white noise process for the population. This would entail very strong density dependence, $D = 1/\hat{\mu}$ (the inverse of the adult annual mortality rate), despite the regression explaining none of the total variance, $R^2 = 0$. The tufted duck approaches this situation, having a low R^2 (table 4.1). Thus, statistical significance of autoregression coefficients is not a valid criterion for the detection of density dependence.

Total density dependence in the life history was significant in four out of the six species. Comparing the strength of total density dependence, D, between species requires correcting the estimates of $\hat{\mu}D$ in table 4.1, through division by the adult annual mortality rate, $\hat{\mu}$. In conjunction with the estimates of $\hat{\mu}D$ derived from autoregression analyses, adult annual survival rates $\hat{s}(=1-\hat{\mu})$ obtained from life-history studies allow estimation of the total density dependence, D, in each of the populations (table 4.1). Strong density dependence occurs in each of the four populations in which significant estimates were obtained. Density dependence for the grey heron and the mute swan, which had the longest time-series, is relatively weak and not significant. The average value of the total density dependence in the six species is $\bar{D} = 1.16$. Thus, on average, a given proportional increase in adult population density, N, produces roughly the same proportional decrease in multiplicative growth rate of the population per generation, λ^T. Because the expected annual rate of return to the equilibrium population size is $\gamma = D/T$, stochastic perturbations from the equilibrium population size for the bird species analysed here all appear to have $0 < \gamma < 1$ and thus to be undercompensated (Begon *et al.* 1996c, pp. 239–240), since even for species with age of maturity equal to one year the generation time is generally larger than D.

Autocorrelation of physical and biotic environments has been discussed as a cause of autocorrelated noise (Williams & Liebhold 1995; Berryman & Turchin 1997). The present theory reveals that environmental covariance in vital rates, operating at different time-lags, creates another source of autocorrelated noise, even in the absence of environmental autocorrelation – see equation (4.3.3). Observed correlations among vital rates (Sæther & Bakke 2000) may be caused both by environmental covariances and by density dependence in the vital rates. Long-term life-history studies of vertebrate species often show that estimates of recruitment of yearlings (reproduction multiplied by first-year survival) are much more variable among years than estimates of adult mortality (Gaillard *et al.* 1998, 2000; Sæther & Bakke 2000), as observed in the tufted duck (Blums *et al.* 1996), grey heron (North & Morgan 1979) and mute swan (Cramp 1972; Bacon & Perrins 1991). This would tend to reduce the environmental covariance of vital rates in the stage-structured model – see equation (4.3.3). The present autoregression analysis, like previous studies (see, amongst others, Turchin 1990, 1995; Royama 1992; Turchin & Taylor 1992; Zeng *et al.* 1998), assumes no autocorrelation of the noise. Residuals from the autoregressions showed no significant autocorrelations, suggesting not only a negligible environmental autocorrelation, but also that environmental covariance of vital rates operating at different time-lags is small. Time-series of at least an order of magnitude longer than the number of autoregression coefficients are required to estimate accurately the total density dependence.

Our results illustrate the advantages of applying demographic theory both to define the total density dependence quantitatively in a life history and to estimate it from population time-series.

We are grateful to André Dhondt, Jenny De Laet and Frank Adriaensen for providing the great tit and blue tit data, the British Trust for Ornithology for providing the grey heron data and Henri Weimerskirch for providing the South Polar skua data. This work was supported by National Science Foundation grant DEB 0096018 to R.L. and by grants to B.-E.S. and S.E. from the Research Council of Norway.

5

Pattern of variation in avian population growth rates

5.1. Introduction

Studies on population growth rates (i.e. how fast population size changes) go back several hundred years, at least to the 16th century when the potential for exponential growth of populations was realized (Caswell 2001). Since then, factors affecting variation in population growth rates have been the major focus for human demography and a major part of population ecology. Although ecologists have dealt with this problem for a long time, surprisingly few generalizations have appeared that allow us to predict variation in population growth rates within and among natural populations. A major reason for this may be difficulties in separating out the relative contribution of density-dependent and density-independent factors on the population growth rate (see reviews in Sinclair 1989; Caughley 1994; Turchin 1995).

The purpose of the present study is to summarize how stochastic effects affect the long-term growth rate of populations with no age structure. We will then extend some recent work (Sæther *et al.* 2002), using data on fluctuations of bird populations, where stochastic factors as well as parameters characterizing the expected dynamics are being separately estimated. We suggest many characteristics of avian population dynamics are closely associated to variation in the specific population growth rate because of the presence of covariation between differences in the expected dynamics and the stochastic component of the fluctuations in population size. Finally, we strongly emphasize that reliable population projections, for example for use in population viability analysis, will not only require estimates and modelling of the expected dynamics as well as the stochastic components, but also assessment of the effects on the predictions of uncertainties in parameter estimates.

5.2.   Stochastic population growth rates

To illustrate the importance of stochastic effects on population dynamics, we consider a simple model describing density-independent growth in a random environment $N(t + 1) = \lambda(t)N(t)$ (Lewontin & Cohen 1969). Here $N(t)$ is population size in year t, and $\lambda(t)$, the population growth rate in year t, is assumed to have the same probability distribution each year with mean $\bar{\lambda}$ and variance σ_λ^2. If the population size is sufficiently large to ignore demographic stochasticity, the environmental stochasticity is $\sigma_e^2 = \sigma_\lambda^2 = \text{var}(\Delta N/N \mid N)$, where ΔN is the change in population size between t and $t + 1$. Population ecologists commonly analyse population fluctuations on the natural-log scale (Royama 1992) so that $X(t + 1) = X(t) + r(t)$, where $X(t) = \ln N(t)$ and $r(t) = \ln \lambda(t)$. If E denotes the expectation, the growth rate during a period of time t, $(X(t) - X(0))/t$, has mean $s = \text{E}r$ and variance $\text{var}(r)/t = \sigma_r^2/t$. For long time intervals the mean slope of the trajectories on log scale all approaches the constant s, which is called the stochastic population growth rate. This leads to the approximation $s \approx \ln \bar{\lambda} - \sigma_r^2/2$ and $\sigma_r^2 \approx \sigma_r^2/\bar{\lambda}^2 = \sigma_e^2/\bar{\lambda}^2$ for the mean and the variance, respectively (Lande *et al.* 2003). As expected, we see by simulating this model that all sample paths are after some time below the trajectory for the deterministic model (figure 5.1). Stochastic effects reduce the mean growth rate of a population on the logarithmic scale by *ca.* $\sigma_r^2/2$, compared with one in a constant environment.

Demographic stochasticity is caused by random fluctuations in individual fitness, which are independent among individuals, giving $\text{var}(\Delta N/N \mid N) = \sigma_e^2 + \sigma_d^2/N$ (Engen *et al.* 1998; Sæther *et al.* 1998), where σ_d^2 is called the demographic variance. It produces a similar reduction in the stochastic growth rate s as the environmental variance σ_e^2, given by

$$s \approx \ln \bar{\lambda} - \frac{\sigma_e^2}{2} - \frac{\sigma_d^2}{2N} \tag{5.2.1}$$

(Lande 1998).

We see that at small population sizes demographic variance creates a deterministic decrease in the long-term growth rate in addition to causing random fluctuations in population size so that $\text{var}(\Delta N \mid N) = \sigma_e^2 N^2 + \sigma_d^2 N$. However, for population sizes of $N \gg \sigma_d^2/\sigma_e^2$ environmental

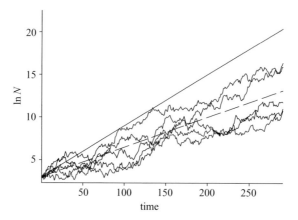

Figure 5.1. Simulation of five populations growing density independently in a random environment, with parameters $r = 0.06$, $\sigma_r^2 = 0.05$ and initial population size $N_0 = 20$. Solid line shows the population growth in a constant environment and the dashed line the expected long-run growth.

stochasticity constitutes the major contribution to stochastic variation in the rate of population growth.

One of the few general patterns in ecology is the dependence of population growth rates on population size (Turchin 1995). In the density-dependent case, the population growth rate at population size N is $r_N = E(\Delta N/N \,|\, N) = E[(\lambda - 1)|\,N)]$ on an absolute scale, and $s_N = E(\Delta \ln N \,|\, N) = E(r \,|\, N)$ on a logarithmic scale. To quantify the effects of density dependence on the population growth rate we use the theta-logistic model (Gilpin & Ayala 1973),

$$E\left(\frac{\Delta N}{N} \,|\, N\right) = r_0 \left[1 - \left(\frac{N}{K}\right)^{\theta}\right],\tag{5.2.2a}$$

where r_0 is the mean specific growth rate of the population at $N = 0$, K is the carrying capacity and θ specifies the density regulation. We can also formulate (5.2.2a) as

$$E\left(\frac{\Delta N}{N} \,|\, N\right) = r_1 \left[1 - \frac{N^{\theta} - 1}{K^{\theta} - 1}\right]\tag{5.2.2b}$$

(Sæther *et al.* 2000a), where r_1 is the mean specific growth rate of the population at $N = 1$. We can then describe a wide variety of density regulation functions by varying only one parameter θ (figure 5.2). For instance, for $\theta = 0$ we get the Gompertz form of density regulation, whereas $\theta = 1$ gives the logistic model (figure 5.2).

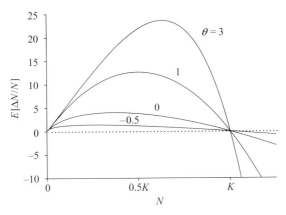

Figure 5.2. The expected change in population size N from time t to $t + 1$, conditioned on N ($E(\Delta N \mid N)$) as function of relative population size (to the carrying capacity K) for different values of θ in the theta-logistic model. Other parameters are $r_1 = 0.1$ and $K = 100$.

The dynamic consequences of variation in r_1 are strongly dependent on the density regulation function (figure 5.3a,b). For a given r_1, the variability decreases with increasing θ. Similarly, an increase in r_1 also reduces population variability if θ is kept constant. This is due to the inverse of the rate of return to the equilibrium at the carrying capacity being

$$\gamma = r_0\theta = r_1\theta/(1 - K^{-\theta}).$$ (5.2.3)

Thus, for a given θ, γ will increase with r_1 and the population will return to equilibrium more rapidly (May 1981b).

One advantage of the theta-logistic model is that it has relatively well understood statistical properties (Gilpin *et al.* 1976; Sæther *et al.* 1996, 2000a; Diserud & Engen 2000). This allows us to calculate approximately the stationary distribution of population size. The variance of this distribution using the diffusion approximation (Karlin & Taylor 1981) is

$$\sigma_N^2 = \frac{K^2\Gamma[(\alpha + 2)/\theta]}{[(\alpha + 1)/\theta]^{2/\theta}\Gamma(\alpha/\theta)}$$ (5.2.4)

where $\alpha = 2r_1/\sigma_e^2(1 - K^{-\theta}) - 1$, and Γ denotes the gamma function (Diserud & Engen 2000). The variance σ_N^2 is strongly influenced by σ_e^2 and θ (figure 5.3c). Furthermore, for small values of θ variation in r_1 strongly affects the stationary distribution of population sizes (figure 5.4), whereas smaller effects on σ_N^2 of variation in r_1 occur for larger θ.

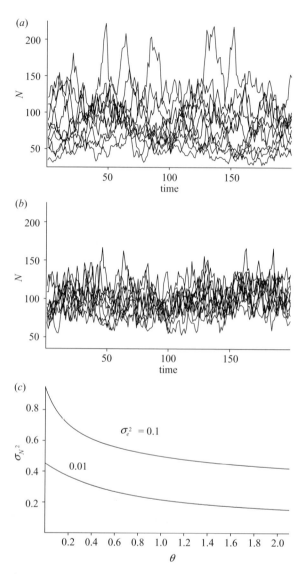

Figure 5.3. Ten trajectories of the population fluctuations during a period of 200 years in the theta-logistic model for (a) $\theta = 0.1$ and (b) $\theta = 1.5$, assuming $\sigma_e^2 = 0.01$. (c) The variance in the stationary distribution of population size (σ_N^2) in relation to θ for $\sigma_e^2 = 0.01$ and $\sigma_e^2 = 0.1$.

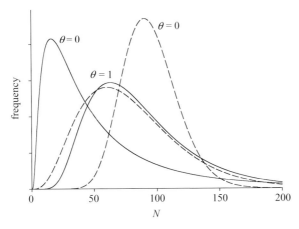

Figure 5.4. The stationary distribution of population sizes in the theta-logistic model – see equation (2.2*a,b*) – for the Gompertz ($\theta = 0$) and logistic ($\theta = 1$) type of density regulation for $r_1 = 0.025$ (solid lines) and $r_1 = 0.1$ (broken lines). Other parameters are $K = 100$ and $\sigma_e^2 = 0.01$.

When demographic stochasticity is present, there is no stationary distribution. For $\sigma_d^2 > 0$, we may compute approximately the quasi-stationary distribution, again using the diffusion approximation (Lande *et al.* 2003). A larger proportion of the distribution lies closer to zero in the quasi-stationary than in the stationary distribution obtained assuming $\sigma_d^2 = 0$ (figure 5.5). Ignoring demographic stochasticity may thus cause serious underestimation of the time to extinction.

Although variation in population growth rates strongly affects the characteristics of the population dynamics, surprisingly little is known about how, for example, r_1 varies among or within species. This may be related to difficulties in estimating r_1 due to the fact that reliable estimates of population growth rates require precise estimates of population size. Uncertain population estimates will lead to over-estimates of r_1 (Solow 1998) as well as the environmental variance. Furthermore, unbiased estimates of population growth rates are also difficult to obtain because many populations fluctuate around K so the population is rarely found at such small population sizes that the growth is close to r_1 (Aanes *et al.* 2002). In addition, large uncertainties will be present in the estimates of K in many time-series that in a statistical sense are short (Myers *et al.* 2001). This makes it even more difficult to obtain reliable unbiased estimates of r_1 because r_1 and K act together on E($\Delta N/N|N$) in (5.2.2*b*).

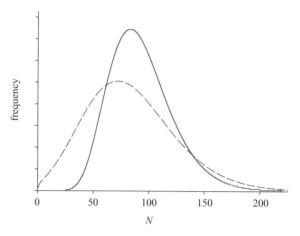

Figure 5.5. The estimated quasi-stationary (broken line) and stationary (solid line) ($\sigma_e^2 = 0$) distributions of the theta-logistic model. The population parameters are $\theta = 0.5$, $r_1 = 0.1$, $K = 100$, $\sigma_e^2 = 0.01$ and $\sigma_e^2 = 1$ (for the quasi-stationary distribution).

5.3.   Density-independent population growth

First, we consider some populations that have strictly declined or increased in size over a period of at least 10 years and where the population counts are assumed to be relatively precise (table 5.1). We estimate the stochastic population growth rate s and σ_r^2 according to Engen & Sæther (2000), following Dennis *et al.* (1991).

When comparing only declining or increasing species, we still find large interspecific variation in s (table 5.1). The most dramatic decline (table 5.1) was recorded in a population of *Pluvialis apricaria* in Scotland (Parr 1992). By contrast, *Branta leucopsis* at the Island of Gotland in the Baltic Sea was the most rapidly increasing population in the dataset, with $\hat{s} = 0.285$ (table 5.1).

There was a positive relationship between the specific growth rate r and $\ln \sigma_r^2$ among the species in the dataset (figure 5.6, correlation coefficient = 0.48, $p = 0.029$, $n = 21$). However, this relationship may be strongly influenced by sampling errors in the parameter estimates. To account for this we performed a meta-analysis (Hunter & Schmidt 1990), a technique that is currently becoming increasingly popular among comparative biologists (e.g. Arnqvist & Wooster 1995). We assume that r and $\ln \sigma_r^2$ are binormally distributed among species. For populations that are not density regulated, it follows from standard normal theory (Kendall &

Table 5.1. *Characteristics of strictly declining or increasing species included in the present study.*
(s *is the stochastic growth rate and* $\sigma_r^2 = \mathrm{Var}[\ln \lambda(t)]$, *where* $\lambda(t)$ *is the population growth rate in year t.*)

Species	Locality	Period	s	σ_r^2	Source
Anas platyrhynchos	Engure Lake, Latvia	1958–1993	0.094	0.152	Blums *et al.* (1993)
Aptenodytes forestri	Terre Adélie, Antarctica	1952–1999	−0.014	0.026	H. Weimerskirch (personal communication)
Branta leucopsis	Gotland, Sweden	1976–2001	0.285	0.065	K. Larsson (personal communication)
Bubulcus ibis	Camargue, France	1967–1998	0.267	0.375	H. Hafner (personal communication)
Ciconia ciconia	Böhmen and Mähren, Czech Republic	1930–1984	0.071	0.034	Hladik (1986)
Ciconia ciconia	Estonia	1939–1981	0.034	0.015	Veromann (1986)
Ciconia ciconia	Denmark	1952–1971	−0.074	0.012	Skov (1986)
Ciconia ciconia	Netherlands	1929–1981	−0.068	0.090	Jonkers (1986)
Ciconia ciconia	Baden-Württemberg, Germany	1950–1965	−0.054	0.017	Engen & Sæther (2000) (from Bairlein & Zink 1979)
Circus aeruginosus	Britain	1927–1982	0.057	0.294	Underhill-Day (1984)
Diomedea exulans	Bird Island, South Georgia	1962–1996	−0.006	0.010	Croxall *et al.* (1997)
Gymnogyps californianus	California, USA	1965–1980	−0.077	0.120	Dennis *et al.* (1991)
Haliaeetus albicilla	Germany	1976–1997	0.044	0.003	Hauff (1998)
Hirundo rustica	Jutland, Denmark	1984–1999	−0.076	0.024	Engen *et al.* (2001)
Milvus milvus	central Wales	1951–1980	0.040	0.008	Davis & Newton (1981)
Otis tarda	Brandenburg, Germany	1975–1995	−0.076	0.012	Litzbarski & Litzbarski (1996)
Pandion haliaetus	Scotland	1958–1994	0.126	0.027	Dennis (1995)
Perdix perdix	Sussex, England	1970–2000	−0.058	0.069	Aebischer (personal communication)
Phalacrocorax aristotelis	Isle of May, Scotland	1946–1992	0.146	0.118	Harris *et al.* (1994)
Phalacrocorax carbo	Vorsø, Denmark	1978–1993	0.116	0.011	Van Eerden & Gregersen (1995)
Phalacrocorax carbo	Ormsö, Denmark	1978–1993	0.206	0.055	Van Eerden & Gregersen (1995)
Pluvialis apricaria	northeast Scotland	1973–1989	−0.176	0.134	Parr (1992)
Recurvirostra avosetta	North Sea coast, Germany	1951–1994	0.091	0.068	Halterlein & Sdbeck (1996)
Sula bassanus	Alisa Craig, Ayrshire, Scotland	1936–1976	0.0269	0.004	Nelson (1978)

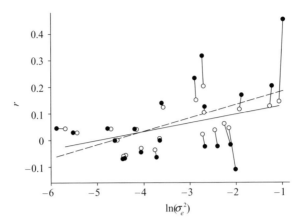

Figure 5.6. The logarithm of the environmental variance σ_r^2 in relation to the specific growth rate r, assuming no density regulation. Solid lines and circles represent the initial analysis, broken line and open circles the meta-analysis that takes into account uncertainties in the parameter estimates (see text). For sources, see table 5.1.

Stuart 1977) that the sampling errors in these two parameters are independent. The uncertainty in the estimate of r must be estimated from the data, whereas the variance in the estimate of $\ln \sigma_r^2$ only depends on the sample size and can be calculated theoretically using well-known properties of the χ^2-distribution. Finally, assuming that the sampling errors are approximately normally distributed, we calculated the conditional expectation of the values for each single species, conditioning on the raw estimates. The results of these adjustments of the estimates are shown in figure 5.6. The estimated correlation in the bivariate between-species model was 0.49, with significance level $p = 0.0580$ calculated by parametric bootstrapping. This therefore suggests the presence of an interaction between the expected dynamics and the stochastic component of the population fluctuations.

Overall, a large proportion of the variation in the stochastic population growth rate s was explained by differences in r (correlation coefficient $= 0.95, p < 0.001, n = 21$). However, the stochastic contribution to s differed between declining and increasing populations. In declining populations, σ_r^2 contributed significantly to the variation in s (correlation coefficient $= -0.66, p = 0.039, n = 10$), whereas no such significant effect was present in increasing species (correlation coefficient $= 0.40, p > 0.2, n = 11$). This shows that not only the expected dynamics, but also quantifying the stochastic effects are important for predicting the rate of decline in such decreasing populations.

Variation in s among populations was not significantly ($p > 0.2$) related to differences in life history (clutch size or adult survival rate), either in declining or in increasing species.

5.4. Density-dependent population growth

Traditionally, population ecologists have analysed population dynamics by fitting autoregressive models to time series of fluctuations in population size (Bjørnstad & Grenfell 2001). Linearity at a logarithmic scale is often assumed (Royama 1992). Here, we take an alternative approach by estimating the parameters in the theta-logistic model (equation (5.2.2a,b)) by maximum-likelihood methods (see Sæther *et al.* (2000a) and Sæther & Engen (2002b) for a description of the methods). This enables us to estimate the form of the density regulation (see figure 5.3c).

In 11 populations where $\theta > 0$, the estimates of θ ranged from 0.15 to 11.17 (table 5.2). This implies that the assumption of log-linearity ($\theta = 0$) is not necessarily fulfilled in time-series of fluctuations of bird populations (e.g. Sæther *et al.* 2002). Large uncertainty was found in most estimates of r_1 with wide confidence intervals of $\lambda_1 = \exp(r_1)$ in several species (figure 5.7). This shows that r_1 (or λ_1) is often difficult to estimate even in comparatively long time-series of population fluctuations.

Even though we find large uncertainties in the estimates, a pattern of covariation appeared among several parameters of the theta-logistic model. As in the density-independent cases (figure 5.6), the environmental variance increased with r_1 (figure 5.8a, correlation coefficient = 0.75, $n = 11$, $p = 0.008$). Furthermore, the logarithm of θ decreased with r_1 (figure 5.8b, correlation coefficient = −0.69, $n = 11$, $p = 0.018$), showing that maximum density regulation occurs at lower relative (to K) population sizes in species with higher population growth rates. The estimates of γ, the strength of density regulation at K (equation (5.2.3)) also increased with σ_e^2 (figure 5.8c, correlation coefficient = 0.61, $n = 11$, $p = 0.048$).

We then examined how variation in the parameters affected interspecific variation in the variance of the stationary distribution (see equation (5.2.3)). Differences in σ_N^2 were only significantly correlated with σ_e^2 (correlation coefficient = 0.62, $p = 0.043$, $n = 11$).

The interspecific variation in r_1 was also correlated with life-history differences. Lower growth rates were found in species with high survival

Table 5.2. *Estimates of parameters in density regulated bird populations.*
(r, specific growth rate; θ, the format density regulation; σ_ε², environmental stochasticity; γ, strength of density dependence at the carrying capacity, K.)

Species	Locality	Period	r_1	θ	σ_ϵ^2	γ	Source
Anser caerulescens	Perouse Bay, Canada	1970–1987	0.09	2.05	0.041	0.58	Cooch & Cooke (1991)
Aythya ferina	Engure Lake, Latvia	1970–1987	0.82	0.15	0.276	0.82	Blums et al. (1993)
Ciconia ciconia	Böhmen and Mähren, Czech Republic	1962–1984	0.09	3.74	0.014	0.37	Hladik (1986)
Haematopus ostralegus	Mellum, Germany	1946–1968	0.74	0.65	0.065	0.56	Schnakenwinkel (1970)
Lagopus lagopus	Kerloch, Scotland	1963–1977	1.50	0.47	0.411	1.94	Moss & Watson (1991)
Melospiza melodia	Mandarte Island, Canada	1975–1998	0.99	1.09	0.437	2.25	Sæther et al. (2000a)
Parus caeruleus	Braunschweig, Germany	1964–1993	1.16	0.54	0.114	1.50	Winkel (1996)
Phalacrocorax aristotelis	Isle of May, Scotland	1946–1992	0.55	0.33	0.072	0.36	Harris et al. (1994)
Phalacrocorax carbo	Vorsö, Denmark	1978–1993	0.17	11.17	0.005	2.10	Van Eerden & Gregersen (1995)
Recurvirostra avosetta	Havergate, England	1947–1986	1.04	0.25	0.104	0.39	Hill (1988)
Rissa tridactyla	North Shields, England	1949–1984	0.17	2.05	0.014	0.37	Porter & Coulson (1987)

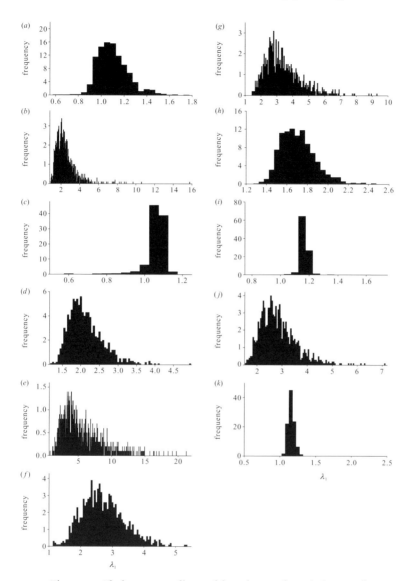

Figure 5.7. The bootstrap replicates of the estimates of population growth rate $\lambda_1 = \exp(r_1)$ for the populations listed in table 5.2. (*a*) *Anser caerulescens*; (*b*) *Aythya ferina*; (*c*) *Cicconia cicconia*; (*d*) *Haematopus ostralegus*; (*e*) *Lagopus lagopus*; (*f*) *Melospiza melodia*; (*g*) *Parus caeruleus*; (*h*) *Phalacrocorac aristotelis*; (*i*) *Phalacrocorax carbo*; (*j*) *Recurvirostra avosetta*; (*k*) *Rissa tridactyla*.

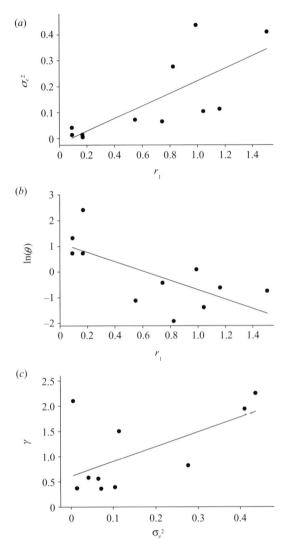

Figure 5.8. (*a*) Environmental variance σ_e^2, (*b*) the logarithm of θ in relation to the specific population growth rate r_1, and (*c*) strength of density regulation at the carrying capacity γ in relation to σ_e^2. For sources, see table 5.2.

rates (figure 5.9*a*, correlation coefficient $= -0.71$, $n = 9$, $p = 0.031$) and small clutch sizes (figure 5.9*b*, correlation coefficient $= 0.68$, $n = 11$, $p = 0.021$) than in species at the other end of this 'slow–fast continuum' (Sæther & Bakke 2000).

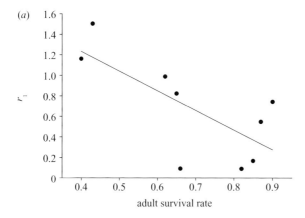

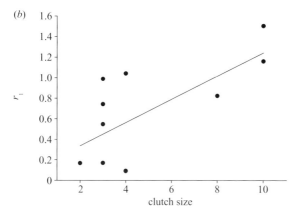

Figure 5.9. The specific population growth rate r_1 in relation to (*a*) adult survival rate and (*b*) clutch size in density regulated bird populations. For sources, see table 5.2.

5.5. Predicting population fluctuations

Population viability analysis has during the last decades become an important tool in the management of threatened or vulnerable species (see reviews in Beissinger & Westphal 1998; Groom & Pascual 1998) because it provides a quantitative assessment of the probability for a population to decline to (quasi-)extinction. Our analyses show that large uncertainties are often found in the estimates of important parameters for population viability analysis – analyses such as the form of the density regulation (Sæther *et al.* 2000*a*) and r_1 (figure 5.7), even in time-series of a length that are rarely available in populations of endangered and threatened species. Predictions from population viability analyses must therefore take into account these uncertainties. Elsewhere, we have suggested (Sæther *et al.* 2000*a*; Engen & Sæther 2000; Engen *et al.* 2001; Sæther & Engen 2002*b*) that the concept of population prediction interval can be useful in such analyses, embracing the effects of the expected dynamics and stochastic factors as well as uncertainties in parameter estimates on future population trajectories. The population prediction interval is defined as a stochastic interval that includes the unknown population size with a given probability $(1 - \alpha)$. We define extinction to occur when $N = 1$. We then predict extinction in a sexual population to occur after the smallest time at which this interval includes the extinction barrier. The width of the population prediction interval increases with the process variance (Heyde & Cohen 1985) and the estimation error. It is important to notice that the uncertainty in the parameters does not change the extinction risk of the population, but it will affect the confidence we have in the population predictions, including the probability of extinction.

We will illustrate our approach by an analysis of factors affecting the width of the population prediction interval in two populations of passerine birds. Many species of birds breeding in areas of Europe with a highly intensified agriculture have declined in numbers (Pain & Pienkowski 1997; May 2000; Møller 2001). In a study of *Hirundo rustica* on Jutland, Engen *et al.* (2001) predicted the time to extinction of such a population that declined from 184 in 1984 to 59 pairs in 1999. The estimates of the population parameters were $\hat{s} = -0.078$, $\hat{\sigma}_d^2 = 0.18$ and $\hat{\sigma}_e^2 = 0.024$, respectively. Ignoring uncertainties in the estimates, the width of the 90% population prediction interval included the extinction barrier after 28 years (figure 5.10*a*). Including uncertainties in the parameter estimates, this time was decreased by 25% to 21 years (figure 5.10*b*). Furthermore, in studies of time to extinction the demographic variance is assumed to be zero. Ignoring the demographic variance, the upper 90% population prediction interval included extinction first after 33 years (figure 5.10*c*). This occurs because the lack of demographic variance fails to account for the acceleration of the final decline to extinction produced by demographic stochasticity (Lande *et al.* 2003).

Similarly, in a widely fluctuating small population of *Melospiza melodia* in Mandarte Island accounting for uncertainty produced a more precautionary assessment of extinction than under the assumption of exactly known parameters (Sæther *et al.* 2000*a*). Again, using the upper 90%

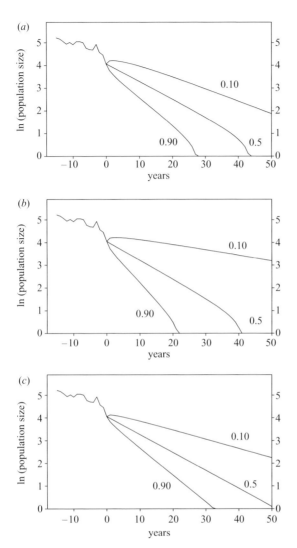

Figure 5.10. The upper bounds of population prediction intervals for a declining population of *Hirundo rustica* in Jutland, Denmark, when (*a*) assuming no uncertainty in the parameters, (*b*) including demographic and environmental stochasticity as well as uncertainties in the parameters, and (*c*) assuming no demographic stochasticity, $\sigma_e^2 = 0$.

population prediction interval the shortest time in which the population prediction interval included the extinction boundary $n = 1$ was 17 years whereas with no uncertainty in parameter estimates this interval was increased to 30 years.

5.6.   Discussion

This study demonstrates a strong covariation among the specific growth rate r_1 and other parameters characterizing fluctuations of bird populations. Both in density-independent and in density-dependent models the population growth rate increased with the environmental variance σ_e^2 (figures 5.6 and 5.8a). Large uncertainties were found in most estimates of r_1 in the density-dependent populations (figure 5.7). However, a relationship still appeared between r_1 and the form of the density regulation, expressed by θ in the theta-logistic model (figure 5.8b). Thus, maximum density regulation occurred at larger densities (relative to K) in populations with a small r_1 than in populations with larger values of r_1. A similar pattern has previously been noted by Fowler (1981, 1988). Furthermore, populations most rapidly growing at smaller densities were more subject to random fluctuations in the environment than populations with smaller values of r_1 (figure 5.8a).

A strong pattern of covariation occurs in avian life-history traits (Sæther 1988; Sæther & Bakke 2000; Bennett & Owens 2002), which can be considered as representing a 'slow–fast continuum' of life-history adaptations. At one end of this dimension we find species that are reproducing at a high rate, but with a short life expectancy at birth. At the other end, species that produce a small number of offspring, mature late in life but have a high adult survival rate, are found. This study shows that in density regulated populations a large proportion of the variance in the specific growth rate r_1 can be explained by the position of the species along this 'slow–fast continuum' (figure 5.9). The specific growth rate was higher in populations with large clutch sizes and low adult survival rates than in less fecund species with long life expectancy at maturity. This is in contrast to fishes, where Myers et al. (1999) failed to identify any relationship between r_1 and any life-history characteristics. The difference among these two studies may be related to larger uncertainties in the estimates of r_1 in fishes than in birds. Furthermore, relative smaller interspecific variation in r_1 may also be present in fishes than in birds.

The covariation among the specific growth rate r_1, environmental variance σ_e^2 and life-history variation (figures 5.8*a* and 5.9) suggests the presence of some common underlying demographic process affecting the expected as well as the stochastic effects on the population dynamics of birds. In solitary birds, we have suggested that spatial limitation of access to breeding size may provide such a general mechanism (Sæther *et al.* 2002). In territorial long-lived species recruitment will be limited by the access to vacant territories. A negative relationship between annual variation in the number of recruits and adult recapture rate is expected in species with such a survival-restricted demography. By contrast, in short-lived species, a positive relationship is expected because high survival is also likely to give a high number of recruits because of common effects of the environment and small age-specific variation in density-dependent effects. Such fluxes of new recruits are likely to affect the population growth rate even at small densities, giving small values of θ. By contrast, in long-lived territorial species, variation in the number of recruits at small densities will only add to the pool of surplus individuals or to occupation of new territories, which probably has a small effect on the population growth. Around K adding extra birds probably increases mortality and rapidly leads to population decline, thus giving a large θ. As a consequence, θ was negatively related to the correlation coefficient between annual variation in the number of recruits and adult recapture rate.

Even though interspecific variation in r_1 was related to other population dynamical parameters (figure 5.8), differences in variability of the density regulated populations were not correlated to r_1. The interspecific differences in the variance of the stationary distribution (standardized in relation to K) of the populations included in this dataset were only significantly related to the environmental variance σ_e^2. Thus, environmental stochasticity rather than variation in the expected dynamics was the major contribution to the interspecific differences in population variability recorded in the present study. This emphasizes the need for estimating and modelling stochastic effects for the understanding of avian population dynamics. Whether this relationship can explain variation in population dynamics in other taxa as well remains to be seen.

An important future challenge when conducting different environmental impact analyses will be to predict the consequences of different management actions on the future dynamics of the population in concern. Such an ability to correctly predict future population fluctuations is a prerequisite for conducting a population viability analysis of threatened and endangered populations. The results presented here (figure 5.10) suggest that development of population projections beyond a short time-span will be difficult for many species because of strong stochastic effects on the population dynamics as well as large uncertainties in the estimates of population parameters such as r_1 (figure 5.7). We strongly recommend that uncertainties in the population projections are evaluated, for instance by using the concept of the population prediction interval, when reporting the results of such analyses.

We are grateful to N. Aebischer, H. Hafner, K. Larsson and H. Weimerskirch for providing unpublished data. The study was financed by grants from the European Union (project METABIRD) and the Research Council of Norway.

6

Determinants of human population growth

6.1. Factors driving previous population growth

The broad outlines of the history of global population growth are by now familiar. At the dawn of the agricultural revolution (8000 years before present), total population was about 250 000 (Cook 1962). It took all of human history (until 1800) for global population to reach one billion – roughly today's population of Europe and North America combined. It took 130 years (until 1930) to reach two billion. It took only 60 more years (1960) to reach three billion. The fourth billion was reached between 1960 and 1975, the five billion mark was passed in 1987 and the six billion mark was reached in 1999.

Less well appreciated are the facts that both the annual growth rate and the annual absolute increment of world population have passed their peaks and are expected to continue to decline. The growth rate peaked at 2.1% per year in the late 1960s and fell to 1.35% by 2000 (see table 6.1), and the annual absolute increment to population peaked at about 87 million per year in the late 1980s and was about 81 million at the end of the 20th century. This does not mean, of course, that little further population growth is to be expected; most mid-range population projections foresee future population rising to 8–10 billion by the end of the 21st century.

As shown in table 6.1, the TFR (average number of children per woman under a period perspective) declined modestly in most parts of the world from 1950–1955 to 1970–1975, then declined over the following 25 years with a rapidity that was unimaginable in the 1960s. This second period of decline was especially pronounced in Asia, where TFR fell by more than two children per woman (a statistic that is, however, heavily influenced by a dramatic fertility decline in China during the 1970s). One exception

Table 6.1. *Demographic trends in the world since 1950 by main regions.*
(From United Nations 2001.)

Region	Total population size (millions)			Growth rate (%)			Life expectancy (both sexes, in years)			Total fertility rate		
	1950	1970	2000	1950–1955	1970–1975	1995–2000	1950–1955	1970–1975	1995–2000	1950–1955	1970–1975	1995–2000
world	2519	3691	6057	1.79	1.93	1.35	47	58	65	5.0	4.5	2.8
MDCs	814	1008	1191	1.20	0.78	0.30	66	71	75	2.8	2.1	1.6
LDCs	1706	2683	4865	2.06	2.35	1.62	41	55	63	6.2	5.4	3.1
Africa	221	356	794	2.17	2.61	2.41	38	46	51	6.7	6.7	5.3
Asia	1399	2142	3672	1.93	2.25	1.41	41	56	66	5.9	5.1	2.7
Europe	548	657	727	0.99	0.59	−0.04	63	67	69	2.7	2.2	1.4
Latin America and Caribbean	167	285	518	2.65	2.45	1.56	51	61	69	5.9	5.0	2.7
North America	172	231	314	1.70	0.97	1.04	69	72	77	3.5	2.0	2.0
Oceania	13	19	31	2.18	2.08	1.37	61	66	74	3.9	3.2	2.4

has been Africa, where fertility rates remained well above six children per woman on average through the late 1980s; since then, the beginnings of a fertility decline have become apparent. Meanwhile, regions such as Europe and North America that had already achieved very low fertility by 1970–1975 saw these rates persist or fall further.

During the 1950s and 1960s, reductions in mortality resulting from the spread of modern hygiene and medicine were even more significant than fertility declines. During the period 1950–1955 (the first period for which estimates are available), life expectancy was lowest in Africa (38 years) and Asia (41 years), while it had already improved significantly in Latin America (51 years). Over the following 20 years life expectancy increased impressively in all parts of the world. In Asia, by far the most populous continent of the world, it increased by 15 years over this short period. In Africa, it improved by 8 years, although this increase was below the world average. Improvements continued over the next 25 years to 1995–2000, but at a somewhat slower speed, with Asia, Latin America and Africa (even with AIDS) seeing substantial improvements.

These trends in fertility and mortality resulted in different patterns of population growth in different parts of the world. In fact, the dominant feature of the global demographic landscape has been the contrast between the well-off populations of Europe, North America and Japan and the poorer populations of Asia, Africa, the Middle East and Latin America. The population of the MDCs is relatively small (about 1.2 billion in 2000) and expanding very slowly (0.3% per year) following a 46% increase since 1950 (see table 6.1). That of the LDCs is large (*ca.* 4.9 billion in 2000) and expanding rapidly (1.6% per year) after increasing by a factor of 2.9 since 1950. As a consequence, the share of today's industrialized countries in the world population decreased from 32% in 1950 to 20% in 2000 and is likely to decrease much more in the future. In addition, despite the rapid changes in most LDCs, inhabitants of MDCs on average live significantly longer (life expectancy at birth for both sexes combined is *ca.* 75 years, versus 63 years in LDCs) and have fewer children (TFR is 1.6, versus 3.1 in LDCs).

The widely varying historical experiences of the different regions of the world have also left a strong imprint on the age structure of their populations. In Africa, age distribution is typical of a rapidly growing population, showing larger and larger cohorts in the young age groups. There are more than twice as many children under age 5 than adults aged 20–25, four times more than those aged 40–45 and 10 times more than the elderly aged 65–70. In Western Europe the pattern is completely different: the number of women aged 60–65 approximately equals the number of children under age 5, while the largest age groups are those between 30 and 40. The age pyramid is narrower at the bottom due to the very low levels of fertility since the 1970s; at the same time, declining mortality rates have widened the top by increasing the size of older age cohorts.

The narrowing of population pyramids at the bottom (from low fertility) and widening at the top (due to extended longevity) is called 'population ageing'. The two components are referred to as ageing 'from the bottom' and 'from the top'. Population ageing is an enormously important social phenomenon, especially in relation to the uncertain future of pension and health care systems. Ageing will continue in MDCs and has already started in LDCs. Just as the speed of mortality improvements accentuated the implications of demographic transition for population growth rates, the speed of LDC fertility decline will accentuate the ageing phenomenon. Both trends are part of the secular change called demographic transition.

The demographic transition began in MDCs in the late 18th century and spread to LDCs in the last half of the 20th (Notestein 1945; Davis 1954, 1991; Coale 1973). The conventional 'theory' of demographic transition predicts that, as living standards rise and health conditions improve, first mortality rates decline and then, somewhat later, fertility rates decline. Demographic transition 'theory' has evolved as a generalization of the typical sequence of events in what are now MDCs, where mortality rates declined comparatively gradually beginning in the late 1700s and then more rapidly in the late 1800s and where, after a lag of 75 to 100 years, fertility rates declined as well. Different societies experienced transition in different ways and today, various regions of the world are following distinctive paths (Tabah 1989). Nonetheless, the broad result was, and is, a gradual transition from a small, slowly growing population with high mortality and high fertility to a large, slowly growing population with low mortality and low fertility rates. During the transition itself, population growth accelerates because the decline in death rates precedes the decline in birth rates.

On a theoretical level there are two different ways to explain demographic transition. One views the fertility decline as a direct response to the mortality decline. This so-called 'homeostasis argument' stresses that societies tend to find an equilibrium between births and deaths. When death rates decline due to progress in medicine and better living

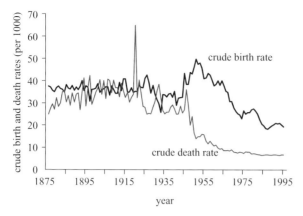

Figure 6.1. Birth and death rates in Mauritius since 1871. Grey line, crude death rate; black line, crude birth rate. (Source: Mauritius Central Statistical Office.)

conditions, the equilibrium is disturbed and the population grows unless birth rates adjust to the new mortality conditions and also start to decline. The fact that fertility tends to decline many years after mortality may be explained by a perception lag. The other view assumes that modernization of society acts as a joint driving force of declining mortality and fertility. Fertility decline lags mortality decline, according to this view, because fertility is more strongly embedded in the system of cultural norms and therefore changes more slowly than mortality relevant behaviour. The historical record of Europe – where fertility sometimes declined simultaneously with mortality and population growth was generally much lower than in today's high fertility countries (Coale & Treadway 1979; Coale & Watkins 1986) – gives more support to the second explanation. But the two arguments are not necessarily mutually exclusive.

Figure 6.1 illustrates the demographic transition in Mauritius, a developing country that has good records for birth and death rates for more than a century. Up to around World War II, birth and death rates show a pattern of strong annual fluctuations, due mostly to diseases and changing weather conditions, which are typical for 'pre-modern' societies. Whenever birth rates are consistently above death rates, the population grows, as was the case in Mauritius during the late 19th century. After World War II, death rates on Mauritius declined precipitously due to malaria eradication and the introduction of European medical technology. Birth rates, on the other hand, remained high or even increased somewhat due to the better health status of women (a typical phenomenon in the early phase of demographic transition). By 1950 this had resulted in a population growth rate of more than 3% per year, one of the highest at that time. Later, birth rates declined, with the bulk of the transition occurring during the late 1960s and early 1970s when TFR declined from more than six to less than three children per woman within only 7 years, probably the world's most rapid national fertility decline. It happened on a strictly voluntary basis and was a result of high female educational status together with successful family planning programmes (Lutz 1994). Because of the still very young age structure of the Mauritian population, current birth rates are still higher than death rates and the population is growing by *ca.* 1% per year despite fertility around replacement level (i.e. two surviving children or TFR somewhat above 2.0, depending on mortality conditions).

Empirically observed trends in all parts of the world have overwhelmingly confirmed the relevance of the concept of demographic transition to LDCs (Tabah 1989; Cleland 1996; Westoff 1996; United Nations 2001). With the exception of pockets where religious or cultural beliefs are strongly pro-natalist, fertility decline is well advanced in all regions except sub-Saharan Africa, and even in that region many signs of a fertility transition can be perceived. In South East Asia and many countries in Latin America, fertility rates are on par with those in MDCs only several decades ago, and in several countries such as China, Taiwan and Korea, fertility is at sub-replacement levels.

The biggest difference between the demographic transition process in what are now MDCs and LDCs has been the speed of mortality decline. Mortality decline in Europe, North America and Japan came about over the course of two centuries as a result of reduced variability in the food supply, better housing, improved sanitation and, finally, progress in preventive and curative medicine. Mortality decline in LDCs, by contrast, occurred very quickly as a result of the application of Western medical and public health technology to infectious, parasitic and diarrhoeal diseases since World War II. Life expectancy in Europe rose gradually from about 35 years in 1800 to about 50 years in 1900, 66.5 years at the end of World War II and 74.4 years in 1995. In LDCs, it shot up from 40.9 years at the end of World War II to 63 years in 2000. The increase that took MDCs about one and a half centuries to achieve came to pass in LDCs in less than half a century. As a result of the speed of the mortality decline, populations in LDCs are growing three times faster today than did the populations of the present MDCs at the comparable stage of their own demographic transition.

Studies of the factors influencing changes in fertility must begin with the proximate determinants of fertility: (i) age at marriage (or beginning of sexual activity); (ii) prevalence and effectiveness of contraception; (iii) prevalence of induced abortion; and (iv) duration of postpartum infecundability, especially due to breast feeding (Bongaarts & Potter 1983). Fertility decline must come through changes in one or more of these four proximate determinants.

The adoption of contraception has been the principal source of fertility decline in LDCs. However, how couples adopt contraceptive practices is a function of many influences. The spread of contraceptive practice is a diffusion process consisting of stages of awareness, information, evaluation, trial and adoption. All of these stages consist of actions undertaken in social networks, leading to path dependence and the persistence of heterogeneity between sub-populations (Kohler 1997). Coale (1973) lists three 'preconditions' required for fertility decline. First, fertility must be regarded as being within the realm of conscious choice. Often, this marks a fundamental change in the way individuals view their lives and their families (Lockwood 1997; Van de Walle 1992); for example, people may change from having a fatalistic attitude toward fertility to making procreation an object of their life-course planning. Yet, in most demographic transitions, fertility regulation was already practised during the pre-transition phase, albeit more for spacing than for limiting the final number of children (Mason 1997). Therefore, second, there must be objective advantages to lower fertility. Third, acceptable means of fertility reduction must be at hand. These three preconditions for a lasting fertility decline suggest three parallel strategies to foster the transition from high to low fertility.

(i) Emphasize universal basic education to bring fertility increasingly into the realm of conscious choice. Modern mass media may also exert an important influence. These strategies are also likely to bring about attitudinal and cultural change.

(ii) Pursue changes in socioeconomic variables, mostly neoclassical economic costs and benefits arising from variables such as child labour, female participation in the modern-sector labour force, support in old age, etc. Changes in the 'value' of children also impact on couples' desired family size.

(iii) Invest in reproductive health and the availability of family planning services, including maternal and child health programmes that reduce infant mortality. Help women match their desired and actual number of children by focusing on the unmet need for family planning.

This framework suggests that if two of the three preconditions are already met, the introduction of the third may trigger a rapid fertility decline. In the previously described case of the rapid Mauritian fertility decline, the young female population was already literate and large families were increasingly perceived as an economic burden. The strong and strictly voluntary family planning campaign that strengthened the negative perception of high fertility and provided efficient family planning services that were even supported by the influential Roman Catholic church (supporting only the 'natural' ones) then triggered the precipitous fertility decline. In some other countries huge investments in family planning were virtually without effect because one of the other two preconditions was not met.

6.2.   Expected future population growth

The human population can be projected for several decades with rather high accuracy because most of the people who will be alive in 20–30 years have already been born, and we know their cohort size. All long-term global population projections employ the cohort-component method. Initial populations for countries or regions are grouped into cohorts defined by age and sex, and the projection proceeds by updating the population of each age- and sex-specific group according to assumptions about three components of population change: fertility, mortality and migration.

Development of this approach was the major innovation in the evolution of projection methodology. It was first proposed by the English economist Edwin Cannan (1895) and was then reintroduced by Whelpton (1936), formalized in mathematical terms by Leslie (1945) and first employed in producing a global population projection by Notestein (1945). Prior to the mid-20th century, the few global population projections that had been made were based on extrapolations of population growth rate applied to estimates of the total population of the world (Frejka 1994). Since Notestein's 1945 projection, the cohort-component method has become the dominant means of projecting population and has remained essentially unchanged. The real work in producing projections lies not in refining the mechanics of the model itself, but in estimating population size and age structure in the base period and in forecasting future trends in fertility, mortality and migration.

Fertility has the greatest effect on population growth because of its multiplier effect – children born today will have children in the future, and so on. Both the projected pace of fertility decline and the assumed eventual fertility level are important in determining trends in population size and age structure. The two factors also interact – the lower the assumed eventual fertility level, the more important the pace of fertility decline becomes to projected population size (O'Neill *et al.* 1999). Mortality projections are typically based on projecting life expectancy at birth. Projections of mortality must specify how the distribution of mortality over different age and sex groups may change over time. Changes in mortality at different ages have different consequences for population growth and age structure. When child and infant mortality decline, for example, a greater proportion of babies will survive to adulthood to have their own children and contribute to future growth. Mortality declines among the older population have a more short-term effect on population growth because the survivors are already past reproductive age.

Future international migration is more difficult to project than fertility or mortality. Migration flows often reflect short-term changes in economic, social or political factors, which are impossible to predict. And, since no single, compelling theory of migration exists, projections are generally based on past trends and current policies, which may not be relevant in the future.

Projection results can be produced in one of two forms: as a set of scenarios or, more recently, as probability distributions. Population projections according to alternative scenarios, called variants in some cases, show what the future population would be if fertility, mortality and migration follow certain paths. The common practice by many statistical offices and the United Nations (UN) Population Division is to define high and low variants (in addition to the median variant), which are based on alternative fertility assumptions and are supposed to cover a 'plausible range' of future population trends. Such a definition of variants is not only imprecise but also disregards the significant uncertainties associated with future mortality and migration trends, and is inconsistent when inferences are being made from national ranges to a global range (National Research Council 2000). While the ranges given by the high and low variants are therefore not very useful, the medium variants are usually taken as best-guess forecasts; they reflect the current thinking about the most likely future trends.

In 1996, the first probabilistic forecasts of the world's population using stochastic birth, death and interregional migration rates were published, based on information as of 1990–1995 (Lutz *et al.* 1996, 1997). These forecasts have recently been updated with the most recent data and

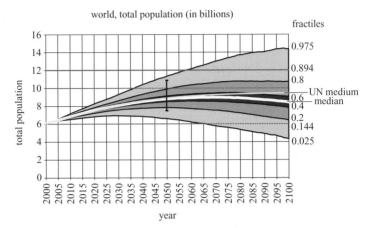

Figure 6.2. Forecast distributions of world population sizes. The error bar refers to the 95% interval as given by the National Research Council (2000) on the basis of an ex-post error analysis.

analysis (National Research Council 2000) as well as an improved methodology (Lutz *et al.* 2001). Owing to the recent acceleration of fertility declines and worsening HIV/AIDS conditions, the new forecasts are lower and show a higher probability that the world's population will stop growing over the course of this century.

Figure 6.2 shows the distribution of simulated world population sizes over time for these newest forecasts produced by IIASA. The median value of these projections reaches a peak around 2070 at 9.0 billion people and then slowly decreases. In 2100, the median value of the projections is 8.4 billion people with the 80% prediction interval bounded by 5.6 and 12.1 billion. There is about a 60% chance that the world population will not reach 10 billion before 2100. There is around an 85% chance that the world's population will stop growing before the end of the century. In 2100, there is around a 15% chance that the world's population will be lower than it is today.

How do these results compare with other recent forecasts? Up to 2045 the IIASA median trajectory is almost identical to the forecasts of the World Bank (2000), the US Bureau of the Census (2000), and the medium variant of the UN (2001). Only the UN long-range projections provide forecasts to the end of the century. The most recent UN medium variant (United Nations 2001) is inserted in figure 6.2 as a thin white line. After an almost identical trend to 2050, the UN forecasts show a virtually constant world population, while the IIASA median begins to decline during the second half of the century. The difference is essentially due to the UN's assumption that after 2050 fertility in all countries will be at replacement level, even in countries where it is already significantly below that level. It also assumes that countries which are still above replacement will never go below replacement fertility. The IIASA projections do not share this assumption, as discussed below.

If we define the end of population growth slightly less literally and take it to correspond to population growth of 0.1% per annum or less, then the UN medium population projection also shows the end of population growth during the second half of the century. Their medium scenario shows world population growth first falling below 0.1% around 2075. Their timing of the end of world population growth is consistent with IIASA's.

Therefore, the general point that the world's population growth is coming to an end does not depend on whether or not the world's fertility rate falls below or remains at the replacement level. Either way, at the end of the 21st century world population growth will probably be over. But whether the population will reach a peak and decrease or just remain virtually stationary does depend on it.

The key determinant of the timing of the peak in population size is the assumed speed of fertility decline in the parts of the world that still have high fertility. On this issue there is a broad consensus that fertility transitions are likely to be completed in the next few decades (National Research Council 2000). For the eventual size of the population and the question of whether or not world population will begin a decline by the end of this century, the key variable is the assumed level of post-transitional fertility. The thorough review of the literature on that subject by the National Research Council (2000) states that 'fertility in countries that have not completed transition should eventually reach levels similar to those now observed in low fertility countries' (p. 106). The IIASA assumptions of long-term sub-replacement fertility are consistent with this view.

Table 6.2 shows the median population sizes and associated 80% prediction intervals for the world and its 13 regions, as defined in Lutz (1996, pp. 437–439). It indicates major regional differences in the paths of population growth. While over the next two decades the medians are already declining in Eastern Europe and the European portion of the former Soviet Union, the populations of north Africa and sub-Saharan Africa are likely to double.

Table 6.2. *Forecasted median world and regional population sizes with 80% prediction intervals (in parentheses) for the world and its 13 regions. (Figures are in millions. From Lutz et al. 2001.)*

Region	Year				
	2000	2025	2050	2075	2100
world	6055	7827 (7219–8459)	8797 (7347–10 443)	8951 (6636–11 652)	8414 (5577–12 123)
1 north Africa	173	257 (228–285)	311 (249–378)	336 (238–443)	333 (215–484)
2 Sub-Saharan Africa	611	976 (856–1100)	1319 (1010–1701)	1522 (1021–2194)	1500 (878–2450)
3 North America	314	379 (351–410)	422 (358–498)	441 (343–565)	454 (313–631)
4 Latin America	515	709 (643–775)	840 (679–1005)	904 (647–1202)	934 (585–1383)
5 central Asia	56	81 (73–90)	100 (80–121)	107 (76–145)	106 (66–159)
6 Middle East	172	285 (252–318)	368 (301–445)	413 (296–544)	413 (259–597)
7 South Asia	1367	1940 (1735–2154)	2249 (1795–2776)	2244 (1528–3085)	1958 (1186–3035)
8 China region	1408	1608 (1494–1714)	1580 (1305–1849)	1422 (1003–1884)	1250 (765–1870)
9 Pacific Asia	476	625 (569–682)	702 (575–842)	702 (509–937)	654 (410–949)
10 Pacific Organisation for Economic Co-operation and Development	150	155 (144–165)	148 (125–174)	135 (100–175)	123 (79–173)
11 Western Europe	456	478 (445–508)	470 (399–549)	433 (321–562)	392 (257–568)
12 Eastern Europe	121	117 (109–125)	104 (86–124)	87 (61–118)	74 (44–115)
13 European part of the Former Soviet Union	236	218 (203–234)	187 (154–225)	159 (110–216)	141 (85–218)

In Western Europe (including Turkey) and North America, future changes will depend not only on fertility and mortality but also significantly on migration volumes. This adds to the uncertainty ranges of future population sizes, with the median starting to decline in Western Europe over the next two decades and continuing to increase in North America.

The projections show that the China region and the South Asia region, which in 2000 have approximately the same population sizes, are likely to follow very different trends. Owing to an earlier fertility decline, the China region is likely to have around 700 million fewer people than the South Asia region by the middle of the century. This absolute difference in population sizes is likely to be maintained over the entire second half of the century and illustrates the strong impact of the timing of fertility decline on eventual population size (O'Neill *et al.* 1999).

6.3.   Population density and human fertility

In the demographic literature there has been surprisingly little systematic analysis on the question of the relationship between population growth rate and human fertility level on the one hand, and population density on the other. As described in the previous section, the study of fertility determinants has largely been focusing on social, economic and even cultural factors influencing reproductive behaviour. Since human fertility, especially under the condition of conscious family planning during the later parts of the demographic transition, is seen primarily as socially determined, ecological factors (such as population density), which are prominent in animal ecology, have played little or no role in demographic analysis. However, population density need not operate only through direct biological mechanisms; perceived population density may also be an important psychological determinant of fertility.

Of the few studies addressing this question in very different settings, most scholars found a significant negative relationship between human population density and the birth rate. In cross-national studies, Adelman (1963), Beaver (1975), Cutright & Kelly (1978), Heer (1966) and Janowitz (1971) found such relationships after controlling the level of urbanization, economic development and other background variables. The negative relationship between density and fertility also has been observed for smaller units, i.e. Ohio counties in 1850 (Leet 1977), townships in Taiwan (Collver *et al.* 1967) and rural areas in Mexico (Hicks 1974). Hermalin & Lavely (1979) described the evolution of Taiwan's agriculture and fertility from a variety of sources using historical information and data, and concluded that the observations were reasonably consistent with theories of rural fertility that propose tenure and inheritance, land availability and capital-output ratios as crucial variables. Firebaugh (1982) studied 22 farm villages in Punjab, India, between 1961 and 1972. Correlation coefficients for pooled data show crude birth rates to be inversely correlated with density, agricultural production, female literacy, the percentage of certain castes and trend variables. Estimated density effects are statistically significant although not very large. Yasuba (1962) analysed the fertility ratios of states in the USA for the period 1800 to 1860. His principal result was that the most important factor associated with fertility differences and trends was population density – the higher the density, the lower the fertility ratio.

Against the background of this rather diverse set of studies, most of them several decades old, the following analysis is, to our knowledge, the first systematic study on the basis of international time-series and also using different density measures.

This analysis is also of particular relevance to the assumptions of long-term fertility levels in different parts of the world in the above-described IIASA population projections. As an alternative to the substantively unfounded assumption of universal convergence toward replacement fertility, those projections assume that long-term fertility in a region within a given range of fertility rates will depend on the population density in that region. The following analysis provides a broader empirical basis for this assumption.

When studying the possible association between population growth rates, fertility and population density, it is not immediately clear which density measure should be applied. The usual measure of dividing the total population of a country by its total area gives a measure of general space, of 'elbow room' so to speak, but it also includes areas of desert and tundra that are not appropriate for agriculture; therefore, one may be interested in considering the potentially arable land of a country rather than the total surface area. In the following analysis we study both density indicators independently, named 'population density' and 'population density (arable land)' in the tables and figures. For the latter case we excluded

Figure 6.3. Bivariate relationship between population growth rate (average annual growth rate calculated from five-year intervals; United Nations 2001) and population density in five groups, according to 1960 density, 1960–2000. (Note: population growth rate lagged by five years.) Each line corresponds to the time-series of one country; numbers at the end of x-axis labels give numbers of countries.

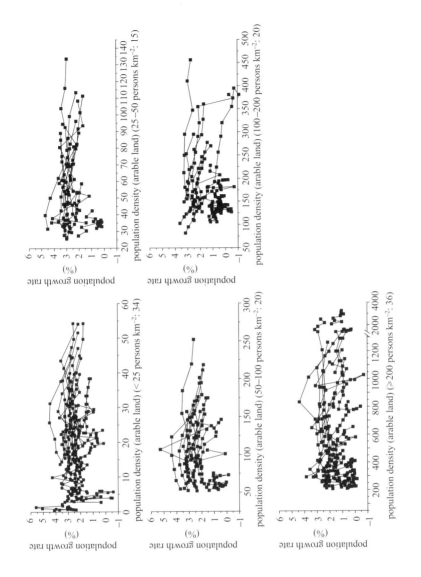

six Middle Eastern desert states because they greatly distort the picture with almost no arable land and rather significant populations mostly due to oil revenues. As the following results show, the two different measures of population density do not produce qualitatively much different patterns of association. The best density variable that one would like to measure in its impact on fertility is perceived density based on perceived living space as it influences behaviour, but we do not know of any data on this. The two density measures applied here cover presumably two important determinants of this perceived density and therefore in combination seem to be an acceptable proxy.

For this analysis, time-series data from 1960 to 2000 (in five-year steps) have been collected for 187 countries mostly derived from international sources (World Bank 2001; United Nations 2001; and see FAO statistical databases at http://apps.fao.org/subscriber/). These data include population size, the different population densities, annual population growth rates, total fertility rates, as well as female labour force participation rates, female literacy rates, urban proportions, GDP per capita in constant US$ and a food production index. Figure 6.3 depicts these data on the bivariate relationship between population density (arable land) (on the horizontal axis) and population growth rates (on the vertical axis). The lines connect the data of individual countries over time. In order to be able to discover some possible patterns among this massive amount of data, the figure sorts the countries according to their population density in 1960 into five groups ranging from the lowest with less than 25 persons per square kilometre to the highest of above 200 persons per square kilometre. Aside from some country-specific peculiarities, the graph does not show any clear bivariate association between the two variables within each of the five groups, neither cross-sectionally nor over time. By comparing across groups, however, it is evident that the average population growth rate is lower for countries with higher density.

Figure 6.4 plots the same country grouping of the time-series but replaces population growth rates with total fertility rates on the vertical axis. Here, a much clearer pattern of association appears. As population density increases over time the mean number of children tends to become

Figure 6.4. Bivariate relationship between total fertility rate and population density in five groups, according to 1960 density, 1960–2000. (Note: a five-year perception lag for TFR has been assumed.) Each line corresponds to the time-series of one country; numbers at the end of x-axis labels give numbers of countries.

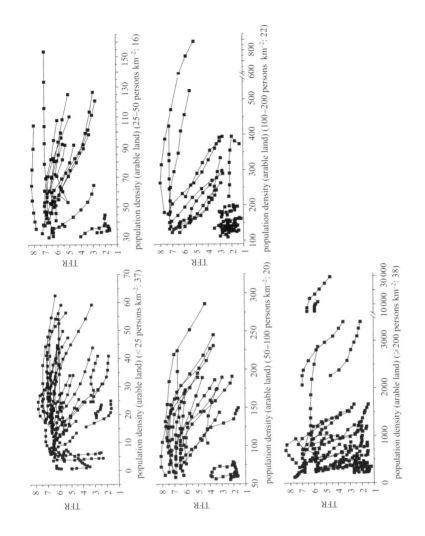

lower. Also, when comparing across groups, it is quite apparent that countries with higher population density have on average much lower fertility rates. Why is the picture so different with respect to the fertility rates than with respect to population growth rates? To interpret this we have to be aware of the fact that even in a population closed to migration, population growth rate is determined by three factors: fertility, mortality and population age structure. Of these three, only fertility is directly a consequence of changing individual behaviour; therefore, only fertility can reflect possible psychological reactions to increasing population density. During the process of demographic transition, mortality is typically positively correlated with fertility. As mortality rates go down and life expectancy increases, fertility rates also go down. This mortality decline counteracts the negative impacts of a fertility decline on population growth rate because more people stay alive, thus contributing to a higher population size. It is also worth noting that since the advent of modern preventive medicine and hygiene, human population density does not seem to have a positive association with the level of mortality as one might infer from animal ecology and considerations of carrying capacity. If there is an association, it seems to be a negative one, with urban areas almost universally showing lower mortality rates than rural areas. This even seems to hold in some of the most polluted megacities because the generally much better access to health facilities in urban areas seems to outweigh the negative environmental impacts.

It is worthwhile to have a closer look at this apparently strong bivariate relationship between density and fertility because it might not really reflect a causal relationship, but rather could be due to some other developmental variables in the background, such as level of income or level of education that might simultaneously lead to lower fertility and make higher population densities possible. For this reason tables 6.3, 6.4 and 6.5 give sets of multiple regressions that study the relationship of population growth and fertility to population density while controlling some of the other social and economic variables measured. To get a more differentiated picture and to avoid serial autocorrelation, the regressions are given separately for the seven points in time and separately for the subset of developing countries, in order to rule out the possibility that the bipolarity between developed and developing countries dominates the appearing pattern. A perception lag of five years has been assumed between the explanatory and the dependent variables, i.e. the independent variables listed for 1960 are being related to fertility and population growth in 1965, and so on. The calculations shown here are based on giving equal weight

Table 6.3. *Multiple linear regressions of several variables on the annual population growth rate (lagged by five years) for 187 countries, 1960–1990.*

Variable	1960 Standardized coefficients	1965 Standardized coefficients	1970 Standardized coefficients	1975 Standardized coefficients	1980 Standardized coefficients	1985 Standardized coefficients	1990 Standardized coefficients
all countries (n = 187)							
female labour force participation rate	−0.041	0.121	0.109	−0.079	0.057	−0.029	−0.208
population density	0.266*	0.277*	0.203	0.082	0.135	0.002	−0.064
female literacy rate	0.139	0.102	0.079	−0.196	−0.105	−0.127	−0.172
population urban	−0.056	0.273	0.079	0.108	0.165	−0.044	−0.126
GDP per capita[a]	−0.076	−0.319	−0.171	−0.214	−0.289*	−0.195	−0.098
food production index[b]	−0.246	−0.122	−0.172	0.088	−0.257*	−0.246*	−0.179
R²	0.142	0.122	0.075	0.067	0.141	0.163	0.134
LDCs[c] (n = 143)							
female labour force participation rate	−0.105	0.086	−0.041	−0.139	0.074	0.004	−0.217
population density	0.228	0.268*	0.185	0.070	0.124	−0.013	−0.094
female literacy rate	0.131	0.111	0.076	−0.180	−0.082	−0.099	−0.179
population urban	−0.087	0.224	0.018	0.049	0.155	−0.034	−0.088
GDP per capita[a]	−0.127	−0.252	−0.138	−0.196	−0.294*	−0.233	−0.210
food production index[b]	−0.239*	−0.102	−0.160	0.110	−0.262*	−0.296*	−0.174
R²	0.146	0.110	0.068	0.068	0.133	0.172	0.177

[a] Constant 1995 US$ for GDP per capita.

[b] 1989–1991 = 100 as reference for food production index.

[c] Comprises all regions of Africa, Asia (excluding Japan), Latin America and the Caribbean plus French Polynesia.

* $p < 0.05$; ** $p < 0.01$; *** $p < 0.001$.

Table 6.4. *Multiple linear regressions of several variables on the TFR (lagged by five years) for 187 countries, 1960–1990.*

Variable	1960 Standardized coefficients	1965 Standardized coefficients	1970 Standardized coefficients	1975 Standardized coefficients	1980 Standardized coefficients	1985 Standardized coefficients	1990 Standardized coefficients
all countries ($n = 187$)							
female labour force participation rate	−0.167	−0.096	−0.085	−0.144	−0.153	−0.117	−0.068
population density	−0.177*	−0.191*	−0.196**	−0.226**	−0.236***	−0.239***	−0.248***
female literacy rate	−0.387**	−0.508***	−0.601***	−0.694***	−0.701***	−0.629***	−0.618***
population urban	−0.024	−0.051	−0.106	−0.218	−0.245*	−0.275*	−0.261*
GDP per capita[a]	−0.514***	−0.389**	−0.244*	−0.051	−0.018	−0.043	−0.048
food production index[b]	−0.133	−0.088	−0.019	0.015	0.028	−0.026	0.137*
R^2	0.655	0.703	0.710	0.714	0.734	0.712	0.698
LDCs[c] ($n = 143$)							
female labour force participation rate	−0.249	−0.146	−0.104	−0.122	−0.102	−0.069	−0.025
population density	−0.193*	−0.225*	−0.229**	−0.245**	−0.264**	−0.281**	−0.296***
female literacy rate	−0.336**	−0.510***	−0.603***	−0.671***	−0.694***	−0.607***	−0.590***
population urban	−0.243	−0.274	−0.299	−0.270*	−0.266*	−0.384*	−0.340*
GDP per capita[a]	−0.399*	−0.180	−0.011	0.033	0.074	0.087	0.076
food production index[b]	−0.145	−0.062	0.011	0.060	0.074	−0.008	0.124
R^2	0.567	0.612	0.637	0.660	0.684	0.664	0.637

[a] Constant 1995 US$ for GDP per capita.
[b] 1989–1991 = 100 as reference for food production index.
[c] Comprises all regions of Africa, Asia (excluding Japan), Latin America and the Caribbean plus French Polynesia.
$* p < 0.05; ** p < 0.01; *** p < 0.001.$

Table 6.5. *Multiple linear regressions of several variables (using density as defined by arable land) on the TFR (lagged by five years) for 181 countries, 1960–1990.*

Variable	1960 Standardized coefficients	1965 Standardized coefficients	1970 Standardized coefficients	1975 Standardized coefficients	1980 Standardized coefficients	1985 Standardized coefficients	1990 Standardized coefficients
all countries (n = 181)							
female labour force participation rate	−0.183	−0.100	−0.056	−0.141	−0.160	−0.102	−0.057
population density (arable land)[a]	−0.184*	−0.186*	−0.162*	−0.156*	−0.153*	−0.173**	−0.203**
female literacy rate	−0.404**	−0.518***	−0.603***	−0.696***	−0.710***	−0.643***	−0.622***
population urban	0.305	0.034	0.014	−0.060	−0.123	−0.169	−0.201*
GDP per capita[b]	−0.591***	−0.492***	−0.372**	−0.257*	−0.210*	−0.196*	−0.153
food production index[c]	−0.126	−0.075	0.014	0.023	0.057	0.007	0.135*
R^2	0.671	0.725	0.727	0.727	0.738	0.722	0.704
LDCs[d] (n = 137)							
female labour force participation rate	−0.280	−0.166	−0.109	−0.164	−0.130	−0.068	−0.006
population density (arable land)[a]	−0.276**	−0.266**	−0.226*	−0.205*	−0.193*	−0.221**	−0.250**
female literacy rate	−0.352**	−0.534***	−0.626***	−0.715***	−0.734***	−0.654***	−0.607***
population urban	−0.204	−0.181	−0.192	−0.219	−0.216	−0.259	−0.237
GDP per capita[b]	−0.469*	−0.289	−0.123	−0.037	−0.009	−0.004	−0.033
food production index[c]	−0.154	−0.073	0.013	0.042	0.087	0.015	0.136
R^2	0.615	0.648	0.648	0.650	0.666	0.659	0.639

[a] Includes land currently used for other purposes such as grassland, forests, protected areas, buildings and infrastructure, etc.

[b] Constant 1995 US$ for GDP per capita.

[c] 1989–1991 = 100 as reference for food production index.

[d] Comprises all regions of Africa, Asia (excluding Japan), Latin America and the Caribbean plus French Polynesia.

* $p < 0.05$; ** $p < 0.01$; *** $p < 0.001$.

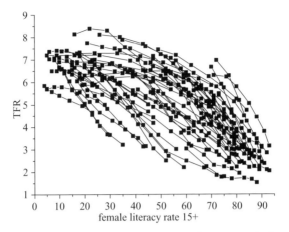

Figure 6.5. Bivariate relationship between literacy rates for females aged 15+ and total fertility rates for time-series of 65 developing countries, 1960–2000. Each line corresponds to the time-series of one country.

to all countries. Additional calculations based on a weighting of the countries by their population size yielded qualitatively similar results and are given in electronic Appendix A available on The Royal Society's Publications Web site.

The results of these 28 multiple regressions cannot be discussed in detail here, but a few general conclusions can be drawn. In almost all regressions for the fertility rate, female literacy seems to be the single most important factor. This is consistent with the large body of literature on fertility determinants and with the theoretical foundations of the process of demographic transition described above. The tables show that the relationship of female literacy to the total fertility rate is more pronounced than that to the growth rates across all points in time. The urban proportion also has a consistent negative association with fertility (the higher the degree of urbanization, the lower fertility) but is not always statistically significant. GDP per capita only shows a significant negative coefficient with fertility during the 1960s on the global level, while it is insignificant with a variable sign in all the other regressions. With respect to population growth rates (table 6.3), the pattern and even the signs of the coefficients are much less consistent over time and are statistically insignificant in general. This has to do with the fact that changes in total population size are also influenced by mortality and migration, which tend to have less consistent associations with density. As a piece of background information, figures 6.5 and 6.6 plot the bivariate relationships of income and

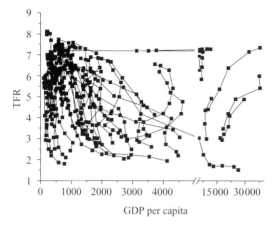

Figure 6.6. Bivariate relationship between GDP per capita (constant 1995 US$) and total fertility rate in 55 developing countries (same countries as in figure 6.5 with available income data), 1960–2000. Each line corresponds to the time-series of one country.

female literacy to the total fertility rate. The comparison of the two figures impressively confirms the view that female literacy is a much more straightforward and almost linear covariate (and determinant) of declining fertility than GDP per capita, where the picture is very mixed.

How does population density – under both definitions used here – come out as an explanatory variable in this multivariate setting? Again the relationship is much stronger and more statistically significant in the case of fertility as the dependent variable, although the signs are consistently negative for both fertility and the growth rate. When explaining the level of fertility, population density comes out second in importance after female literacy, yet still well ahead of the traditionally studied factors: female labour force participation, income, urbanization and food security. This strong negative effect of population density on the level of fertility five years later is statistically significant in almost all years, both at the global level and among the sub-group of developing countries. When comparing the results for the two definitions of density (see tables 6.4 and 6.5), the one based on arable land turns out to be slightly less significant than the one based on total area.

In order to understand better the possible effects of population density on human fertility, more research is needed in terms of studying both these associations and sub-national scales, and in terms of understanding better the possible mechanisms of causation. For the former, table 6.6 gives a simple correlation analysis for the 30 provinces of China, which

Table 6.6. *Correlation coefficients between population density (under three different definitions) and the TFR as well as population growth rate (lagged by five years) in China's 30 provinces (numbers of provinces in parentheses), 1970–1990.*

(Sources of data: Yin Hua & Lin Xiaohong 1996; Population Census Office under the State Council et al. 2001; Fischer et al. 1998.)

Variable	1970	1975	1980	1985
population density–population growth rate	−0.764** (30)	−0.294 (30)	−0.103 (30)	−0.346* (30)
population density–TFR	−0.581** (28)	−0.587** (28)	−0.556*** (30)	−0.529** (30)
population density (potential cultivated land)–population growth rate	−0.629** (29)	−0.160 (29)	−0.004 (29)	−0.339* (29)
population density (potential cultivated land)–TFR	−0.474** (28)	−0.460** (28)	−0.477** (29)	−0.479** (29)
population density (currently cultivated land)–population growth rate	−0.746** (29)	−0.254 (29)	−0.019 (29)	−0.316* (29)
population density (currently cultivated land)–TFR	−0.532** (28)	−0.536** (28)	−0.501** (29)	−0.522** (29)

*$p < 0.05$; ** $p < 0.01$.

also confirms the above-described associations at a sub-national level for the world's most populous country, which has seen dramatic fertility declines over the past three decades. While the correlations between density and fertility are consistently high over time, the relationship to the growth rate is also affected by changing patterns of inter-provincial migration.

As to the possible mechanisms of causation, direct biological factors such as decreasing fecundability due to 'density stress' are rather unlikely candidates for the human population, especially in a technologically advanced stage of development. Instead psychological factors, such as perceived living space, may play a role. An earlier study (W. Lutz, personal communication) identified a clear 'island factor' in the onset of fertility declines, i.e. the fact that in otherwise comparable socioeconomic settings, small islands – where the spatial limitations are obvious – began their fertility transitions earlier. But even with respect to contemporary European fertility levels, it is conspicuous that the very-low-density regions of northern Scandinavia have significantly higher fertility than the high-density areas of central and southern Europe. This clearly needs further investigation.

In conclusion, we have shown that the process of demographic transition has led to unprecedented growth in the human population, but will also lead to significant population ageing and the likely end of world population growth. What will determine human population growth in the very long run, once the momentum of the demographic transition-induced population growth comes to an end? This is an open question at this point. Biological and ecological factors will clearly be very important for the future human life span and health, but they may also play an increasingly important role with respect to human fertility.

The section on past population growth draws partly from O'Neill *et al.* (2001). The section on future trends draws partly from Lutz *et al.* (2001).

7
_____

Two complementary paradigms for analysing population dynamics

7.1. Introduction

For more than 100 years, ecologists have been estimating populations of animals, beginning with those of economic value, and have tried to make sense of the resulting data. How to make sense of quantitative population data is not immediately clear. Once an ecologist has two successive estimates of population size, he or she follows the first law of quantitative ecology, which is to divide one number by the other, producing the finite population growth rate (λ) that Sibly & Hone (Chapter 2) described. However, what to do next?

This is the critical step. Being good scientists, most ecologists would wish to predict the size of the population growth rate and would proceed in one of two directions to do this. First, they could adopt the density paradigm of Sibly & Hone (Chapter 2) and plot population growth rate against population density. (The concept of a paradigm as promulgated by Thomas Kuhn (1970) has been used in many ways, and one might argue that the paradigms discussed here are better labelled as 'conceptual approaches'. I have no quarrel with this comment and I use the term 'paradigm' as shorthand for what ecologists do (cf. den Boer & Reddingius (1996).) Alternatively, they could adopt the mechanistic paradigm and plot population growth rate against an ecological factor, such as the amount of food available per capita, which may explain the change. What are the problems and what are the advantages of going in one direction rather than another?

However, let us drop back for a moment to consider a whole set of assumptions that we have already made about our population. Many of these assumptions are discussed in other papers of this issue.

(i)   We assume that we can define a population unambiguously. This can be a problem with open populations.

(ii)   We assume that we can measure population size accurately and can convert this to absolute population density. This is more difficult than many ecologists think.

(iii) We assume that we have defined a biologically relevant time-step over which to measure the population growth rate. The time-step is not always obvious (Lewellen & Vessey 1998).

(iv)   We assume, at least initially, that all of the individuals in the population have equal impact, regardless of sex, age and genetic composition. We can relax this assumption later.

(v)   We assume a uniformity of nature, such that whatever variable we can find to predict population growth rate will be the critical variable at other times and in other places. This assumption of repeatability is rarely tested.

(vi)   We assume that we can substitute time for space, or space for time, so that there is a uniform predictive function.

All of these assumptions operate within the equilibrium paradigm, and all of them are, potentially, hazardous if we assume a non-equilibrium world view in which transient dynamics are the rule rather than the exception. In this paper, I discuss primarily assumptions (iv)–(vi).

Given that population ecologists must start somewhere, we admit to these assumptions for the moment and ask which direction to follow.

7.2.   Density paradigm

The density paradigm instructs our ecologist to plot population growth rate against population density. At this point, our ecologist might become suspicious because the same variable appears in both the *x*- and the *y*-axis. However, we are assured by some biometricians that this is not a problem (Griffiths 1998) so we disregard this potential problem. If the density data are a time-series of one or more plots, much now depends on the trend shown by the data. If density is monotonically falling (or rising), it will not be possible to estimate the equilibrium point, except by extrapolation. If the population does not vary much in density, the relationship may well look like a shotgun pattern.

The decision tree (figure 7.1) illustrates how to proceed. If there is a negative relationship between population growth rate and density, the next question is, which of the demographic components drive this relationship? Given that data are available to answer this question, the

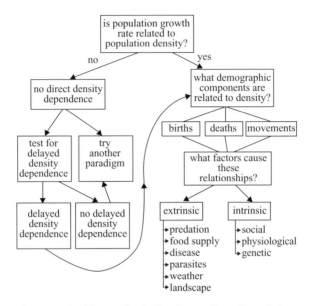

Figure 7.1. Decision tree for the density paradigm of population regulation.

next step is to find out which factors, or combinations of factors, cause these changes in births, deaths or movements (if the population is not closed). All of this is what I will call the standard analysis procedure of the density paradigm. What happens if there is no pattern in the plot of growth rate against density?

We are assured by both theoreticians and empiricists (e.g. Nicholson 1933; Sinclair 1989; Turchin 1999) that there must be a negative relationship between population growth rate and density. If this is true, it raises an interesting question in respect of the relationship of theory in ecology to empirical data. If there must be a relationship, the problem of the field ecologist is to describe this relationship in terms of its slope and intercept. The problem is not to ask if indeed such a relationship exists (Murray 1999, 2000). There is no alternative hypothesis to test.

The first strategy that is adopted after finding that there is no relationship between population growth rate and population density is to invoke delayed density dependence (Turchin 1990). This is a reasonable strategy because virtually every interaction in population ecology involves some time delays. However, this strategy opens a Pandora's Box because data analysis begins to take on the form of data dredging since we have no *a priori* way of knowing what the critical time delays might be. There are elegant methods of time-series analysis that can be applied to population data to estimate the integrated time-lags in a series of density estimates (Stenseth *et al.* 1998), but it is far from clear how to translate these estimated time-lags into ecological understanding. Do predators respond to changes in prey abundance instantly, via movements (e.g. Korpimäki 1994) or more slowly via recruitment processes (e.g. O'Donoghue *et al.* 1997; Eberhardt & Peterson 1999)?

If delayed density dependence can be identified in a time-series of population densities, we can proceed in the same manner as the standard analysis procedure of the density paradigm and try to determine what causes these time-lags. The remaining problem is what to do with cases in which no direct or delayed density dependence can be identified in a time-series. In theory, this situation cannot occur, but it seems to arise frequently enough to cause endless arguments in the literature about the means of testing for direct and delayed density dependence (den Boer & Reddingius 1989; Dennis & Taper 1994). Most ecologists in this situation would not give up studying population regulation, but would switch to the second paradigm discussed by Sibly & Hone (Chapter 2), the mechanistic paradigm.

7.3.   Mechanistic paradigm

The mechanistic paradigm can be viewed in two ways. Sibly & Hone (Chapter 2) consider it an elaboration of the density paradigm, as shown in figure 7.1, and indicate that one can proceed to this level of analysis for populations that are well studied in a reductionist manner. Krebs (1995), by contrast, viewed the mechanistic paradigm as an alternative to the conventional approach through the density paradigm. The mechanistic paradigm short-circuited the search for density dependence, on the assumption that no predictive science of population dynamics could be founded on describing relationships between vital rates and population density without specifying the ecological mechanisms driving these rates.

The key question seems to be whether any density-dependent relationship is repeatable in time or space. I have been able to find few ecologists who have asked this question. The most well-studied groups in this regard might be commercial fishes, birds and large mammals. The Pacific salmon fisheries of western North America are managed partly on the basis of Ricker curves, which plot stock versus recruitment and are another form of a plot for density dependence. The clear conclusion from much research

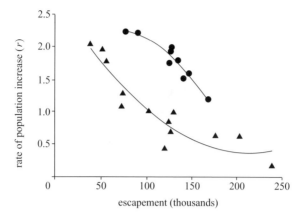

Figure 7.2. Non-repeatability of the relationship between population density and rate of population growth for Columbia River chinook salmon (*Oncorynchus tshawytscha*) over time. You could not manage this fishery in the 1950s using the relationship from the 1940s; this was because both oceanic and freshwater environments had changed. Upper curve illustrates data for the period 1938–1946; lower curve illustrates data for the period 1947–1959. (Data from Van Hyning 1974.)

work is that these Ricker curves cannot be specified as a fixed relationship either temporally, in the same river system, or spatially, between different rivers (Walters 1987). Figure 7.2 gives one illustration for a Chinook salmon stock from the Columbia River system. The Ricker curve for this salmon stock has changed over time, which is not surprising since there has been so much human influence on this river system that many extrinsic environmental factors, as well as intrinsic factors (Ricker 1982), have changed over time.

Considerable work on bird populations allows us to test whether density-dependent relationships are repeatable over time and space. Both (2000) reviewed studies on density dependence in clutch size in passerine birds and found that, for the great tit (*Parus major*) in Europe, only 12 out of 24 long-term studies showed significant density dependence in clutch size. So even within the same species, there is no consistency of density dependence among different populations. Moreover, even in those areas with density-dependent clutch size, no consistent relationship applied to all areas (figure 7.3). This means that one cannot use the data from one area to predict what to expect in another area – density dependence is area-specific. The conclusion is that density-dependent relationships occur often but are not repeatable and are an unreliable basis for a predictive ecology.

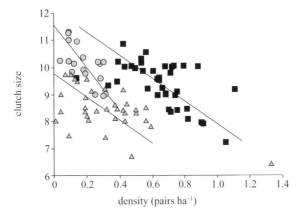

Figure 7.3. Non-repeatability of the density-dependent relationship between clutch size and population density for great tits (*Parus major*) in three woodlands in the Netherlands. You cannot use the density-dependent relationship from one area to predict clutch size in another area. Circles, Hoge Velue A; triangles, Vlieland; squares, Hoge Velue B. (Data courtesy of Both 1998.)

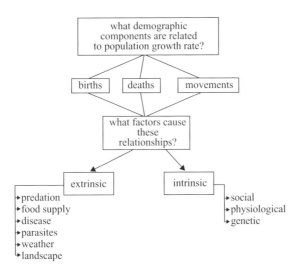

Figure 7.4. Decision tree for the mechanistic paradigm of population regulation. The key difference from figure 7.1 is that we ask what demographic factors are related to population growth rate, not population density.

Figure 7.4 illustrates the flow diagram for the mechanistic paradigm. It looks identical to figure 7.1, but has one very significant difference: instead of asking what demographic components are related to population density, it asks which are related to population growth rate. In cases in which density is closely related to population growth rate, there will be no difference between these two approaches. However, in every non-equilibrial system, the differences can be very large. The critical assumption again depends on whether there is an equilibrium point for the system under study. The mechanistic paradigm is best adapted to short-term considerations in which questions about ultimate equilibrium states are not particularly relevant. It is closely related to the approach to population dynamics typified by the Leslie matrix (Caswell 1989).

The mechanistic paradigm asks how individual animals are influenced by the factors affecting density and recognizes that individuals vary in their responses to predators, food supplies, parasites and weather, as well as in their social standing within the population. Behavioural ecology has made a particularly strong contribution to our understanding of individual differences and is pushing strongly to utilize this understanding to enrich population dynamics.

Let us consider four case studies in order to contrast the density paradigm with the mechanistic paradigm.

(a)   Fire ants

The fire ant (*Solenopsis invicta*) is an introduced pest in the southern United States. It occurs in two forms, a monogyne form with a single queen and a polygyne form with multiple queens per nest. Monogyne fire ants are territorial, whereas polygyne fire ants are non-territorial and reach much higher average densities (Tschinkel 1998). Adams & Tschinkel (2001) carried out a removal experiment in an area of Florida occupied by the monogyne form. They removed all fire ant colonies from a circular core area with a radius of 18 m and then followed the recolonization for a period of five years (figure 7.5). Recolonization was rapid and ant biomass returned to control (equilibrium) values within two years, illustrating a density compensation driven by territoriality. Adjacent ant colonies expanded and new colonies arose from the dispersal of new queens. Population biomass in this area varied slightly from year to year (coefficient of variation of density 13%), but was on average quite stable. This experiment illustrates very well the standard analysis procedure of the density paradigm, which works well in this fire ant system. It also illustrates the mechanistic paradigm because the population carrying capacity was set by territoriality among colonies.

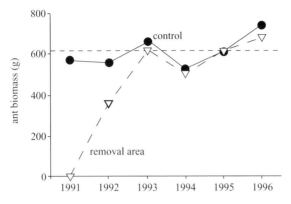

Figure 7.5. Density convergence experiment on the monogynous (territorial) form of the fire ant *Solenopsis invicta* in Florida. All colonies in core areas of 1018 m² were removed from six plots in the spring of 1991. Recolonization was followed by measuring spring biomass of ants in each of the next five years. Convergence was 60% after one year and complete after two years, demonstrating density regulation back to the average control density of 613 g per 1018 m² measured on six unmanipulated plots. The dotted line shows the average control ant biomass. Control biomass showed a coefficient of variation of 13%. (Data from Adams & Tschinkel 2001.)

(b)   Song sparrows

The song sparrow (*Melospiza melodia*) on Mandarte Island, British Columbia, has been the subject of a long-term study since 1962 and has been reported by Smith & Arcese (1986), Arcese & Smith (1988), Smith (1988) and many others. Figure 7.6 illustrates the population density changes in the song sparrow on this 6 ha island since 1975. The population trend consists of periods of three to four years of population growth followed by a catastrophic 1-year decline, and this has been repeated three times in the last 25 years. The first two of these population declines were correlated with severe winter weather; the third was not. Arcese & Smith (1988) showed that fledgling production declines at high density in this population, and these demographic symptoms could be relieved by adding food to territories. This population shows a clear difference between the density and the mechanistic paradigms. If we ask what prevents population increase, we answer that reproductive output is reduced as density increases and the mechanism limiting reproductive output is food shortage. If we ask what causes the largest changes in population growth rates, we answer that the major or key factor is severe winter mortality. The population trace of this species is the net result of negative

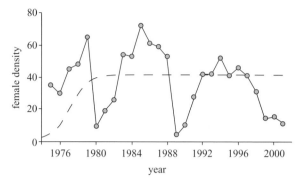

Figure 7.6. Density of song sparrow (*Melospiza melodia*) females on Mandarte Island, British Columbia, 1975–2001. Data courtesy of J. N. M. Smith and P. Arcese. The dashed line is the logistic equation fitted to these data by Sæther *et al.* (2000*a*) and is clearly a very poor descriptor of the population trace. Female density is given per 6 ha.

feedback of high density on reproductive output and occasional major winter mortality. Does this population have an equilibrium density? We can ask what would happen to this population if there were no winter losses. The answer to this question is hypothetical and problematic because none of the identified density-dependent relationships does more than slow down the rate of population increase; they do not set it to zero (Arcese & Smith 1988). To complicate the matter more, this island is part of a metapopulation of song sparrows in the general region and while immigration is rare, it is critical for the maintenance of genetic diversity and for recovery from low numbers (Smith *et al.* 1996). Recent analyses (P. Arcese and J. N. M. Smith, personal communication) suggest that immigrants strongly affect the population growth rate because their outbred offspring have much higher survival and reproductive rates compared with birds with no immigrant genes in their lineage. The key to winter losses seems to be the genetic quality of individual birds. To summarize, the song sparrow on Mandarte Island is a very well-studied bird population and we have a good understanding of its population dynamics, which can be well described by both the density and the mechanistic paradigms. If, for some reason, we had to manage this population, we would try to manipulate the level of outbreeding to maintain high individual quality. Density dependence in this population does not prevent instability.

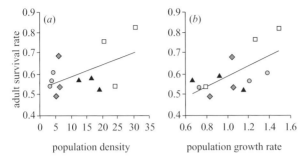

Figure 7.7. Relationship of annual adult survival rate to (*a*) population density and (*b*) population growth rate (λ) for house sparrows (*Passer domesticus*) on four islands in Hegeland, north Norway, 1993–1996 (circles, Gjærøy; squares, Infre Kvarøy; triangles, Ytre Kvarøy; diamonds, Hestmannøy). The density paradigm would expect adult survival rate to fall as population density rose. However, the opposite was observed. (Data from Sæther *et al.* 1999.)

(c)   House sparrow

Sæther *et al.* (1999) have analysed the demography of the house sparrow (*Passer domesticus*) on four islands off north Norway over a period of four years. They were particularly interested in metapopulation dynamics, but their detailed studies allow us to ask how well we can understand their results with the approach suggested by the density paradigm. Sæther *et al.* (1999) measured breeding population density by direct enumeration, and, from 1993 to 1996, estimated reproductive success (number of off-spring fledged per female), juvenile survival over the first year and annual adult survival rates. None of these vital rates was negatively related to population density (figure 7.7). All three variables – reproductive success ($r = 0.40$), juvenile survival rate ($r = 0.51$) and adult survival ($r = 0.65$) – were positively related to population growth rates, and jointly determined whether or not a particular island was a source or a sink population in any given year. There was no correlation between population density and population growth rate. Sæther *et al.* (1999) suggested that the large year-to-year variations in the rates of increase on the different islands were associated with weather variation, but it is not clear if the impact of weather was direct or through changes in food supplies. Thus, the exact mechanisms causing change are not known for these populations. The suggestion for these house sparrow populations is that the density paradigm does not work, and the mechanistic paradigm is not sufficiently evaluated to show whether a predictive model based on particular mechanisms could be defined.

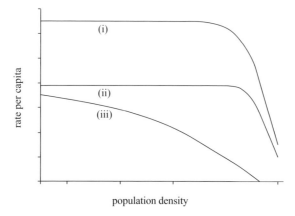

Figure 7.8. The density paradigm model for large mammals as articulated by Fowler (1987b). In this model reproductive output and adult survival rates are not affected by density until the population almost reaches carrying capacity. Graph (i) represents reproduction; graph (ii) represents adult survival and graph (iii) juvenile survival.

(d)   Large mammals

The density paradigm is particularly well defined for large mammals (Gaillard *et al.* 1998). Figure 7.8 illustrates the curvilinear pattern of density dependence postulated for large mammals by Fowler (1987b). Juvenile survival is predicted to be most sensitive to population density, while adult survival and reproductive rates are predicted to begin to decline only at high densities. This paradigm is well accepted by many large-mammal ecologists (Huff & Varley 1999) and we can use the extensive data from North American elk to test this paradigm.

Native ungulates in North American national parks have been subjected to a variety of management policies during the last 100 years (Houston 1982). In 1968, Yellowstone National Park instituted a new management policy ('natural-regulation' management) which was a hands-off policy that permitted ungulates to reach an unmanipulated population level. The natural-regulation management policy assumes first that density-dependent changes in birth and death rates will occur as ungulates increase and reach a dynamic stable equilibrium and second, that this equilibrium will be reached without extensive impacts on vegetation, soils or other species of animals in the community (Singer *et al.* 1998). The Northern Yellowstone elk population has been particularly well studied (Houston 1982). As elk populations increased after control by shooting in the Park was stopped in 1968, Houston (1982) found a slight reduction in pregnancy rates and an increase in the age at sexual maturity at high elk densities, as well as a major density-dependent decline in calf survival during the first year of life. These density-dependent processes would tend to move the elk population toward an equilibrium density.

Wolves, a major predator until Europeans arrived, were missing from the Yellowstone ecosystem after the 1920s until they were reintroduced in the early 1990s. Mechanisms of population limitation for large mammals are simplified if major predators are missing from a system and the major candidate mechanisms remaining are disease, food shortage, weather and social factors. Ungulates are rarely candidates for social regulation (Wolff 1997) and we are left with only three potential factors to consider. Brucellosis is of minor importance to this elk herd (Cheville *et al.* 1998) and we are left with only two potential mechanisms to drive demography.

Can we describe the population dynamics of the Northern Yellowstone elk herd by the density paradigm? Data on elk populations (figure 7.9) do not fit the simple density-dependent paradigm of Fowler (1987b). Eberhardt *et al.* (1996) followed the population growth of an elk population in eastern Washington and found a density-dependent decline in calf recruitment at very low elk densities. This decline was not due to a reproductive failure, since all adult females were pregnant, and they suggested that calf mortality in the first few weeks of life might be the mechanism behind the density-dependent response shown in figure 7.9*a*. Predation might be the mechanism of loss but no detailed studies were possible to test this speculation. In contrast to this view of elk density dependence, Singer *et al.* (1997) found that summer calf mortality was only loosely related to elk numbers ($r^2 = 0.29$, $n = 17$ years), but winter calf losses were much higher at high densities ($r^2 = 0.65$) (figure 7.10).

There has been only a vague relationship between population size and rate of population growth in the Northern Yellowstone elk populations for the past 20 years (figure 7.11). The reason for this is that two climatic variables have a strong influence on demographic parameters (Huff & Varley 1999). Severe winter weather increases calf mortality rates (Houston 1982). A combined regression model with elk density and winter weather severity can explain, statistically, 73% of the variability in winter calf losses (Coughenour & Singer 1996). Summer precipitation has a strong impact on summer plant production and, consequently, on calf summer survival. The overall picture for the North Yellowstone elk population is of a population with vague density dependence buffeted by variable weather

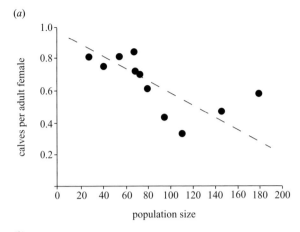

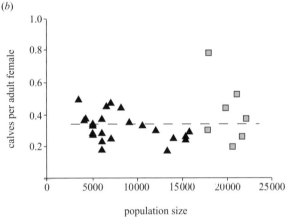

Figure 7.9. Non-repeatability of density-dependent relationships in North American elk from (*a*) Eastern Washington ($r^2 = 0.44$) and (*b*) Northern Yellowstone ($r^2 = 0.03$; squares represent 1985–1991, triangles represent 1951–1979). The ratio of calves per adult female measures the combination of natality and juvenile survival during the first six months of life. Data from Eberhardt *et al.* 1996; Singer *et al.* 1997; Houston 1982. The results are not consistent with the density paradigm model for large mammals (Fowler 1987b).

conditions that impact on summer grazing conditions and winter snow levels. The recent introduction of wolves to the Yellowstone ecosystem is predicted to reduce elk numbers, but the predicted reduction in equilibrium density varies from *ca.* 10–20% reduction (Mack & Singer 1993), to a 50–66% reduction (Gasaway *et al.* 1992). It is clear that the northern Yellowstone elk population has not yet reached its equilibrium and may never get there because of stochastic variation in weather as well as human influences in the areas surrounding Yellowstone.

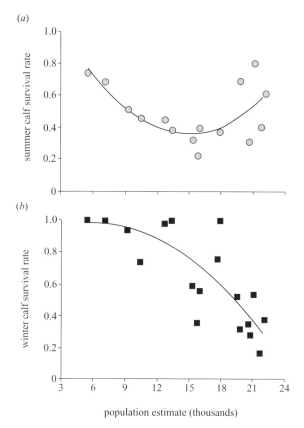

Figure 7.10. (*a*) Summer and (*b*) winter elk calf survival in the Northern Yellowstone national park, 1968–1990, in relation to the number of elk in the herd. Sample size is 16 years for summer, and 19 years for winter data. (Data from Singer *et al*. 1997.)

My perception is that the density paradigm for large mammals does not work well (Peterson 1999), in spite of the common belief that it does. Bison populations in Yellowstone show no clear evidence of density dependence (Singer *et al*. 1998). The moose–wolf interaction on Isle Royale in Lake Superior has not provided a good fit to the large-mammal paradigm illustrated in figure 7.8 (Peterson 1999).

(e) Two contrary views

There are two points of view, that argue strongly against the mechanistic paradigm. The multiple factor hypothesis of population regulation (Holmes 1995) argues that there are no necessary conditions or predictable

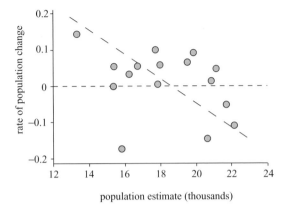

Figure 7.11. Northern Yellowstone elk population, 1975–1992, rate of population change (r) versus estimated population size, $r^2 = 0.12$, $n = 16$. Data corrected for hunting removals. Data from Cheville *et al.* 1998; Singer *et al.* 1997. The horizontal line divides increasing populations above from declining populations below. A negative trend is apparent but with great variability.

relationships between ecological factors like predation and disease and changes in population density. What happens in one population in a given year cannot be predicted from its density, from what happened last year, or from any set of mechanistic relationships. The multiple factor viewpoint, in its extreme form, is not consistent with the density paradigm, which expects population growth rate to fall in a predictable way with population density. If the multiple factor hypothesis is correct, it explains the failure of ecologists to achieve a predictive theory of population dynamics – there can be no such theory. Such a view would appear to condemn ecologists to *a posteriori* descriptions of population changes.

A different type of multiple factor hypothesis suggests that several factors will interact to determine population changes (Lidicker 1994). This view appears to be stated as a polar opposite to the single-factor hypothesis and seems to be based on a confusion of the distinction between necessary and sufficient conditions. Most hypotheses in population regulation do not clearly state whether the proposed mechanism is necessary or sufficient or both, so this confusion is understandable. The problem reduces to what factors are assumed constantly present as background sources of mortality for a population, and this is confounded even more when no distinction is made between additive and compensatory forms of loss.

The multiple factor hypothesis of population regulation is an important viewpoint and we do not yet have enough data on populations of the same species in different environments to know how general our explanations might be. In the bad-case scenario, predictive relationships for one population will not apply to another in a different region. In the worst-case scenario, predictive relationships for one population will not even apply to the same population in later years, so that every population is unique. If this turns out to be correct, ecologists will become environmental historians instead of scientists, charting how populations change with no predictive insights. At the heart of both the density and the mechanistic paradigms is the faith that, although there are many variables that impact on a population, the major controlling variables will show strong signals through the noise of contingent events. We do not know at this time whether this belief is well placed.

A second contrary view is that climate change will invalidate all of the relationships that we ecologists can establish between populations and their ecological agents of control. This view argues that we are now in a state of transient dynamics with no possible predictability of future trends or outcomes. Again, in this case we have no way of determining whether this belief is correct or not, and all we can try to do is falsify it by achieving the goals of the density paradigm and the mechanistic paradigm.

7.4. Conclusions

If not all population ecologists can agree that populations are regulated (Murray 2000), we might, at least, hope to find that there are predictive relationships between ecological mechanisms and population growth rates. We can demonstrate these kinds of predictions for only a few population systems, and our goal should be to increase the breadth and variety of case studies of mechanistic population regulation. A first cut can be to distinguish populations whose growth rate is limited top–down by predators and diseases from those whose growth rate is limited bottom–up by nutrients or food supplies (Kay 1998; Power 1992).

My suggestions here are parallel to those of Chitty (1996), who has argued for the view that comparative studies can untangle the Gordian knot of density dependence by searching for mechanistic differences between experimental and control populations. The key here is to use the experimental approach, particularly manipulative experiments where they are possible, and to consider at all times multiple working hypotheses. Our experiments ought to be designed to evaluate several alternative hypotheses, not just our favourite one.

One of the enigmas of the study of population dynamics is how, historically, it has become so entwined with the ideas of density dependence. The major thrust of ecology over the past 50 years has been to show that population density can be decomposed into sets of individuals with variable traits and interesting ecological interactions. This study of individual differences has brought behavioural ecology into the limelight over the past 20 years, and is now doing the same for disease ecology. Not all individuals are the same, as George Orwell told us long ago, and yet we must aggregate these individuals into a density if we are to use density dependence as a central pillar in our theory of population dynamics. My plea here is to concentrate our efforts on finding out in the short-term why population growth rate is positive or negative. In doing this, we can abandon the worries about equilibrium that have caused so much controversy and put more interesting experimental biology into population dynamics. By concentrating on what factors affect population growth rate, we can provide a science that will be useful to decision makers and managers of the diversity of populations on our planet.

The author thanks The Royal Society and the Novartis Foundation for their support. The author also thanks Jim Hone and Richard Sibly for the invitation to attend The Royal Society meeting and the challenge to clarify thinking on population dynamics. The author thanks Dennis Chitty, Alice Kenney, Peter Hudson and Peter Calow for comments on the manuscript, Tony Sinclair for clarifying ideas on population regulation and graduate classes for endless challenges to points of view about population regulation.

8
_____

Complex numerical responses to top-down and bottom-up processes in vertebrate populations

8.1. Introduction

The intrinsic rate of growth of animal populations (r_{max}) is a species-specific character that is determined by a trade-off between reproductive capacity and survival. In simple form, given a finite amount of resources such as food and time, a species can evolve adaptations that either enhance reproduction and result in lower survival, or increase survival at the cost of lower per capita reproduction. These life-history features are related to body size in a wide range of animal species from protozoa to mammals, with r_{max} negatively related to body size (Blueweiss *et al.* 1978; Caughley & Krebs 1983; Sinclair 1996).

The species-specific adaptation, r_{max} determines how species respond to environmental impacts. In a given environment, both large and small species experience the same negative environmental effects, and the degree to which the species are adapted to resist decline or tolerate them is reflected by r_{max}. Body size buffers large mammals against environmental disturbance compared with smaller mammals, and this contributes to the greater apparent stability of large-mammal populations. Therefore, in mammals, population variability is inversely related to body size when considered over absolute time. However, when corrected for generation length, there is no relationship between population variability and body size. This implies that all species show the same intrinsic degree of population variability. Thus, when lifespan is taken into account, small species do not experience any more severe extrinsic perturbations than larger species (Sinclair 1996).

Population variability is measured by changes in the observed instantaneous rate of increase through time, *r*. Such variability, although constrained by the intrinsic features of $r_{\max}$, is determined by a variety of extrinsic factors. This paper explores the underlying factors that change r and the several ways in which the basic pattern is modified by (i) influences from higher trophic levels, and (ii) chance effects from environmental disturbances.

8.2.   Bottom-up effects on rate of increase

We start with the basic proposition that for all populations the rate of growth, r, is determined by the food supply per capita, a bottom-up process. As the population increases, there is less food per capita and the growth rate declines, as seen, for example (figure 8.1a), in the Serengeti wildebeest (*Connochaetes taurinus*). Thus, the per capita rate of increase (r) of a population is negatively related to population density (figure 8.1b). The decrease in r occurs through either an increase in per capita mortality or a decrease in per capita natality. Although this relationship of r to per capita food must apply to all populations, we usually detect it only when there is a relatively constant rate of renewal in the food supply. When food fluctuates widely through large random variations in weather, we may not see a relationship between these two factors or even between r and population size (Choquenot & Parkes 2001).

Populations experiencing a negative feedback mechanism are 'regulated' (Murdoch 1994). The simplest model assumes that r declines uniformly with density, coincident negative feedbacks having constant effects at all population sizes. However, observations of large mammals indicate that density dependence is more likely to be nonlinear, with stronger effects occurring at high population numbers and weak or no effects taking place at low population numbers. Although such curvilinear density dependence can lead to increased stability at high numbers, as in the wildebeest of Serengeti, in other circumstances it can lead to overcompensation, while at low numbers it can lead to instability. By contrast, small-mammal populations regulated more by resources mediated through social behaviour may have a more linear relationship between r and density. Indeed, smaller vertebrates may exhibit the opposite nonlinear effects with stronger density dependence at low population size (Fowler 1987a; Chapters 2 and 5). This relationship is more likely to be produced by predator regulation and causes instability at high population levels. We shall return to these aspects later (see §8.4).

(*a*)

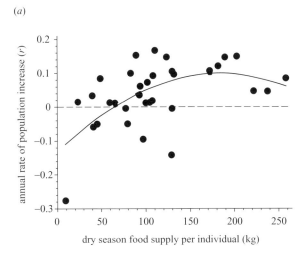

(*b*)

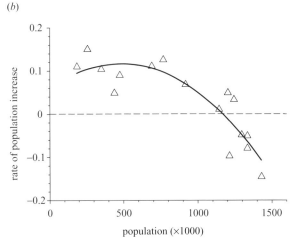

Figure 8.1. (*a*) Annual instantaneous rates of increase for the Serengeti wildebeest population as a function of dry season food supply per individual (kg dry season grass individual^{-1} km^{-2}). (*b*) Wildebeest rate of population increase plotted against population size (data from Mduma *et al.* 1999).

The rate of increase can also be expressed as a function of the total food available to the population. Figure 8.2 illustrates such a relationship for western grey kangaroos (*Macropus fulginosus*) that feed on grasslands in eastern Australia (Bayliss 1987): at low food availability, *r* is negative but increasing monotonically to a positive asymptote at high food

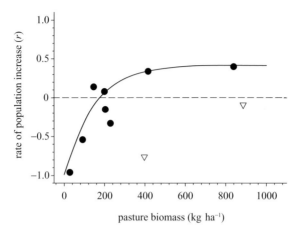

Figure 8.2. Caughley's numerical response expressed as the rate of increase relative to the total food available to the population. Rate of increase of western grey kangaroos in Kinchega National Park, Australia in relation to pasture biomass (kg ha⁻¹) (from Bayliss 1987).

abundance. Caughley *et al.* (1987) termed this relationship the 'numerical response', more recently renamed the 'demographic numerical response' (Choquenot & Parkes 2001).

Holling (1965) provided a different definition for his numerical response, in this case the relationship between population size and food available (figure 8.3), an example of which is seen in the convex relationship between wolf numbers (*Canis lupus*) and their prey, the moose (*Alces alces*) (Messier 1994). There is an important distinction between the two approaches. Holling recognized that predator numbers do not continue to increase as available food increases (as implied in figure 8.2), but rather that they reach an asymptote because of other factors, such as lack of space and intraspecific interference, which limit the population. In particular, vertebrate predators tend to be territorial and territories limit the eventual size of the predator population. Invertebrates that are not territorial, such as parasitoid hymenoptera, are also limited through mutual interference and avoidance at high food levels. Thus, the population of predators stabilizes initially at different levels of food (P₁ and P₂ in figure 8.3). However, when space becomes limiting the population remains at a similar density (P₃) with respect to food. Holling's numerical response is in effect an isocline where $r = 0$, and so has been called the 'isocline numerical response' (Choquenot & Parkes 2001). Thus, the family of curves in figure 8.4 represents different points along the isocline in figure 8.3. Some

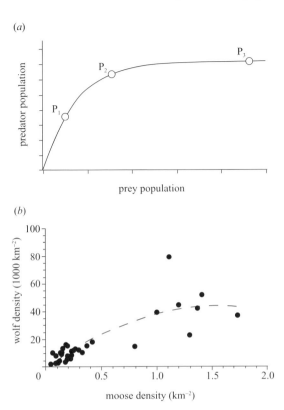

Figure 8.3. Holling's (1965) numerical response relationship between population size and available food. (*a*) Over a range of prey values, the predator population stabilizes at different levels as a function of food supply (e.g. P_1 and P_2). When space or other resources become limiting, (P_3) the predator population reaches a ceiling. (*b*) Observed numerical response of wolves to moose in North America. (From Messier 1994).

herbivores are also territorial and would exhibit the Holling type of response but many other species are not territorial and tolerate high levels of crowding. These species would exhibit the Caughley type of numerical response. Alternatively, we might view Caughley-type populations as those that never reach the density levels above P_2 in figure 8.3.

8.3.   Top-down effects on rate of increase

Predator responses to increasing prey numbers not only involve the numerical changes discussed above, but also changes in the behaviour of

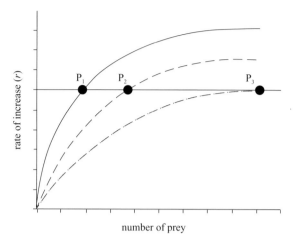

Figure 8.4. The family of numerical response curves represents different points on the isocline in figure 8.3. P_1 is the stable point when there are no intraspecific interference effects on predator numbers, P_2 represents intermediate interference effects and P_3 occurs when predator numbers are limited by intraspecific interference and not by food supplies. Horizontal line is the predator zero isocline.

individual predators, the functional response (Solomon 1949). Differences in the way that predators search for and catch their prey ultimately affect the rate of population increase. The two basic behaviour patterns are identified as the type II and type III functional responses (Holling 1959, 1965). In the type II response, the predator eats more as prey density increases but this relationship curves monotonically to an asymptote due to satiation and the effect that available time has on the maximum rate of prey offtake. The proportional effect on the prey population is uniformly inversely density-dependent (depensatory). The type III response has a theoretical S-shape where prey are avoided at low density but are then actively sought at higher density. The proportional effect on the prey increases at low prey densities (density dependent) and decreases, as in type II, at high prey densities. Operationally, these two types can be distinguished by whether the predation curve starts near the origin and reaches an asymptote quickly (type II), or whether the predation curve starts at a higher prey density (type III) because the predator is either ignoring prey at low density or avoiding that habitat. In practice, it may be difficult with empirical data to distinguish between type II and type III curves.

The total response of predators is the total mortality imposed by predators expressed as the proportion of prey killed (i.e. the per capita

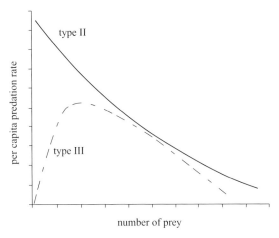

Figure 8.5. Total response curves for predators expressed as per capita mortality of prey, without density dependence in functional and numerical responses (type II) and with density dependence at low prey densities (type III).

mortality rate). It is the product of Holling's functional and numerical responses. If there is no density dependence in either functional or numerical response, then the proportional effect of the total response (the predation curve) is uniformly inversely density-dependent and is of type II form (figure 8.5). If there is density dependence then the shape of the total response is of type III form and shows density dependence at low prey densities while remaining depensatory at high prey densities (Ricklefs 1979; Sinclair 1989; Sinclair *et al.* 1998).

The effect of the two types of total response (figure 8.5) on the prey rate of increase depends on their relationship to the per capita net recruitment curve of the prey in the absence of predators. The difference between net recruitment and predation provides the instantaneous rate of increase realized by the prey population (figure 8.6). Figure 8.6*a* shows the prey rate of increase relative to population size (number of prey) for a family of curves that differ in the magnitude of type II total predation. Figure 8.6*b* shows the equivalent curves for type III total predation. In figure 8.6*a*, there is one stable point (C) at high prey density if total predation is low (curve 1). At higher predation levels (curve 2), the rate of increase is positive at prey densities of between B and C. Point B is an unstable boundary and predators can drive the prey to extinction if numbers fall below B. Above B, the number of prey can increase to C. At even higher predation levels (curve 3), there are no equilibria and predators can cause extinction

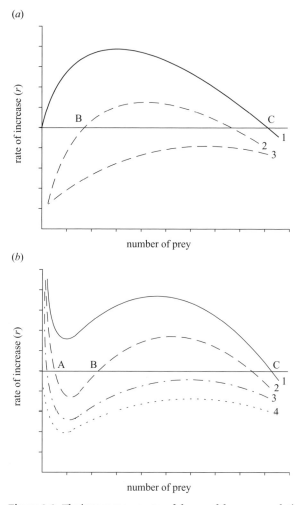

Figure 8.6. The instantaneous rates of change of the prey population experiencing different levels of (*a*) type II and (*b*) type III predation. Point A represents a stable point from regulation by predators, point C a stable point due to regulation from food with predation not regulating, and point B is an unstable threshold. Curves 1–4 represent different intensities of predation: 1, lowest predation level; 4, highest predation level.

at all prey densities. The curves in figure 8.6*a* represent those where prey are secondary and predators depend on some other primary prey.

Where the total response is type III and shows density dependence (figure 8.6*b*), there are several possible outcomes of predation. At low predation (figure 8.6*b*, curve 1), there is a single equilibrium (C), similar to

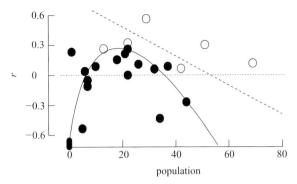

Figure 8.7. Instantaneous rates of increase per year (*r*) for black-footed rock-wallaby (*Petrogale lateralis*) in Western Australia with type II predation (closed circles) and without predation (open circles). Zero rate of increase is given by the dotted line. There is no lower stable prey density with predators. (Data from Kinnear *et al.* (1988), after Sinclair *et al.* (1998)). Lines fitted by eye represent trends in the data.

that in figure 8.6*a*. At progressively higher predation levels, there are two stable points (A, C) (curve 2), a single stable point (A) at low prey density (curve 3) where predators regulate prey and finally a predation rate too high to allow the persistence of prey (curve 4). These curves show that stability rather than extinction can occur at low prey densities provided that predation is not too high. Whether the prey is held at A or C is determined by both the magnitude of predation and the presence of disturbances that switch prey numbers between these points. We address the effect of disturbance in §8.5.

(a) Type II interactions where predators and other (primary) prey determine the prey rate of increase

Black-footed rock-wallabies (*Petrogale lateralis*) live in small rocky outcrops in Western Australia (Kinnear *et al.* 1988, 1998). In two of these outcrops, foxes were removed after an initial 4-year study period, while in the remaining three areas foxes were allowed to persist. Rates of change per year for these rock wallabies (figure 8.7) from the five populations before foxes were removed show a curve similar to figure 8.6*a* curve 2, with a higher stable equilibrium lying in the range of 20–40 animals and a lower instability boundary in the region of 5–10 animals. Below this level, the population heads to extinction. The four values of *r* from the two fox-absent populations are all greater than those when foxes were present, and extrapolation from these indicate a carrying capacity in the region of

(a)

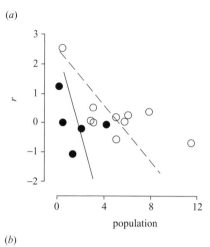

(b)

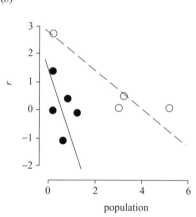

(c)

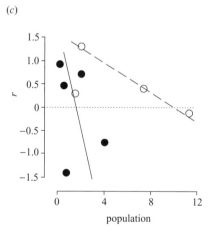

70–80 animals, doubled when foxes are present. These data suggest that foxes were responding to rock-wallabies in a type II fashion, treating the prey as secondary to some more abundant and persistent primary prey. As such, predators can cause extinction of the prey.

(b) Examples of type III interactions where predators and habitat refuge determine the prey rate of increase

The numbat (*Myrmecobius fasciatus*) is a 300 g marsupial termite eater. Once commonplace, they are now confined to two small populations in Western Australia. The population at Dryandra Woodland Reserve has been monitored periodically since 1955 (Friend & Thomas 1994) and fox removal has been instituted since 1982. Instantaneous rates of increase for this population before and after fox removal show two stable states, one with predators at a density index of 1.4 and the other without predators at a density index of 5.9 (figure 8.8*a*).

In large tracts of Eucalypt forest in Western Australia, several small marsupials have coexisted with red foxes (Morris *et al.* 1995). Populations with foxes have been monitored since 1985. In part of the forest, foxes were removed for 4 years (1990–1993). Growth rates of western quoll (*Dasyurus geoffroii*), a 1.5 kg carnivore, and brush-tailed bettong (*Bettongia penicillata*), a 1.3 kg herbivorous macropod, are illustrated in figure 8.8*b,c*. In both species, lower and higher stable values are seen for *r*, predicted from type III responses, with the higher points some six times those of the lower (Sinclair *et al.* 1998). All of these examples indicate a change in the magnitude of predation from that in curve 3 to that in curve 1 (figure 8.6*b*). The lower state is so low, however, that it would be exposed to extinction due to random events.

8.4. Disturbance and predation on prey rate of increase

Whether predators keep prey at low or high densities depends on both the efficiency of the predator at catching prey and the ability of the prey to

Figure 8.8. Instantaneous rates of increase per year (*r*) for prey with type III predation (closed circles, solid lines) and without red fox predation (open circles, dotted lines) in Western Australia. (*a*) Numbat (*Myrmecobius fasciatus*) observed 100 km^{-1} of transect, (*b*) western quoll (*Dasyurus geoffroii*) percentage trap success and (*c*) brush-tailed bettong (*Bettongia penicillata*) percentage trap success. Zero rate of increase is given by the horizontal dotted line. There is a lower stable prey density with predators and an upper one without predators. (Data from Friend & Thomas (1994) and Morris *et al.* (1995), after Sinclair *et al.* (1998)).

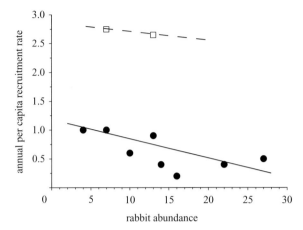

Figure 8.9. In northern New South Wales, Australia, net recruitment of European rabbits declined with rabbit density due to density-dependent fox predation in 1981 and 1982, years of normal rainfall (circles). In 1979, high rainfall produced increased food supplies and very high reproductive rates (squares), allowing rabbits to escape predator regulation and outbreak to plague levels. (Data from Pech *et al.* 1992)

avoid predation. Disturbance events can change these biological abilities of predator and prey.

The feeding rate of a predator depends on its ability both to find and to capture prey. For example, predatory fish may be unable to see their prey in turbid water and so the survival rate of the prey increases (McQueen 1998). Alternatively, disturbance might increase the reproductive rate of prey. In Australia, pulses of high rainfall cause rapid increases in the availability of high-quality food. This produces high rates of reproduction in rabbits (*Oryctolagus cuniculus*) and house mice (*Mus domesticus*) that outpace the depredatory effects of carnivores and raptors. Consequently, outbreaks of rabbits and mice follow periods of high rainfall (Sinclair *et al.* 1990; Pech *et al.* 1992; figure 8.9). In both of these examples, extrinsic disturbances improve the prey rate of population increase so that the prey population escapes predator regulation (point A, figure 8.6*b*) and outbreaks to the higher level (point C, figure 8.6*b*).

By contrast, environmental disturbance might reduce the ability of prey to escape from predators and so become more vulnerable to predation. In 1985, wolves reappeared in Banff National Park having been extirpated there in the 1930s. The wolves hunted the large and expanding elk (*Cervus elaphus*) population living throughout the Bow valley. This valley exhibited a gradient of snow depth in winter, with deep snow at the top and shallow snow towards the lower end. Elk ran much slower in deep snow when chased by wolves and were heavily depredated in comparison with those in the shallow snow. By the early 1990s, elk numbers had declined and the population is now confined to the shallow snow areas (Huggard 1993) and limited by predators. A similar dynamic is apparently being played out in the Yellowstone National Park, USA, where wolves were reintroduced in 1995 (Singer & Mack 1999).

8.5.   Multiple states, meta-stability and disturbance

Multiple states are special cases where a population can exist at two levels under the same conditions (Holling 1973; May 1977; Scheffer *et al.* 2001; figure 8.6*b*, curve 2). Movement between the two levels requires some form of disturbance, such as an environmental event or a temporary increase in predation, human harvesting or habitat loss. A boundary between states exists if a system disturbed from one state to another does not return to its original state once the cause of the disturbance returns to its original value. A second factor holds the system in the second state. There are now several known examples of multiple states from lake systems, tundra, savannah and deserts (Walker *et al.* 1981; Dublin *et al.* 1990; Hik *et al.* 1992).

In the Allegheny National Forest, Pennsylvania, USA, white-tailed deer were maintained experimentally at five densities and hardwood tree seedlings were monitored in both clearcut and uncut (control) forest plots (Tilghman 1989; Schmitz & Sinclair 1997). In uncut forest, even low deer numbers kept seedling numbers down because light and root competition depressed seedling recruitment (figure 8.10*a*). However, in clearcuts (areas with all trees removed) with high productivity, seedling density remained high under the same deer densities. It is only when deer numbers exceed 80/259 ha that seedling densities are expected to drop to those in uncut forests, while deer numbers must drop below 10/259 ha before seedling densities can increase in uncut forest. In this example, there are two densities of seedlings with the same density of deer. The existence of the two states within a given area of forest depends on its stage structure.

A second example is seen in the interaction of elephants feeding on *Acacia* trees in Serengeti (figure 8.10*b*). In the first half of the 20th century, there was high tree density (30% cover). Elephant numbers were low as a result of heavy elephant hunting in the 1800s, but they gradually increased throughout the 1950s and 1960s, feeding on mature trees but

(*a*)

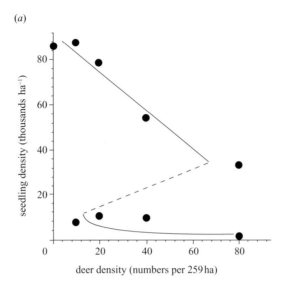

(*b*)

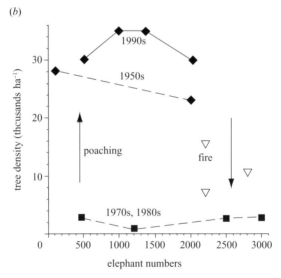

Figure 8.10. Examples of multiple states. (*a*) Tree seedlings in North American eastern hardwood forest can exist at two densities under the same density of white-tailed deer herbivores. The two states depend on the stage structure of the rest of the forest that is disturbed by clear cutting (from Schmitz & Sinclair 1997). (*b*) In Serengeti, there are two levels of *Acacia* tree cover. Elephants do not reduce high levels but do maintain low levels of tree cover. Fire reduced tree cover to low levels and poaching reduced elephant numbers to low levels (data from Norton-Griffiths 1979; Dublin *et al.* 1990; Dublin 1995; A. R. E. Sinclair, unpublished data).

not substantially reducing their number. Widespread and frequent hot grass fires throughout this same period resulted in little recruitment of juvenile trees, senescence of old trees and eventual precipitous decline in tree density at the end of the 1960s (Norton-Griffiths 1979). With low tree densities, elephants were able to prevent tree regeneration by their intense de-predation of seedlings and thus they maintained a grassland state throughout the late 1970s and early 1980s (Dublin *et al.* 1990; Dublin 1995). This state was maintained in Serengeti despite reduction of elephants through rampant poaching in the 1980s until numbers dropped to some 20% of their original density. At this stage, tree recruitment escaped elephant predation and a new period of high tree cover has returned in the 1990s and 2000s. This high tree density is persisting despite an exponential increase in elephants since poaching was stopped (Sinclair 1995; A. R. E. Sinclair, unpublished data). In essence, two states existed in tree cover and external perturbations from fire and poaching were required to change the state from high to low and from low to high cover, respectively.

8.6.   Predator–prey cycles and combined top-down and bottom-up effects

Systems that have both bottom-up and top-down regulation can produce lag effects from predation that result in population cycles of all three trophic levels. This is illustrated by the snowshoe hare (*Lepus americanus*) cycle in Canada (figure 8.11) (Krebs *et al.* 2001a). During the increase phase of the hare population cycle, *r* declines as predicted by resources (see figure 8.1). During the decrease phase, *r* remains negative and does not follow the same path as that for the increase phase due to delayed density-dependent effects of the specialist predators, lynx (*Lynx canadensis*), coyotes (*Canis latrans*) and great-horned owls (*Bubo virginianus*). Consequently, the hare population collapses to about one tenth of the highest density. At this point, *r* jumps to a higher level again as hares respond to abundant food resources. Hence, we see two growth curves, due in this case to interacting bottom-up and top-down processes.

8.7.   Rate of increase at very low population sizes

(a)   Disturbance
Disturbance is particularly important at low population sizes when density-dependent effects are curvilinear and so are weak at low density.

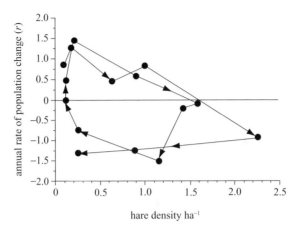

Figure 8.11. Rate of population increase (*r*) in relation to snowshoe hare density at Kluane, Yukon, from 1986 to 2001. These data comprise the better part of two hare cycles. The rate of increase for the decline phase of the hare cycle follows a different track from that for the increase phase due to time delays in both food and predator effects (Data from Krebs *et al*. 1999; C. J. Krebs, unpublished data).

We have already illustrated such curvilinear density dependence in the wildebeest in §8.2, figure 8.1. Other examples are European bison (*Bison bonasus*), northern fur seals (*Callorhinus ursinus*) and other species illustrated in Fowler & Baker (1991). Under these conditions of curvilinear density dependence, there is little or no compensation for stochastic disturbances and populations can exhibit a random walk towards extinction (Mangel & Tier 1994) or simply be held at low levels. The heath hen (*Tympanuchus cupido*) drifted to extinction once it dropped below a certain threshold (Allee 1938). Rates of increase relative to population size then exhibit a hyperbolic curve, as seen for American bison and pronghorn antelope in Yellowstone National Park (figure 8.12*a,b*).

Stochastic effects can occur in two ways. First, environmental disturbances from weather (cold, drought, storms) are random. They reduce *r* and so they can eliminate small populations (Tuljapurkar 1997; Efford 2001). The population of 'ou' (*Psittirostra psittacea*), an endemic bird in the forests of Kauai, Hawaiian islands, numbering fewer than 20, was in danger of being extirpated by the hurricane that devastated the island in 1992 (Pratt *et al*. 1987). Other chance effects are the appearance of predators, or epidemic diseases, whose impacts become destabilizing for very small prey populations. The Vancouver Island marmot (*Marmota vancouverensis*) in British Columbia numbers fewer than 30 (Bryant & Janz 1996). A single

(a)

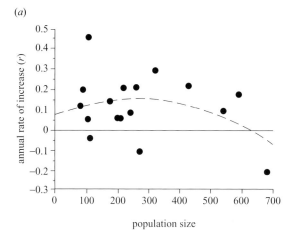

(b)

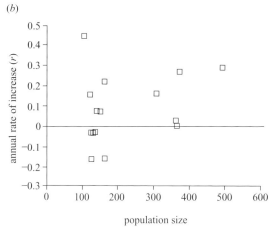

Figure 8.12. Per capita rates of increase (*r*) that are depressed at low densities in large-mammal populations: (*a*) Yellowstone plains bison, and (*b*) Yellowstone pronghorn antelope. (Data from Singer & Norland (1994), after Sinclair (1996)).

wolf killed several individuals in a few days in 2001, causing a further decrease in the population and increasing the probability of extinction (A. Bryant, personal communication). The Atlantic northern right whale (*Eubalaena glacialis*) population has been unable to increase above 300 animals in the past 80 years due to persistent mortality from ships, fishing gear and climate change (Fujiwara & Caswell 2001). By contrast, environmental stochasticity may enhance rather than depress *r* in plants that have a storage life-stage (Higgins *et al.* 2000, 2001).

Second, stochastic effects that reduce *r* at very low population sizes can take place from chance distortions of demography (Lande 1998). This can occur through a sex ratio that drifts towards too many males, as occurred in the flightless Kakapo parrot (*Strigops habroptilus*) on the south island of New Zealand or in the Chatham Island robin (*Petroica traversi*). In the former case, all females died (though other island populations are just surviving); in the latter case, two remaining females have rescued the population (Morris & Smith 1988; Clout & Merton 1998).

(b)   Social effects on the rate of increase
(i)   The Allee affect
Allee (1931, 1941) described the advantages of group-living, first, in overcoming hazards of the environment (such as extreme temperatures, radiation etc.) and avoiding predation that single individuals experience. Second, at very low population levels, widely dispersed individuals might have difficulty in finding mates and hence their reproduction would be lower than those living in groups at the same population size (Dennis 1989; Courchamp *et al.* 1999; Stephens & Sutherland 1999). Third, very small populations might experience deleterious genetic effects from inbreeding that would both reduce reproductive capacity and survival of progeny. In general, therefore, these social and genetic features of a species (the intrinsic effects) result in a decreased *r* at very low population sizes, exacerbating other environmental and demographic stochastic effects.

(c)   Trophic level effects
McNaughton (1979a,b, 1983) has shown that plants sometimes respond to moderate levels of herbivory by growing more than they would at very low levels of herbivory. This is observed in the regrowth of willow and birch twigs that are fed upon by snowshoe hares in winter (Krebs *et al.* 2001b). This suggests that, up to a point, a growing population can promote recovery in its own per capita availability of food. Most probably, this effect pertains to herbivores rather than to higher trophic interactions. The consequence of compensatory growth in plants is that herbivore populations at extremely low densities may have reduced *r* relative to populations at somewhat higher density.

In summary, the combined effects of stochastic, social and trophic processes both reduce the mean per capita growth rate at very small population size and increase the variance around *r*. The population level where some measure of the range of variability of *r* rises above zero could represent a MSP that should be achieved for conservation (figure 8.13). We

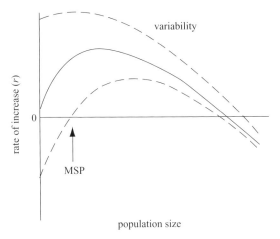

Figure 8.13. The combined effects of stochastic, social and trophic processes reduce the mean per capita growth rate and increase the variance at very small populations. The population level where the variability in r is above zero represents the MSP that should be achieved for conservation.

note this only applies where we can detect a consistent relationship between r and population size. Although all these effects have at times been included in the 'Allee effect' (Fowler & Baker 1991; Courchamp *et al.* 1999), we find it useful to distinguish the different mechanisms that reduce r at very small populations because they have different management implications for the restoration of declining populations.

8.8.   Other complex effects on the rate of increase

(a)   Dispersal behaviour

Even if bottom-up regulation, driven by food supply, is the primary process affecting r, both social behaviour and dispersal are major processes that can alter the relationship between r and population size. If dispersal does not occur in a linear density-dependent way but rather when thresholds of density are reached, then a population will grow in an eruptive fashion. Pulses of range expansion are followed by density increases within patches. Growth rate will show cycles in relation to overall density. This was postulated for Himalayan thar (*Hemitragus jemlahicus*) in New Zealand (Caughley 1970b). Recently pulsed population growth has been demonstrated for elk in Yellowstone National Park (Lemke *et al.* 1998), muskox (*Ovibos moschatus*) on the tundra (Reynolds 1998) and wood bison (*Bison bison*) in Canada (figure 8.14) (Larter *et al.* 2000).

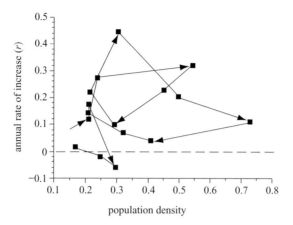

Figure 8.14. Rate of population growth (r, yr^{-1}) of the wood bison in the Mackenzie Bison Sanctuary, Northwest Territories, Canada, after their introduction in 1964 (starting arrow) until 1998. Pulsed dispersal in the wood bison results in cycles of r relative to population size. At each dispersal event, a new area is incorporated so that density drops but the population remains close to stationary. The new area then allows a rapid increase in r after a lag of 2–3 years, followed by an increase in density again. (After Larter *et al.* 2000).

In some cases, animals can disperse at low density because there is vacant space available to them, but are prevented from doing so at a high density because other space is already filled. In these cases, dispersal is inversely density-dependent, allowing overall population growth at low density but not at high density because dispersal is prevented by a 'social fence' (Hestbeck 1982, 1987). Social interactions replace dispersal in limiting population growth at the local level by affecting other demographic rates either through reproduction or mortality.

Dispersal is also relevant to metapopulations through 'rescuing' small local populations that go extinct (Gotelli 1991). Thus, dispersal allows the metapopulation to become more stable than any of the local populations (Harrison 1991; Hanski 1998). In essence, population growth rate for any single population is influenced by immigration and emigration and not just by births and deaths.

(b)   Climate change

Slow change in one direction in an ecosystem, such as an increase in plant productivity due to global warming, will have nonlinear consequences on growth rates of higher trophic levels due to several of the processes mentioned above. First, top-down regulated systems could switch from low prey densities to high if prey reproductive rates increase through increases in food. Outbreaks of insects could occur this way (Myers 1993). Alternatively, if plant communities change in dominant species from palatable plants to unpalatable plants, then food for herbivores could switch to a low level and this would affect all higher levels. The switch from dominance by green algae in lakes to dominance by unpalatable blue-green algae when nutrients are added is one example of this effect (Smith 1983).

8.9.   Conclusion

Food supply is the primary factor determining population growth rate in animal populations, and we postulate bottom-up control as the universal primary standard. However, bottom-up control can be overridden or severely modified by three secondary processes: top-down processes from predators, social interactions within the species and stochastic disturbances. Interactions between these four controls produce the variety of complex, nonlinear effects on population growth that we see in nature.

We develop generalizations to classify populations. Bird populations are driven by primary food limitations coupled with social interactions over territories. Food supply drives changes in large-mammal populations and top-down processes rarely intervene. Small mammals may be affected more by top-down controls, coupled with social interactions, and rarely seem to have their population growth limited by food supplies. Fish populations and many invertebrates, by contrast, seem to have their population growth affected more by stochastic disturbances affecting recruitment processes through primary food limitation. These generalizations should be considered hypotheses to be tested by studying the comparative dynamics of many populations.

Conservation and management of populations depend critically on what factors drive population growth, and we need to develop universal generalizations that will relieve us from the need to study every single population before we can make recommendations for conservation and management.

We thank The Royal Society and the Novartis Foundation for support. We are grateful to Richard Sibly, Jim Hone, David Choquenot and another referee for helpful comments.

9

The numerical response: rate of increase and food limitation in herbivores and predators

9.1. Introduction

(a) Herbivores and predators: types of consumer–resource systems

The resources used by animal populations are either non-consumable or consumable (Caughley & Sinclair 1994). While the absolute level of non-consumable resources is generally not influenced through its use (e.g. shelter), the level of consumable resources is (e.g. food). The most comprehensive classification of the relationship between resources and animals is that developed for grazing systems by Caughley & Lawton (1981). They accounted for the degree to which herbivores interact with their food resources and interfere with each other's capacity to access those resources. Interactive grazing systems are those in which herbivore consumption influences the rate of renewal of food plants, which in turn influences the dynamics of the herbivore population itself. Interactive grazing systems are further differentiated into interferential systems in which herbivores can affect each others capacity to assimilate food plants, and laissez-faire systems in which they do not. Non-interactive grazing systems are those in which herbivore feeding has no influence on the rate of renewal of food plants and, hence, no reciprocal influence on the dynamics of the herbivore population. Non-interactive grazing systems are differentiated between reactive systems in which rate of change in herbivore abundance is a function of food plants, and non-reactive systems in which herbivore population dynamics are largely independent of food availability. We argue that this classification encompasses the range of mechanisms that link most animal consumer systems to their food resources and so is applicable to both herbivores and predators. Any food

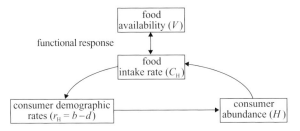

Figure 9.1. A diagram describing a Lotka–Volterra model of interaction between a predator (consumer) and its prey (food). Symbols are those used in equations (9.2.3) and (9.2.4): food availability (V); food intake (C_H); consumer demographic rates (annual rate of increase r_H = births − deaths or $b - d$); and consumer abundance (H).

resource available to an animal population has the potential to elevate average reproduction and/or survival. The availability of food resources to an animal population will be potentially reduced through the use of those resources by the animal population itself (i.e. the negative feedback loop).

(b)   Food availability and consumer abundance (a short history)

Solomon (1949) recognized that an increase in food availability would generally elicit two responses in a consumer population limited by those food resources: a 'functional response' which elevates the per capita rate of food intake, and a consequent 'numerical response' which increases consumer abundance through enhanced reproduction, survival or both. By directly linking food availability and consumer population demography and abundance through the numerical response, Solomon (1949) was generalizing features of more specific models of trophic interaction (primarily Lotka–Volterra predator–prey models) to his central theme of animal population regulation. These models assume that both prey (food) mortality due to predation and predator (consumer) survival are proportional to the product of food (H) and consumer (P) abundance (i.e. bHP and cHP respectively). In effect, this implies that both the functional response of consumers to variation in food availability and the consequent change in consumer demographic rates are linear, indicating that the transfer of biomass from the food to consumer populations is conserved. Perhaps more importantly, the structure of the model drives changes in consumer abundance according to the direct effect that food intake rate has on consumer demography (figure 9.1).

Since Solomon's original definition, two types of numerical response have been defined and used to help elaborate the broad interactive dynamics between consumer populations and their food. These are: (i) a 'demographic' numerical response that links rate of change in consumer abundance to food availability (Caughley 1976; May 1981a); and (ii) an 'isocline' numerical response that links consumer abundance *per se* to food availability (see Holling (1965, 1966) for total predator responses).

In this paper, we review how both approaches to the numerical response have been used to enhance understanding of herbivore and predator population regulation, and attempt to increase their realism and utility by explicitly accounting for the existence of non-equilibrium dynamics due to environmental variability, biological interactions and situations where multiple factors simultaneously limit rates of change in population abundance. The different approaches to describing numerical responses have also been recently reviewed by Sibly & Hone (Chapter 2).

9.2.   Demographic numerical responses

(a)   Single-species logistic models of population growth

The dominant paradigm in large-herbivore ecology proposed that density-dependent mortality regulates population density through food shortage (i.e. the so-called 'food hypothesis' (Sinclair *et al.* 1985)). Most tests of the relevance of this hypothesis to large herbivores have either reduced herbivore population density (or allowed a natural catastrophe to do so), and assessed whether the population returns to its pre-reduction level (Houston 1982; Sinclair *et al.* 1985), or looked for density dependence in *r* or some valid demographic correlate of *r* (i.e. growth, body condition, fecundity or survival) (O'Roke & Hammerston 1948; Woodgerd 1963; Boyd & Jewell 1974; Sinclair 1977; Sauer & Boyce 1983; Skogland 1983, 1985; Messier & Crête 1984; Clutton-Brock *et al.* 1985b; Eberhardt 1987; Fryxell 1987; Choquenot 1991; Messier 1991). Both of these approaches focus on the dynamics of the herbivore population, interpreting any decline in *r* or its index as the population moves towards its hypothetical equilibrium as the effect of declining per capita food availability (May 1981b). Because these tests do not consider food explicitly, they are either implicitly or directly underpinned by single-species models of interaction between herbivores and their food resources (Caughley 1976). The simplest model that is generally applied to herbivore populations is the generalized logistic which has the form

$$r = r_\mathrm{m} \left(1 - \left(\frac{N}{K} \right)^z \right),$$

$$(9.2.1)$$

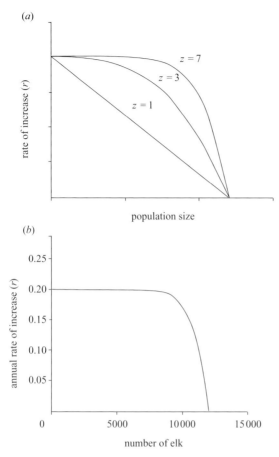

(a)

rate of increase (r)

z = 7

z = 3

z = 1

population size

(b)

annual rate of increase (r)

0.25

0.20

0.15

0.10

0.05

0   5000   10000   15000

number of elk

Figure 9.2. Hypothetical relationships between (a) rate of population increase (r) and population size predicted from the generalized logistic model in which z is varied from 1 to 3 (see lines on graph), and (b) the generalized logistic model for Yellowstone elk estimated from population census data (Eberhardt 1987). The parameter values estimated for elk are $r_m = 0.2$ p.a., carrying capacity $K = 12\,000$ and $z = 11$.

where r_m is the maximum rate of increase, K is the density of the herbivore population where the rate of renewal in food resources is just sufficient to balance reproduction and survival (where $r = 0$), N is prevailing population size and z is a coefficient describing the degree to which the density-dependent decline in r with N is delayed until higher levels of N are attained (Fowler 1981, 1987b; figure 9.2a). The value of z reflects the degree to which the amount of food currently available to herbivores is determined by the number of herbivores currently consuming that food ($z = 1$), or the number that have fed on the food in the past ($z > 1$). Eberhardt (1987) used a fairly high value of $z = 11$ in fitting equation (9.2.1) to population census data for elk (*Cervus elaphus*) in Yellowstone National Park in the western United States, implying that current food availability was heavily dependent on past elk density. Delayed effects of density on r mean that most density dependence is observed at densities near K (figure 9.2b).

The most pressing limitation of single-species models for large herbivores (and hence on tests of the food hypothesis based on single-species models) is that K must be assumed to be relatively constant if the relationship between population density and r is to be consistent (and hence detectable) (Caughley 1976; Choquenot 1998). The importance of this assumption can be illustrated by contrasting the growth trajectories for elk projected from Eberhardt's (1987) model, where K is alternatively stable (1% year-to-year variation in K; figure 9.3a) or unstable (5% year-to-year variation in K; figure 9.3b). While growth towards equilibrium follows a clearly density-dependent trajectory where K is relatively stable, density-dependent population growth is not evident where K is less stable.

(b)   Interactive consumer–resource models

The more explicit demographic numerical response links change in consumer demographic rates to food availability. In contrast to the single-species density-dependent approach above, consumer–resource models are by nature multi-species models, but only in the sense that separate predator and prey components are explicitly modelled and linked. Additionally, whilst only one predator species is usually modelled, prey may involve all food species lumped on one axis or a subset of most important food species. As a step in formulating an interactive plant–herbivore system, Caughley (1976) described a demographic numerical response (after May 1981a) that linked variation in herbivore demographic rates (summarized by the instantaneous rate of population increase, r_H), to the biomass of available food (V):

$$r_H = -a + c_1(1 - e^{-V d_1}), \qquad (9.2.2)$$

where a is the maximum rate at which the population declines in the absence of food, c_1 is a constant describing the difference between the maximum rate at which the population can increase (r_{mH}) and a (i.e. $r_m = c_1 - a$), and d_1 is the demographic efficiency of the population indexing how quickly r changes from being negative to positive as vegetation

(*a*)

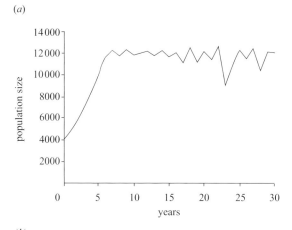

(*b*)

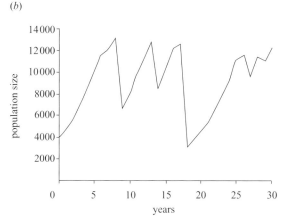

Figure 9.3. Trajectories of growth for elk populations predicted from a generalized logistic model estimated by Eberhardt (1987), with stochastic variation in *K* equivalent to (*a*) 1% of the mean, and (*b*) 5% of the mean value of *K*.

biomass increases. The general form of the response and a diagram of the full interactive model are shown in figure 9.4*a,b*.

The other components of Caughley's interactive model were the growth of ungrazed vegetation and the herbivore functional response. Vegetation growth in the absence of grazing was modelled using a simple density-dependent logistic function to link the instantaneous rate of change in vegetation biomass (r_V) to standing biomass (*V*):

$$r_V = r_{mV} \left(1 - \frac{V}{K} \right), \tag{9.2.3}$$

(a)

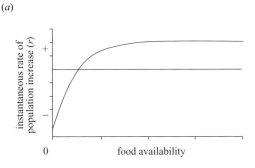

(b)

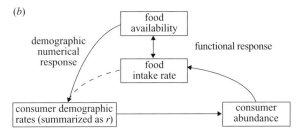

Figure 9.4. (*a*) The general form of a demographic numerical response described by Caughley (1976), and (*b*) the structure of the interactive model within which the response was used. The dashed line in (*b*) indicates a relationship that is explicit in Lotka–Volterra models but is subsumed by the demographic numerical response in Caughley's interactive model.

where r_{mV} is the maximum rate of increase in vegetation biomass and K is vegetation biomass where shading or competition for water or nutrients limits further plant growth. The herbivore functional response, which describes the increase in per capita vegetation offtake by herbivores (C_H) with increasing vegetation biomass, was modelled using the same exponential form as the numerical response:

$$C_H = c_2 \left(1 - e^{-(V - V_g)d_2}\right), \tag{9.2.4}$$

where c_2 is the maximum rate of vegetation intake by each herbivore, V_g is the vegetation biomass where intake by herbivores falls to 0 (i.e. the value of V_g determines whether or not the curve goes through the origin; May 1981*a*, table 9.5), and d_2 is the efficiency of the functional response describing how rapidly vegetation intake increases to its maximum rate with increasing vegetation biomass.

The important differences in the model developed by Caughley and the Lotka–Volterra model described above are: (i) the curvilinear functional and numerical responses (equations (9.2.1) and (9.2.3)); and (ii) the fact that Caughley's numerical response links consumer demography directly to food availability rather than food intake rate. The more complex form of the functional response used in Caughley's model accommodates more sophisticated ideas on how food availability and other environmental factors influence animal foraging behaviour (e.g. Watt 1959; Ivlev 1961; Allden 1962; Holling 1966). In applying the same general form to the demographic numerical response, Caughley's model simply allows the possibility that transfer of biomass between adjacent trophic levels is conserved in the same way as is assumed in Lotka–Volterra models (i.e. maximum reproduction and survival is dependent entirely on the rate of food acquisition). Under these conditions the functional and numerical responses can be parameterized so that maximum rates of increase (r_m) are approached at levels of food availability that produce maximum rates of food intake (C). This would reproduce the linear relationship between the rate of food intake and rate of change in predator abundance used in Lotka–Volterra models. Of course, other forms of this relationship are possible. Crawley (1983) argued that the relationship between food intake rate and r would be curvilinear where (i) maximum reproduction or survival was limited by factors other than food intake or (ii) a threshold rate of food intake was required before reproduction was possible. Different forms for the numerical response would need to be considered if these alternatives were to be accommodated.

While the demographic numerical response used in the interactive model subsumes the direct link between food intake and animal demography, it provides a very powerful summary of the indirect effect that food availability, as a limiting factor, has on an animal population. Demographic numerical responses to food availability have been estimated for a range of herbivores including kangaroos (Bayliss 1987), brush-tailed possums (Bayliss & Choquenot 1998), wild pigs (Choquenot 1998) and wild house mice (Pech et al. 1999). The primary use to which these numerical responses have been put is in the development of simulation models that explore dynamic interactions between herbivore populations and their limiting food resources (Caughley 1976, 1987b; Caughley & Gunn 1993; Bayliss & Choquenot 1998; Choquenot 1998).

Perhaps one the best examples of how demographic numerical responses can be applied to help understand interactions between animal populations and their food resources is the work of Caughley (1987b) and his co-workers. They estimated the components of the interactive model described by equations (9.2.2), (9.2.3) and (9.2.4) for the grazing system comprising red kangaroos and native pastures in Australia's eastern rangelands. This grazing system is highly stochastic, being driven by the vagaries of rainfall, which varies up to 47% from year to year, with low correlation between years and between seasons within years. This highly stochastic variation leads to wide, seemingly random fluctuations in the abundance of kangaroos and the pastures they feed on. Between 1977 and 1985, Caughley and his co-workers exploited these natural fluctuations to estimate the form of density-dependent pasture responses to rainfall (Robertson 1987a,b), and the numerical response of kangaroos to pasture biomass (Bayliss 1985a,b, 1987). During that time, the functional response describing pasture intake by kangaroos to changes in pasture biomass was also estimated in a series of graze-down trials using captive kangaroos held in semi-natural enclosures (Short 1985, 1987).

The vegetation response obtained was modified from that described in equation (9.2.3) to account for empirically estimated effects of variation in rainfall on pasture growth and die-back, over and above the density-dependent effects of pasture biomass. The modelled vegetation response was

$$\Delta V = -55.12 - 0.01535V - 0.00056V^2 + 3.946R, \tag{9.2.5}$$

where ΔV is the pasture growth increment over three months in the absence of grazing, V is pasture biomass at the start of those three months and R is the rainfall in millimetres over that period. The pasture growth increment was taken as a random draw from a normal distribution with mean equal to the solution of equation (9.2.5) and a standard deviation of 52 kg ha^{-1}, equivalent to the variation in pasture growth not accounted for by rainfall and standing biomass (Robertson 1987b). The functional response of red kangaroos (Short 1985) was estimated as

$$C = 86(1 - e^{-V/34}), \tag{9.2.6}$$

(assuming an average body weight of 35 kg), and their demographic numerical response as

$$r_H = -1.6 + 2(1 - e^{-0.007V}). \tag{9.2.7}$$

The dynamics of the grazing system was simulated over 100 years, with seasonal rainfall drawn from normal distributions with means and standard deviations estimated from long-term records. Successive pasture growth increments were estimated from equation (9.2.5), and changes in

(a)

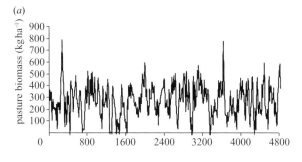

(b)

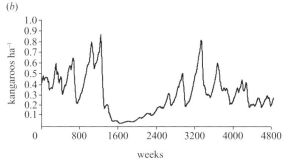

weeks

Figure 9.5. Temporal variation in (*a*) pasture biomass and (*b*) kangaroo density, predicted from a model of interaction between kangaroos and pasture, developed by Caughley (1987*b*).

the per capita rate of pasture consumption and kangaroo density from equations (9.2.6) and (9.2.7). Changes in pasture biomass and kangaroo density were accounted weekly. Figure 9.5*a,b* shows changes in pasture biomass and kangaroo density, respectively, from a typical run of the model.

Caughley (1987*b*) found that despite high season-to-season variation in pasture biomass, and year-to-year variation in kangaroo density, kangaroos persisted indefinitely in the modelled grazing system, neither crashing to extinction nor increasing without limit. Stochastic rainfall variation led to dramatic fluctuations in pasture biomass that were largely independent of kangaroo density. These fluctuations constantly buffeted the grazing system away from its potential equilibrium, creating the rapid oscillation in pasture biomass and large swings in kangaroo density evident in figure 9.5*a,b*. However, despite this constant buffeting, the reciprocal influence that kangaroos and pasture exerted over each other's abundance imparted a sufficiently strong tendency towards equilibrium

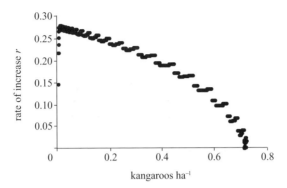

Figure 9.6. Relationship between density and rate of increase for kangaroos in the absence of stochastic variation in rainfall and pasture growth, predicted from a model of interaction between kangaroos and pasture developed by Caughley (1987*b*).

(i.e. 'centripetality'), that kangaroos persisted indefinitely. The tendency that the grazing system has towards equilibrium is driven essentially by density-dependent competition amongst kangaroos for available pasture. The fact that kangaroos compete for pasture is evident in the 43% increase in average predicted pasture biomass that occurs when kangaroos are removed from the model. This indicates that the grazing system achieves centripetality through the same trophic processes that are represented implicitly by the density dependence of single-species models of other herbivore populations (Sinclair 1989). For example, in the absence of density-independent fluctuations in pasture biomass, the pattern of variation in r for kangaroos with density conforms to that of a generalized logistic model (figure 9.6). Hence, the interactive model is the general case for vegetation–herbivore systems, single-species models being a 'short-hand' or 'contracted' version that represents the statistical association of herbivore density and rate of increase that emerges when density-independent perturbation of these systems is low or uncommon.

(c)   Consonance with observation
(i)   Non-equilibrium dynamics (including multiple equilibria):
      environmental variability and kangaroos
Bayliss (1987) developed numerical response models for red and western grey kangaroos in two locations (a national park and a sheep station). Caughley (1987*b*) used a slightly modified version of these functions to simulate overall grazing system dynamics. An Ivlev (1961) function was

fitted (figure 9.7*a*) to the rate of increase versus food availability data for red kangaroos, using maximum likelihood estimation. Results here are for red kangaroos on a national park. Similar patterns were found for both kangaroo species in all locations. The a priori model assumes that food is the major proximate factor that regulates kangaroo population dynamics. Two post-drought outliers were hence excluded from the original analysis because they did not fit the a priori assumption (figure 9.7*a*). However, a time-trace of the rate of increase data (figure 9.7*b*) shows that the population dynamics of kangaroos entering a drought from conditions of high food abundance is quite different from that for populations recovering from conditions of low food abundance during a drought. For populations recovering from drought, rates of increase remain negative despite high levels of food (pasture biomass). This 'hysteresis' or 'lens' pattern (resulting in two equilibria where $r = 0$) may reflect the existence of two alternate system states over a drought cycle, where the transition between each is across a threshold or break point.

A density-dependent (isocline) numerical response model demonstrates the non-equilibrium system properties more clearly (figure 9.7*c*). Results are for western grey kangaroos on Kinchega National Park; however, similar patterns were found for both species in all locations. A time-trace of rate of increase with a six-month time-lag shows clearly the break point or hysteresis between two postulated 'domains of attraction', reflecting drought and non-drought conditions. Although lagged density may confound both extrinsic (pasture food) and intrinsic (spacing behaviour) regulation processes, P. Bayliss and D. Choquenot (unpublished data) argue that the two parallel and negative linear correlations may simply reflect the high and low phases of a stable limit cycle (i.e. an open-ended ellipse). Once again, this graphical analysis points to the existence of two equilibria (where $r = 0$), one at high density and high pasture biomass, the other at low density and low pasture biomass. Both equilibria may be locally stable but globally unstable because pasture biomass is driven largely by stochastic rainfall events (Robertson 1987*a,b*).

Hence, globally, the kangaroo grazing system is a non-equilibrium system but with two postulated local 'domains of attraction' towards stability. The trajectories and positions within this binary system depend critically on initial conditions of pasture biomass (food) and kangaroo density. Populations at high density entering a drought exhibit different population dynamics from low-density populations emerging from a drought (e.g. different sex and age structures, response time-lags and reproductive

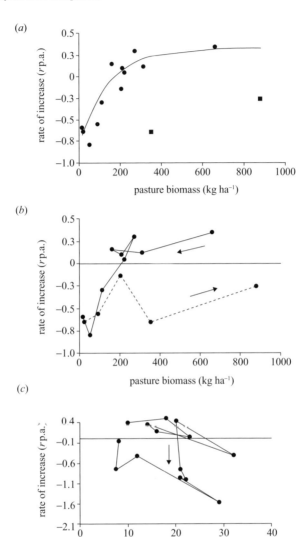

Figure 9.7. (*a*) Demographic numerical response (*r* p.a.) for red kangaroos and their food availability (pasture biomass *V*, kg ha^{-1}). The fitted function is $r = -0.8 + 1.14 \times (1 - e^{-0.007V})$, which excludes two post-drought data (square symbols; Bayliss (1987)); (*b*) time-trace of the same rate of increase data including previously discarded outliers (solid line, high-density populations entering a drought; dotted line, low-density populations leaving a drought); and (*c*) time-trace of western grey kangaroo rate of increase versus lagged density (km^{-2}, six-month time-lag), showing the break-point transition between two domains of attraction (drought and non-drought conditions).

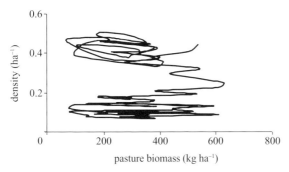

Figure 9.8. Phase plane trajectory plotting isoclines of pasture food (V, kg ha^{-1}) and kangaroo density (D, ha^{-1}) using a demographic numerical response model. Note the apparent existence of two domains of attraction at high (H) and low (L) densities. Each domain is locally stable (centripetal) but globally unstable because of unpredictable changes in rainfall-driven pasture biomass.

condition (see Bayliss 1980; Cairns & Grigg 1993)). The phase plane trajectory (figure 9.8), or time-trace of zero isoclines of kangaroo density (H, ha^{-1}) and food abundance (V, kg ha^{-1}), clearly illustrate the dynamics between alternate periods of very high and low to medium kangaroo densities.

This result is surprising given that the numerical response model is in fact an equilibrium model applied to a stochastic environment. A probable cause may be the asymmetrical relationship between rate of increase and the availability of food, which is in itself driven largely by stochastic rainfall events; for the same amount of rainfall about the annual mean (where $r = 0$), a much greater rate of decrease occurs than a rate of increase. Hence, rainfall variance reduces long-term mean density (Caughley 1987b), but may also create a 'two-state' system (a run of high-density periods followed by a run of low-density periods; see figure 9.5b).

Although not incorporated into the Caughley (1987b) 'structural' interactive grazing model, the composition of pastures is also likely to differ between pre- and post-drought domains, which should add yet another dimension of complexity and stability. Pastures of course respond in two ways to grazing (see McNaughton (1979a) for grassland–herbivore dynamics in the Serengeti), either through changes in biomass and productivity (modelled here), or shifts in pasture composition (not modelled here). Surprisingly, even eastern grey kangaroo populations living in more stable seasonal, temperate environments of eastern Australia exhibit non-equilibrium behaviour. Figure 9.9 shows an apparent 4–5-year

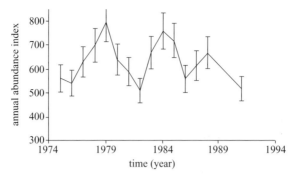

Figure 9.9. Annual trends in an index of abundance (numbers observed km^{-1} ± s.e.) of eastern grey kangaroos on Tidbinbilla Nature Reserve (ACT, Australia), showing apparent 4–5-year stable limit 'cycles' between 1975 and 1991 ($n = 17$ years). Kangaroos were counted at night by spotlight along fixed transects across the reserve.

stable limit cycle for eastern grey kangaroos on Tidbinbilla Nature Reserve, ACT, Australia (P. Bayliss, unpublished data). Although we do not know what causes these apparent cycles (fox predation, competition with rabbits and/or other macropods, disease, El Niño, seasonality effects or a combination of causes), or if in fact they are cycles (coincidence), the system is definitely not an equilibrium system as we would predict from the Caughley (1976, 1987b) interactive kangaroo grazing model applied to more stable environments. Hence, kangaroos appear to be very good examples of non-equilibrium systems because their numbers over time are characteristically unstable; the abundance of all censused populations of kangaroos varies widely.

(ii)   Biological interactions and possums in New Zealand

Caughley & Krebs (1983) identified two categories of regulation in order to explain the apparent dichotomy studies of small and large mammal population dynamics. One is intrinsic (self) regulation, where rate of increase (r) is suppressed by some form of spacing behaviour (or physiological process) as density increases. This type of regulation is generally expressed in terms of the negative prediction between r and instantaneous density (e.g. single-species logistic models). The other is extrinsic regulation, where r is governed by the relationship between the consumer and an external factor (food availability, predation, disease, weather or a combination of factors). However, Erb et al. (2001) found that patterns of population dynamics in small versus large mammals contradict those predicted by the Caughley & Krebs (1983) hypothesis. Nevertheless, their distinction between intrinsic and extrinsic regulatory mechanisms remains an important distinction and is retained here. The term 'density dependent' generally refers to a prediction between r and density, and 'density independent' a lack thereof. However, density per head of population may index a limiting resource (extrinsic regulation) and/or spacing behaviour (intrinsic regulation). Because the population regulatory mechanisms are not explicitly defined, single-species 'density-dependent' models often subsume or hide biological processes rather than expose them.

Although extrinsic and intrinsic regulation are not mutually exclusive they are generally modelled as such. However, some populations may be regulated by both mechanisms (e.g. grazing interference or facilitation may attenuate the numerical response to food availability; see Vessey-Fitzgerald 1968). A theoretical framework for such combined regulatory influences on a species population dynamics already exists (e.g. Caughley & Lawton 1981; Caughley & Krebs 1983), and is encapsulated in a class of numerical response models called interferential models (Caughley & Lawton 1981; Caughley & Krebs 1983; Barlow 1985). Caughley & Lawton (1981) examined a common form of the interferential numerical response, such that

$$r_H = r_m \left(1 - \frac{JH}{V}\right), \tag{9.2.8}$$

where r_H and V are as defined previously, and J is a proportionality constant related to the availability of food needed to sustain consumer H at equilibrium. However, Barlow (1985) argued that, despite its widespread use, this particular type of interferential numerical response model is biologically meaningless. Nevertheless, the addition of an intrinsic density-dependent factor which is unrelated to the extrinsic availability of food is easily expressed by including a density term in laissez-faire numerical response models exemplified by equation (9.2.2). A good example is that provided by Tanner (1975) and explored by Caughley & Krebs (1983), such that

$$r_H = -a + c_1(1 - e^{-Vd_1}) - gD, \tag{9.2.9}$$

where r_H, a, c_1, V and d_1 are as previously defined, with D being instantaneous density and g a coefficient depending on the magnitude of the effect. Ginzburg (1998) suggests that it is preferable to keep separate any terms in the numerical response function which reflect unrelated biological phenomena, and this is the approach adopted by Caughley & Krebs (1983), Bayliss & Choquenot (1998) and Pech *et al.* (1999). This is a more realistic and explicit model of interference or aggregation effects than the model proposed by Caughley & Lawton (1981), as it separates the food intake and self-regulation terms in the consumer numerical response whilst leaving the nature of the self-regulation unspecified, as highlighted by Barlow (1985).

Bayliss & Choquenot (1998) suggested that interferential numerical response models of equation (9.2.9) may provide a more useful framework for understanding population dynamics because the combination of extrinsic (consumer–resource) and intrinsic (animal–density) regulation processes may embody the much broader spectrum of population regulation mechanisms that most probably exist in species. They examined this proposition for the introduced possum in New Zealand forests, which is summarized below.

Population models developed for possums *per se* have curiously been entirely single-species logistic models (e.g. Barlow & Clout 1983) and, hence, ignore plant–herbivore interactions despite the marked impact that possums have on native forests (although Barlow (1991) and Barlow *et al.* (1997) developed comprehensive multi-species possum-disease models as extensions of the earlier single-species logistic model). Hinau (*Elaeocarpus dentatus*) is an endemic New Zealand hardwood tree that lives up to 400 years and is known to mast; fruit production alternates between periods of low and super abundance in cycles of two plus years. Possums are primarily folivorous, but also feed extensively on fruits, flowers and buds of many native trees such as hinau. Hinau fruit is a critical winter food source and may even index all winter food sources (Bell 1981). Bell found that the birth date of possums, the percentage of females with pouch young and body weight all were positively correlated to the annual crop of hinau fruit. Cowan & Waddington (1990) found that hinau fruit production increased dramatically when possums were eradicated, and that fruit production was suppressed again with subsequent recolonization. The ecological relationships between possums and food availability were examined in greater detail in two stages, using hinau fruit to index food availability. First, the physiological relationships between possums and hinau abundance were examined and used to underpin the population-level analyses. Second, an interferential numerical response model was developed using the availability of hinau fruit as an index of food supply overall and 'instantaneous density' to index possible spacing behaviour effects (e.g. feeding interference and/or competition for nest sites) which may be independent of the effects of food. We argue that behavioural density-dependent effects are likely to be instantaneous (i.e. no time-lag) and, by contrast, additional extrinsic effects unrelated to food (e.g. predation or disease) are more likely to be indexed by lagged density effects. Nevertheless, the exact cause of any instantaneous density-dependent effect is unknown and, therefore, additional extrinsic effects (in interaction or in combination) cannot be ruled out.

Possum reproductive and body-condition data (weights and fat storage) were collected during a long-term trap–kill study of possums in the Pararaki Valley ($n = 32$ years; 1965–97), New Zealand (M. Thomas and J. Coleman, unpublished data). Female body fat increased with increasing availability of hinau fruit at Orongorongo Valley (20 km distance from Pararaki Valley; figure 9.10a). The trend is mostly linear although overall significantly nonlinear (quadratic, concave down) due to one point. The proportion of female possums with pouch young increased with increasing body fat condition up to a maximum level (figure 9.10b). An Ivlev curve was fitted by maximum likelihood estimation, explaining a high proportion of variance ($r^2 = 89\%$). The proportion of females with pouch young increased with increasing food availability up to a maximum level (figure 9.10c). A logistic curve was fitted by maximum likelihood estimation, explaining a high proportion of variance ($r^2 = 99.5\%$). There was a negative linear correlation between birth date (arbitrarily estimated as days since 1 January on a logarithmic scale) and hinau food availability (figure 9.10d). More possums were born early when food levels were high compared with more possums being born late when food levels were low. However, only a low proportion of variance was explained ($r^2 = 15\%$) by this relationship.

Long-term ($n = 31$ years; 1966–97; Efford 1998, 2000) population-level data were collected at Orongorongo Valley by mark–recapture and contemporaneously with estimates of the annual crop of hinau fruit (via seedfall traps (Cowan & Waddington 1991; P. E. Cowan, A. Tokeley, D. C. Waddington and M. J. Daniel, unpublished data)). A classic Ivlev curve was fitted a priori by maximum likelihood estimation to the numerical response between rate of increase and food (figure 9.11a), explaining 40% of observed values ($r_{poss} = -0.60 + 0.85 \, (1.0 - e^{-0.037V})$). The estimate of r_m is 0.25 p.a. which compares favourably with the range of estimates (0.22–0.25 p.a.) derived by Hickling & Pekelharing (1989). There was a significant and negative linear correlation between rate of increase and

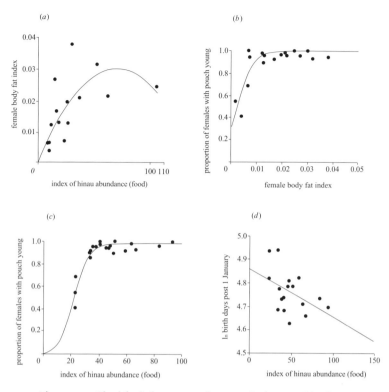

Figure 9.10. Physiological responses of possums in the Pararaki Valley (New Zealand) to winter food supply as indexed by the availability of hinau fruit in the Orongorongo Valley 20 km away (trap–kill data, 1965–97) for: (a) female body fat index versus food (hinau fruit) availability (quadratic polynomial regression: $r^2 = 48\%$, d.f. $= 2/13$, $p < 0.01$); (b) proportion of females with pouch young (PY) versus female body fat index ($r^2 = 89\%$, $n = 17$, $p < 0.01$); (c) proportion of females with pouch young versus food (hinau fruit) availability (logistic model: $r^2 = 99.5\%$, $n = 20$, $p < 0.001$); and (d) timing of births (natural logarithm of arbitrary number of days since 1 January) versus food (hinau fruit) availability (negative linear correlation, $r^2 = 15\%$, $n = 17$, $p < 0.05$). All nonlinear functions were fitted using maximum likelihood estimation.

instantaneous density (figure 9.11b), explaining 35% of observed values ($r_{poss} = 0.65 - 0.084D$). Backwards extrapolation to the y-axis ($D = 0$) predicts an r_m value of 0.65 p.a., significantly higher than that estimated by the demographic numerical response above. However, the linear extrapolation is well outside the range of observed data, and may be an overestimate if the relationship between r and density is nonlinear as expected (because of the interaction between limiting food resources).

The dynamics of hinau fruit production as impacted on by possums, and of possums as influenced by intrinsic and extrinsic regulatory

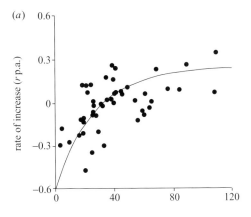

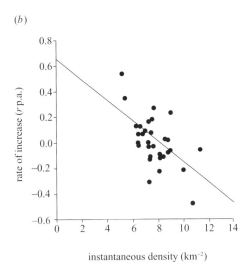

Figure 9.11. The (*a*) extrinsic numerical response (*r* p.a. versus index of food availability *V*: Ivlev curve fitted a priori is $r_{poss} = -0.60 + 0.85\,(1.0 - e^{-0.037V})$, where $r^2 = 40\%$, $n = 50$, $p < 0.001$) and (*b*) intrinsic numerical response (*r* p.a. versus instantaneous density *D*, ha^{-1}: $r_{poss} = 0.65 - 0.084D$, where $r^2 = 35\%$, $n = 31$, $p < 0.001$) of possums in Orongorongo Valley, New Zealand (see Bayliss & Choquenot (1998) for methods). Data are winter and summer annual exponential rates of increase (*r* p.a.), and the index of food availability (hinau seedfall) has a six-month time-lag. The parameters of the Ivlev curve fitted to (*a*) were estimated by maximum likelihood estimation; r_m was estimated at 0.25 p.a.

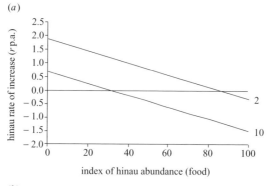

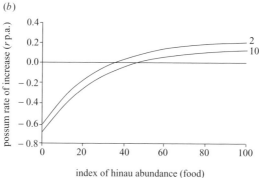

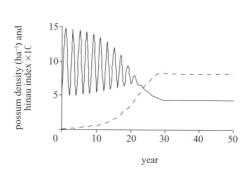

Figure 9.12. Plant–herbivore model for possums. The numerical response (r_{hin} p.a.) of (*a*) hinau fruit production as combined negative functions of the previous year's production V (via a logistics model) and possum density (D_{poss}, ha^{-1}), $r_{hin} = 2.2 (1 - V/100) - 0.15 D_{poss}$ ($r^2 = 68\%$, $p < 0.001$, d.f. $= 1/47$); and (*b*) possums (r_{poss} p.a.) as functions of hinau (food) availability (V) and possum density (D_{poss}, ha^{-1}), $r_{poss} = \{[-0.60 + 0.85 (1 - e^{-0.037 V})] - 0.01 D_{poss}\}$ ($r^2 = 69\%$, d.f. $= 1/47$). A family of numerical response curves exist for both hinau and possums depending on possum density, and are here illustrated with 2 and 10 possums ha^{-1}. (*c*) Simulated equilibration between hinau fruit production (solid line) and possums (dashed line) 50 years after liberation, predicting extinction of the hinau masting cycle.

mechanisms, are characterized by a family of curves or relationships and, hence, multiple equilibria (figure 9.12a,b). Bayliss & Choquenot (1998) developed an interferential numerical response function for possums by statistically combining both extrinsic (r_{poss} versus hinau, V) and intrinsic (r_{poss} versus density, D_{poss} ha^{-1}) numerical responses (figure 9.12b) into a joint multiple regression equation ($r_{poss} = \{[-0.60 + 0.85\,(1 - e^{-0.037V})] - 0.01D_{poss}\}$). Both variables were statistically significant entries into the overall regression equation, which explained 69% of observed values. These two numerical response models are a good example of the contrasting paradigms described by Sibly & Hone (Chapter 2).

The dynamics of hinau fruit production in the presence of consumption by possums is best described by a logistic model with an independent term for the negative impact of possum density (figure 9.12a). The high intrinsic rate of increase of annual hinau fruit production ($r_{hin(m)} > 2.0$) produces stable limit cycles with a periodicity of 2.0 years (see years 1–20; figure 9.12c). Without the impact of possums, hinau would mast every two years (i.e. low one year and high the next), although environmental variability (P. E. Cowan, A. Tokeley, D. C. Waddington and M. J. Daniel, unpublished data) may mask detection of any regular cycles. Figure 9.12c shows a simulated trend in hinau and possum abundance after a liberation of 0.1 possums ha^{-1}. The modelled stable limit cycle of hinau fruit is flattened out after 20 years because the increasing abundance of possums and associated absolute consumption of hinau has an equilibrating effect. The abundance of hinau fruit eventually stabilizes at a level of 44 (cf. an observed mean value of 47). Similarly, possum densities are predicted to equilibrate at 8.4 ha^{-1} in contrast to mean observed densities of 8.2 ha^{-1} (and mean $r_{poss} = 0$, as predicted by the joint regression model). Although this close fit is not an independent test of the model, the outcomes at least concord with observed data. Without the density effect in the possum numerical response function, the model predicts an equilibrium possum density of 10.7 ha^{-1}, 30% above the observed mean, whilst hinau fruit abundance decreases to 33% less than the observed mean.

9.3.   Isocline numerical responses

(a)   Underlying assumptions

An isocline numerical response links changes in the abundance of a consumer population *per se* directly to the availability of its food resources (figure 9.13a). Isocline numerical responses are generally formulated as an asymptotic increase in consumer abundance, indicating that the

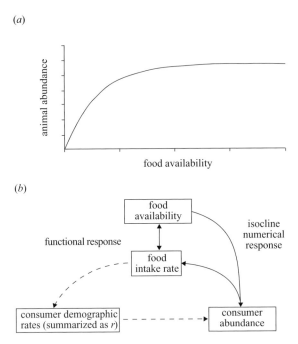

Figure 9.13. (*a*) The general form of an isocline numerical response derived by Rosenzweig & MacArthur (1963), and (*b*) the structure of the predator–prey model within which the response was used. The dashed lines in (*b*) indicate relationships that are explicit in Lotka–Volterra models but subsumed by the isocline numerical response used in Rosenzweig & MacArthur's model.

upper limit to consumer abundance is imposed by some factor which is independent of food availability (e.g. available territories, nest sites or some socially mediated crowding effect). Isocline numerical responses subsume both the effect that food availability has on the rate of food intake through the functional response, and the influence that food intake has on demographic rates of the consumer population (figure 9.13*b*). While the isocline approach greatly simplifies the way in which interaction between food and consumer abundance can be represented, it also assumes that territorial behaviour or interference competition regulates populations at high density, and the accessibility or availability of limiting resources at low density (Choquenot & Parkes 2001).

(b)   Stability properties of the wolf–moose system: graphical analyses

Probably the best-known application of isocline numerical responses is in the graphical analysis of the stability properties of consumer–resource systems (Rosenzweig & MacArthur 1963; Noy-Meir 1975; Sinclair *et al*. 1990; Messier 1994; Caughley & Sinclair 1994). In these analyses, proportional gains and losses to the food population are contrasted over the full range of its potential abundance in order to identify levels of food availability which are relatively stable. Potential increases in the abundance of the food population are usually represented by a generalized-logistic model similar to that shown in figure 9.2*a*. This model assumes that in the absence of the consumer population, the upper limit to food population abundance is imposed by interspecific competition for some limiting resource or through social regulation. Potential losses from the food population are the product of consumer population's abundance (estimated from its isocline numerical response to food availability; figure 9.14*a*) and per capita food intake (estimated from its functional response). When potential loss is expressed as a proportion of the food population it is generally termed the consumer total response. Figure 9.14*b* shows an example of a graphical analysis of the stability properties for a wolf–moose system in North America using predator–prey population parameters estimated by Messier (1994). Our graphical derivation of an equilibrium point between wolves and moose agrees with the conclusion by Eberhardt & Peterson (1999) that a two-state system need not apply. The isocline numerical response underpins calculation of the proportion of the moose population that would be consumed by wolves at given moose densities. These isocline curves are essentially the same as those shown by Sinclair *et al*. (1998) for predation in general.

9.4.   Discussion

(a)   Density dependence and population regulation

While density-dependent, density-independent and inversely density-dependent factors can all limit population density, only density-dependent factors impart a tendency towards an equilibrium. Hence, while any process that affects population density will be a limiting factor, only density dependent factors are also regulating factors (Sinclair 1989).

Density dependence can arise from both intrinsic and extrinsic factors operating on a population (Krebs 1985). Changes in the exponential rate of increase (*r*) of an intrinsically regulated population slows at high density through the effect of some type of spacing behaviour on mortality, fecundity or migration as population density increases. Populations regulated in this way can be thought of as 'self regulating', with rate of change in their abundance determined instantaneously by their prevailing density

(a)

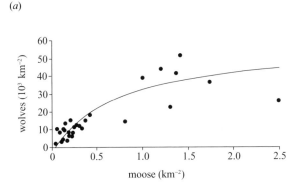

(b)

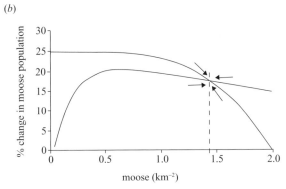

Figure 9.14. The wolf–moose system in North America showing (a) an example of an isocline numerical response linking the density of consumer populations (y, wolves) to the availability of their food resources (x, moose) (Messier 1994) with the fitted function $y = 58.7\,(x - 0.03)\,/\,0.76 + x$, and (b) a graphical stability analysis representing the interaction between moose and wolves based on empirically derived functions in Messier (1994). The lower curve, describing the percentage loss of moose to wolves as a function of prevailing moose density, was calculated from the functional and isocline numerical responses of wolves to moose density (see (a)). The upper curve, describing the gain in moose abundance in the absence of wolves, was derived by fitting a generalized-logistic model to data (r_m = 0.22 p.a., K = 2.0 moose km^{-2} and z = 4.0) presented in Messier (1994). The dashed line indicates the equilibrium moose density predicted by the model.

(Caughley & Krebs 1983). By contrast, r for an extrinsically regulated population is determined by the availability of some environmental resource such as food or nesting sites, or by the effect of some limiting environmental agent such as predators or disease (Caughley & Krebs 1983). Rate of change in the abundance of an extrinsically regulated population is determined instantaneously or cumulatively by the availability of the critical resource or the effect of the critical agent.

(b)   Population regulation: single-species models

The abundance of large-herbivore populations is widely held to be limited by extrinsic factors, most commonly food supply (Caughley 1970, 1987b; Laws *et al.* 1975; Sinclair 1977; Houston 1982; Skogland 1983; Sinclair *et al.* 1985; Fryxell 1987; Choquenot 1991, 1998), predation (Bergerud 1980; Gasaway *et al.* 1983; Messier & Crête 1984; Messier 1991, 1994), or both (Caughley 1976, 1977). Factors that limit the size of large-herbivore populations may or may not also regulate them, depending on whether they operate in a density-dependent fashion. Sinclair (1989) reviewed studies of regulation in large terrestrial mammals and concluded that the majority (including all ungulates) were regulated by density-dependent mortality related to food shortage. Similar conclusions were reached by Fowler (1987) in a review based on many of the same studies. Sinclair (1989) also found that while predator removal experiments have shown predation to be an important limiting factor for large-herbivore populations, there was no empirical evidence that predation could also be a regulating factor for large herbivores. Skogland (1991) and Boutin (1992) concurred with Sinclair (1989), finding no consistent evidence for regulation of ungulate populations by predation. However, Messier (1994) inferred, from a comparative study of interaction between moose (*Alces alces*) and wolves (*Canis lupus*) across North America, that moose populations could be regulated by wolf predation. If predation does not commonly regulate the abundance of large-herbivore populations, processes that could impart density dependence to their dynamics reduce to intrinsic (socially mediated) mechanisms, the debilitating effects of disease or parasites, or food availability.

Several recent reviews (e.g. Gaillard *et al.* 1998) of empirical evidence for density dependence in large-herbivore demography indicate that the role of density-independent variation in the abundance of herbivores and their key food resources has not been fully recognized in tests of the food hypothesis. Caughley & Gunn (1993) argued that in areas with highly variable environments, factors such as unpredictable precipitation could introduce significant degrees of density-independent variation in food availability and herbivore abundance. When combined with lags and overcompensation in vegetation and herbivore responses, density-independent variation can obscure any tendency that herbivore density may have towards equilibrium. Similarly, Putman *et al.* (1996), and Sæther (1997) considered that, even in temperate grazing systems, stable equilibria between large herbivores and their food resources were unlikely because of the direct effects of environmental variation on herbivore demographic rates and overcompensation in herbivore responses to variation in food availability. If stable K cannot be assumed for large-herbivore populations, tests of the food hypothesis which fail to detect density-dependent variation in r (or its correlates) cannot differentiate between perturbation of the system by density-independent limiting factors, and the absence of regulation through density-dependent food shortage. To account for the potential effects that density-independent variation in food availability and herbivore abundance have on the tendency of a herbivore population towards equilibrium, interaction between herbivores and their food resources must be considered in a more explicit framework than can be provided by single-species density-dependent models (Choquenot 1998).

(c)   Population regulation: interactive models

Explicit models of vegetation–herbivore interaction were developed by Noy-Meir (1975) and Caughley (1976), who derived them from the predator–prey models of Rosenzweig & MacArthur (1963) and May (1973). The interactive model described by Caughley (1976) for a deterministic environment is described in detail here as applied to kangaroos in the Australian rangelands, a stochastic environment driven by unpredictable rainfall (Caughley 1987*b*). The three components of the interactive model (growth of ungrazed plants, functional and numerical responses) operate collectively as two negative feedback loops governing the influence that vegetation and herbivores exert over each other's abundance. Density-dependent growth of ungrazed vegetation forms a vegetation biomass feedback loop, reducing vegetation growth at high biomass and keeping vegetation in check regardless of how good the seasonal conditions may be for plant growth, or how low vegetation offtake by herbivores is. The functional and numerical responses of the herbivore form a vegetation–herbivore feedback loop, increasing the number of herbivores and how much vegetation each consumes at high vegetation biomass, and reducing herbivore abundance and their per capita consumption of vegetation at low vegetation biomass. Caughley (1976) combined these feedback loops in two linked differential equations which predict coincident variation in the abundance of herbivores and the vegetation they feed on.

Herbivores reach a stable equilibrium point that is qualitatively similar to that produced by the generalized logistic model described for elk. However, in the interactive model, equilibrium is achieved through the reciprocal influence that vegetation and herbivores exert over each other's abundance, while in the generalized logistic model, equilibrium reflects an algebraic limit to herbivore population growth imposed by the existence of K. This does not mean that the herbivore population is not regulated through essentially density-dependent processes.

Regardless of the density from where the herbivore population starts, its density moves back towards equilibrium through the same series of dampening oscillations. While the processes producing the tendency towards equilibrium are essentially density dependent (i.e. the vector of vegetation and herbivores at any point in time is a direct consequence of past grazing activity, which is a consequence of past herbivore density), the tendency itself is of more importance to the stability of the grazing system than is the attainment of any specific point equilibrium. To reflect this, Caughley (1987b) coined the useful term 'centripetality' to describe the tendency that a vector in vegetation and herbivore abundance has towards equilibrium. Centripetality de-emphasizes the importance of an equilibrium in the dynamics of vegetation–herbivore systems, focusing instead on the stabilizing properties that the potential existence of an equilibrium imparts.

(i)   Stochastic rangelands grazing systems: kangaroos

Caughley (1987b) demonstrated that stochastic rainfall variation in the rangelands of Australia produced high-frequency and high-amplitude fluctuations in pasture biomass with wearying monotony and largely independently of kangaroo density. Hence, the kangaroo grazing system is constantly buffeted away from its potential equilibrium. McLeod (1997) argued that, in this environment, the concept of equilibrium carrying capacity density has no meaning. Nevertheless, the reciprocal influence that kangaroos and pasture exerted over each other's dynamics and, ultimately, abundance, imparted a strong tendency towards equilibrium (i.e. centripetality). Hence, in this model ecosystem kangaroos were able to persist indefinitely. Interactive models may be the general case for vegetation–herbivore systems (and by extension all consumer–resource systems such as predator–prey systems). By contrast, contracted single-species models are a 'short-hand' proxy, often represented by a statistical negative correlation between herbivore rate of increase and density that manifests when density-independent perturbation of these systems is low or uncommon. By contrast, McCarthy (1996) used variable rainfall as a surrogate for pasture biomass (see Bayliss 1985a,b) to examine the combined statistical relationships between red kangaroo rate of increase, food availability and past kangaroo density. This approach falls neatly between the single-species and interactive consumer–resource approach, and could be better modelled with an interferential numerical response model.

(ii)   Habitat effects on food limitation

Like most animals, large herbivores balance their foraging efficiency with exposure to direct sources of mortality or debilitation in order to maximize the rate at which they can utilize their food resources to increase individual fitness (Belovsky 1981, 1984; Stephens & Krebs 1986). Hence, the quality of the various habitats in which large herbivores may elect to spend time will reflect both the availability and quality of food found there, and the degree to which habitat-related constraints affect the rate at which this food can be assimilated to enhance their reproduction or survival. Habitat-related constraints can influence the rate at which herbivores can find and ingest food relative to its availability (foraging constraints), or the degree to which ingested food can be used to enhance reproduction or survival (demographic constraints). For example, increased predation risk in open areas reduces the foraging efficiency of snowshoe hares (*Lepus americanus*) by limiting their access to the food available in these areas (Hik 1995). Alternatively, the demographic efficiency of caribou (*Rangifer tarandus*) is limited by the availability of habitat that affords their calves protection from predators, over and above any effects of food availability (Skogland 1991).

Habitat-related constraints on foraging or demographic efficiency inhibit the potential that a herbivore population has to respond demographically to variation in its food resources. Hence, models of how habitat-related foraging and demographic constraints influence large-herbivore population dynamics need to be formulated within a framework which explicitly represents interaction between the herbivores and their food resources. Caughley's (1976) interactive model links herbivore foraging efficiency to food availability through a functional response, and herbivore demographic efficiency to food availability through a numerical response. These two responses collectively form a vegetation–herbivore feedback loop that controls the interdependent effects that vegetation and herbivores exert over each other's abundance. However, the vegetation–herbivore feedback loop implies that variation in food availability affects herbivore demography (summarized as r) independently of its effect on their rate of food intake. While this simplification is of little consequence where constraints on foraging or demographic efficiency are constant, it compromises the usefulness of the interactive model where these constraints vary between habitats. Hence, where habitat quality is in part determined by constraints on herbivore foraging or demographic efficiency, the general form of the interactive model cannot be used to consider how habitat quality influences herbivore population dynamics. For example, Choquenot & Dexter (1996) hypothesized that wild pigs in the rangelands thermoregulate when radiant heat loads are high by seeking refuge under the more or less continuous cover afforded by riverine woodlands. Behavioural thermoregulation when ambient temperatures are high is obligatory for wild pigs inhabiting other arid and semi-arid environments (Van Vuren 1984; Baber & Coblentz 1986). Choquenot & Dexter (1996) suggested that the thermoregulatory needs of wild pigs when temperatures were high could link the spatial accessibility of riverine woodlands to either (i) their foraging efficiency by restricting the area over which they could forage or (ii) their demographic efficiency by restricting the area within which they could survive and reproduce. In either case, the quality of any particular location to wild pigs will be determined by both habitat composition of the immediate area around the location (i.e. the accessibility of riverine woodlands) and the availability of food in that area.

(d)   Conclusions

The two main reasons why we construct ecological models are to predict and to aid understanding of the system (Caughley 1981). Both modelling functions are essential to the development and implementation of population management goals for whatever objective (conservation, control or harvesting). Simulation of population management scenarios to assess their efficacy is becoming increasingly popular if not necessary because of the general lack of experimentation. However, the success of this approach depends entirely on the ability of the model to capture real ecological processes, but progress in understanding ecological systems has been less than spectacular. A little scrutiny shows that most wildlife populations are still managed by trial and error rather than by scientific knowledge, and that most managers still lack tight criteria for the success or failure of their actions (including doing nothing). Nevertheless, the contributions of consumer–resource models to the research and management of overabundant kangaroos and possums have been substantial. Curiously, though, the modelling approaches adopted in Australia and New Zealand have followed two independent paradigms (or multi-species versus single-species models; or mechanistic versus density; Sibly & Hone, Chapter 2). Our re-examination of current kangaroo and possum models, however, indicates that a more useful framework for understanding how marsupial populations work may be obtained by combining the two modelling approaches. A marriage between extreme extrinsic (animal–resource) and intrinsic (animal–density) regulation models could embody the much broader spectrum of population mechanisms that most likely exist within species. Although more complex population interactions may be exposed, the trade-off may be increased predictive power and, hence, utility. This approach increases the realism and predictive power of existing population models for kangaroos and possums at manageable levels of complexity.

We conclude by reinforcing the axiom that in order to manage populations effectively we need to understand their dynamics. However, research costs can be substantial both in terms of time and money. For example, the possum model was developed a posteriori (no model in mind) with data collected after 32 years of intensive study of a population more or less in equilibrium. By contrast, the kangaroo model was developed a priori (a model in mind) with experimental data collected after 5 years of intensive study of a non-equilibrium grazing system. Hence, one impediment to more widespread use of more useful interactive ecological models is the daunting and costly task of 'parameterizing' such models, especially for populations that exhibit little dynamics within time-scales dictated by funding and career cycles. An adaptive management strategy (Walters 1997), however, may allow a new breed of population models to be developed and tested cost-effectively by integrating focused population-scale management experiments with the modelling process. One such model is the interferential numerical response function, because it may help bridge three major historical dichotomies in population ecology (equilibrium versus non-equilibrium dynamics, extrinsic versus intrinsic regulation and demographic versus isocline numerical responses).

We thank Malcome Thomas, Jim Coleman, Murray Efford, Phil Cowan (New Zealand Landcare Research) and Environment ACT (Lyn Nelson) for access to unpublished data, Environment Australia-eriss (Max Finlayson), all staff of the NT University Key Centre for Tropical Wildlife Management (in particular Barry Brook, Peter Whitehead & David Bowman) and staff of the Arthur Rylah Institute (Victorian DNRE) for support. We also thank Jim Hone, Richard Sibly and Suzi White for organizing the conference and workshop, and The Royal Society and Novartis Foundation for financial support to attend.

STEPHEN A. DAVIS, ROGER P. PECH AND EDWARD A. CATCHPOLE

10

Populations in variable environments: the effect of variability in a species' primary resource

10.1. Introduction

The relationship between the dynamics of animal populations and the variability of their environment is a central concern of applied and theoretical ecology, if only due to the long-standing debate over the roles of density-dependent and density-independent factors in determining animal abundance (e.g. Sæther 1997). There is a gradation between predictable environments and those that are highly variable that is quite apart from the gradation between cold or wet environments and those that are hot or dry. Much of southern and central Australia is characterized by extreme climatic variability. The coefficient of variation in summer rainfall, for example, can be close to unity. In these environments, models of wildlife population dynamics have emphasized the numerical response (Solomon 1949) to variable resources, rather than density-dependent processes. This is the approach that was taken by Caughley (1987a) and subsequently used by others to analyse the population dynamics of both native and introduced herbivores in Australia (Cairns & Grigg 1993; Caley 1993; Choquenot 1998; Pech & Hood 1998; Pech *et al.* 1999; Brown & Singleton 1999; Cairns *et al.* 2000). In each case the authors proposed a numerical response of the herbivore to either rainfall or pasture biomass. Pech & Hood (1998) used a similar form of the numerical response for the interaction of red fox populations with their prey.

The change in the numbers of a predator in response to the density of its prey was termed the 'numerical' response by Solomon (1949). The term has since been used more generally (May 1981a) to describe how the rate of increase of an animal population changes in response to its food supply. The numerical response may be modelled as either an instantaneous rate or as a finite rate over a discrete time interval. Some of the different functional forms that have been used are listed by May (1981a). The numerical response is commonly a component of interactive models that describe the dynamics of two or more species, but it can be used in single-species models. In addition, changes in the rate of increase can be modelled as a response to a climatic variable where this is a good predictor of food supply. For example, in at least some parts of southern Australia there is a tight relationship between lagged rainfall and pasture biomass (Robertson 1987a), and Bayliss (1987) used lagged rainfall as the independent variable in an Ivlev form (Ivlev 1961) of the numerical response of populations of red kangaroos (*Macropus rufus*).

The exponential rate of increase is related to the numbers of individuals in a population by

$$r_t = \ln\left(\frac{N_{t+1}}{N_t}\right), \tag{10.1.1}$$

where N_t represents the abundance (or some index of abundance) at time t. The Ivlev form of the numerical response of a population is given by

$$r_t = -a + c\left(1 - e^{-dV_t}\right), \tag{10.1.2}$$

where a, c and d are positive constants, and V_t represents the abundance of the population's primary resource at time t. Each of the parameters in equation (10.1.2) have been interpreted as representing a biological process (Bayliss 1987): a represents the maximum rate of decrease of the population when food is scarce ($V_t \to 0$), c determines the maximum rate of increase, r_m, when food is abundant ($r_m = c - a$), and d, referred to as the demographic efficiency of the population, reflects the sensitivity of the population to changes in the availability of food.

One fundamental aim of ecology is to understand what determines the abundance and distribution of species. The purpose of this paper is to explore how temporal variability in the resources required by a species affects its distribution and abundance. More specifically, what is the effect of increasingly unreliable rainfall on herbivore populations (and their predators) that respond to rainfall-driven growth and senescence of pasture? These populations are affected by environmental variability quite differently to ungulate species for which climatic variables are known to have a direct effect on vital rates (Sæther 1997; Gaillard *et al.* 1998; Coulson *et al.* 2000). However, it is a form of environmental variability that is faced by a range of species that stretches from mice (*Mus domesticus*) in Australia (Pech *et al.* 1999), to coyotes (*Canis latrans*) in New Mexico (Windberg *et al.* 1997) to greater kudu (*Tragelaphus strepsiceros*) in Africa (Owen-Smith 1990).

The paper is divided into two sections that develop models for the effect of environmental variability on population dynamics. Both sections deal with herbivore populations in which the numerical response of the herbivore to rainfall, or rainfall-driven pasture, is a central component. The models are used to document the role that variability in rainfall plays in determining the abundance, dynamics and persistence of herbivore populations and, in the second section, their predators. Section 10.2 presents a class of single-species models for those herbivore populations that do not have a significant influence on their food supply and where lagged rainfall may be used as an index of pasture. Section 10.3 presents a parameterized model with three trophic levels: pasture, herbivore and predator. In this model the growth and senescence of pasture is determined by rainfall and there is a feedback mechanism such that the abundance of the herbivore affects the amount of pasture available to the herbivore population at a future time.

10.2. Persistence and abundance of populations in variable environments

(a)   Long-term growth rate

A numerical response to rainfall (rather than pasture) implies a grazing system that is non-interactive (Caughley & Lawton 1981), that is, the grazing of the herbivore population does not reduce the amount of food available to either themselves or later generations. If the amount of rainfall over one period is independent of the amount that fell in preceding periods, then the realized values of V_t in equation (10.1.2) may be represented as a set of independent samples of a random variable V. If the distribution of V is specified, then together with equation (10.1.2), a stochastic process is defined for the dynamics of the population. These dynamics may be characterized by the stochastic growth rate (Tuljapurkar 1997, p. 70), representing the mean long-term growth rate of the population, and is defined as

$$\lim_{t \to \infty} \frac{1}{t} E\left[\ln \Lambda_t(\omega)\right], \tag{10.2.1}$$

where

$$\Lambda_t(\omega) = \frac{N_t(\omega)}{N_0}, \tag{10.2.2}$$

and ω denotes a particular sample path of the stochastic process. The expectation (average, E) in equation (10.2.1) is taken over all possible sample paths. A population will not persist if the mean long-term growth rate is negative. In the case of equation (10.1.2) the stochastic growth rate is just the expected value of r. This follows by rewriting $\Lambda_t(\omega)$ as

$$\Lambda_t(\omega) = \frac{N_t(\omega)}{N_{t-1}(\omega)} \frac{N_{t-1}(\omega)}{N_{t-2}(\omega)} \cdots \frac{N_1(\omega)}{N_0}, \tag{10.2.3}$$

and substituting this expression into equation (10.2.1), giving

$$\lim_{t \to \infty} \frac{1}{t} E\left[\ln \Lambda_t(\omega)\right] = \lim_{t \to \infty} \frac{1}{t} E\left[\sum_{i=1}^{t} \ln\left(\frac{N_i(\omega)}{N_{i-1}(\omega)}\right)\right]$$

$$= \lim_{t \to \infty} \frac{1}{t} E\left[\sum_{i=1}^{t} r_i(\omega)\right] = \lim_{t \to \infty} \frac{1}{t} \sum_{i=1}^{t} E\left[r_i(\omega)\right]$$

$$= \lim_{t \to \infty} \frac{1}{t} \sum_{i=1}^{t} E\left[r\right] = E\left[r\right], \tag{10.2.4}$$

because, from equation (10.1.2), r is independent of both time t and the path ω.

For the Ivlev form of the numerical response, equation (10.1.2), the expected value of r is given by

$$E\left[r\right] = -a + c\left(1 - E\left[e^{-dV}\right]\right) = -a + c\left(1 - m_V(-d)\right), \tag{10.2.5}$$

where m_V is the moment generating function of V (Mendenhall *et al.* 1981). For many distributions (such as the normal and gamma distributions) the moment generating function has a closed form. In these cases the stochastic growth rate may be written as a simple function of the parameters of the Ivlev numerical response and of the mean and variance of V.

The numerical response of red kangaroos to rainfall, as an annual rate of increase, was given by Bayliss (1987) and is shown in figure 10.1a. If annual rainfall may be adequately represented by a gamma distribution (with mean μ and variance σ^2) then, as an example of equation (10.2.5), the stochastic growth rate of red kangaroos is given by

$$E\left[r\right] = -a + c\left(1 - \left(1 + \frac{d\sigma^2}{\mu}\right)^{-\mu^2/\sigma^2}\right), \tag{10.2.6}$$

where $a = 0.57$, $c = 1.14$ and $d = 0.004$. This expression is shown as a surface in figure 10.1b. It is an increasing function of the mean rainfall and a decreasing function of the variability of rainfall. At high values of the

(a)

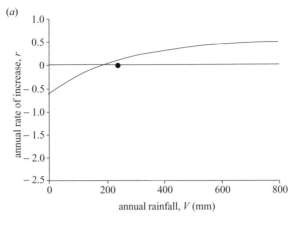

(b)

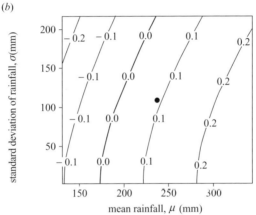

Figure 10.1. (*a*) The numerical response (curved line) of red kangaroos to lagged rainfall (Bayliss 1987) (filled circle, mean rainfall) and (*b*) the stochastic growth rate (contour lines) of red kangaroos as a function of the mean and standard deviation of rainfall. The distribution of rainfall is represented by a gamma distribution. At the site at which Bayliss (1987) measured the numerical response of red kangaroos, the mean annual rainfall between 1883 and 1984 was 236 mm and the standard deviation was 107 mm (shown in (*b*) as a filled circle) and hence the coefficient of variation was 0.45.

standard deviation, the stochastic growth rate is negative implying population decline and eventual extinction. This is perhaps the most interesting feature of equation (10.2.5) as it implies that a population may have a positive rate of increase for the mean value of rainfall but have a negative stochastic growth rate. This means that in a variable environment, the value of V for which $r = 0$ in equation (10.1.2) has less ecological meaning than the contour in figure 10.1*b*, where the mean stochastic growth rate is zero.

The conclusion that variability in *V* reduces the stochastic growth rate of a population does not depend on the distribution assumed for *V*. Intuitively, the stochastic growth rate is reduced by variability because the numerical response is convex: the benefit of a positive deviation from the mean does not compensate for losses incurred by a negative deviation from the mean. The net loss caused by such deviations increases with the magnitude of the deviations. This was illustrated by Caughley (1987*b*) for a symmetrical distribution for *V*, but it is in fact true for any distribution. This can be shown by taking the expectation of equation (10.1.2) and comparing the result with the same equation evaluated at $V = \mathrm{E}\,[V] = \mu$. The latter gives the growth rate of the population if there was no variability at all. Thus,

$$\mathrm{E}\,[r] = -a + c\big(1 - \mathrm{E}\,[e^{-dV}]\big) < -a + c\big(1 - e^{-d\mu}\big) \qquad (10.2.7)$$

because by Jensen's inequality (e.g. Karlin & Taylor 1975, p. 249),

$$\mathrm{E}\,[e^{-dV}] > e^{-d\mu}, \qquad (10.2.8)$$

for any distribution *V*, provided that $c > 0$ and $d > 0$ (which ensure that the Ivlev form of the numerical response is convex).

(b)   The distribution of species

A consequence of equation (10.2.6) is that for a fixed value of the mean of *V*, the sensitivity of the stochastic growth rate to changes in the variance of *V* is proportional to the parameter *c*,

$$\frac{\partial \mathrm{E}\,[r]}{\partial \sigma} \propto -c. \qquad (10.2.9)$$

The parameter *c* represents the difference between the maximum rate of increase, r_{m} and the maximum rate of decrease, *a*. This means that the slope, in the direction of σ, of the surface given by equation (10.2.6) (and illustrated in figures 10.1*b* and 10.2*b*) can be related to the difference between the lower and upper limits of the rate of increase.

Although sensitivity to variability is not wholly determined by *c*, equation (10.2.9) indicates that populations most sensitive to changes in the variance of *V* will have extreme values of both r_{m} and *a*, or at least large values for *a*. The distribution of these species is more likely to be influenced by variability in resources. Red kangaroos on the one hand, have evolved mechanisms, such as a reproductive response to rainfall (Lee & Cockburn 1985), which enable them to persist during periods of low rainfall, and so have a relatively small value of *a*. Feral pigs (*Sus scrofa*) on the other hand, which in Australia are an introduced herbivore, tend to decline rapidly during dry periods (Choquenot 1998). Published estimates of the maximum rate of decrease for red kangaroos (Bayliss 1987; Caughley 1987*b*; Cairns & Grigg 1993) are smaller than those for feral pigs (Caley 1993; Choquenot 1998) regardless of whether estimates are based on a fitted numerical response or simply the maximum rate of decrease observed. The numerical response for feral pigs and the stochastic growth rate predicted by equation (10.2.6) are shown in figure 10.2. Note that the response surface for feral pigs in figure 10.2*b* shows greater sensitivity to both the mean μ and the standard deviation σ of the rainfall than is the case for the red kangaroos in figure 10.1*b*.

(c)   Density dependence

A key assumption for the single-species models presented so far is that population growth rate does not depend on density. This assumption is clearly not valid when a herbivore has a substantial impact on future levels of its food supply. Caughley & Krebs (1983) proposed that Ivlev's form for the numerical response could be modified by adding a linear term in density so as to introduce density dependence. Equation (10.1.2) becomes

$$r_t = -a + c\left(1 - e^{-dV_t}\right) - bN_t, \tag{10.2.10}$$

where *b* is a positive constant. Graphically, the final term in equation (10.2.10) represents a downward vertical shift of the numerical response on the axes shown in figures 10.1*a* and 10.2*a*. As abundance increases, the amount of the primary resource required to maintain positive rates of increase becomes higher such that the population cannot continually increase. This means that either the abundance of the population will tend to zero (this is the case if the stochastic growth rate calculated from equation (10.2.10) with $N_t = 0$ is negative), or abundance will vary about some long-term average. In the latter case, the long-term average can be found by taking the expected value of equation (10.2.10),

$$E\left[r_t\right] = -a + c\left(1 - m_V(-d)\right) - bE\left[N_t\right]. \tag{10.2.11}$$

Setting $E\left[r_t\right] = 0$ (which must be the case if abundance varies about a

(*a*)

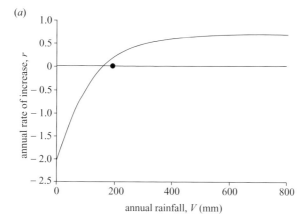

(*b*)

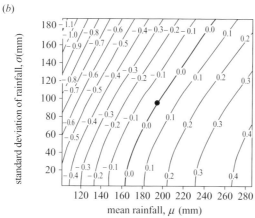

Figure 10.2. (*a*) The numerical response (curved line) of feral pigs to lagged rainfall. Values for r_m and *a* are those given by Choquenot (1998). In the absence of a published estimate, *d* was chosen such that E $[r] = 0$. This means that in the long term, the population is expected to persist but not increase. (Filled circle, mean rainfall.) (*b*) The stochastic growth rate (contour lines) of pigs as a function of the mean and standard deviation of rainfall. The distribution of rainfall is represented by a gamma distribution. Choquenot (1998) gives the mean annual rainfall as 193 mm and the standard deviation as 94 mm (shown in (*b*) as a filled circle) and hence the coefficient of variation was 0.49.

positive long-term average) defines

$$E\left[N_t\right] = \frac{1}{b}\left(-a + c\left(1 - m_V\left(-d\right)\right)\right). \tag{10.2.12}$$

The similarity between equation (10.2.5) and equation (10.2.12) implies that, like the stochastic growth rate, average abundance is a decreasing function of the variance of *V*.

10.3. Trophic interactions and variable environments

(a)   Three trophic levels: foxes, rabbits and pasture

Pech & Hood (1998) developed an interactive model for pasture, rabbits and foxes in semi-arid southern Australia. The model consists of a functional and numerical response for the predator and prey species, and a pasture growth model. The latter is largely determined by rainfall but includes a density-dependent term that prevents the amount of pasture from increasing without limit in the absence of herbivory. The model is discrete and runs on a quarterly basis. The full model given by Pech & Hood (1998) includes the presence of disease and an alternative prey species, and incorporates additional stochasticity in the growth of pasture (apart from that due to variability in rainfall). We reduce the model to the following simple interactions between rainfall and the three trophic levels:

$$\Delta V = -55.12 - 0.0153 V_t - 0.00056 V_t^2 + 2.5 R, \qquad (10.3.1)$$

where ΔV is the growth of pasture in the absence of herbivores, V_t is the pasture biomass at time t and R is the quarterly rainfall over the interval $(t, t+1)$. In the presence of rabbits,

$$V_{t+1} = V_t + \Delta V - C_{t+1}, \qquad (10.3.2)$$

where C_{t+1} represents the pasture removed by rabbits over the interval $(t, t+1)$, determined by the number of rabbits and their functional response to pasture. The number of rabbits, denoted by N_t, is determined by their numerical response to pasture, r_N, and the total predation rate by foxes per rabbit per quarter, G_P,

$$N_{t+1} = N_t \exp(r_N - G_P). \qquad (10.3.3)$$

The predation rate G_P is related to the functional response of foxes, denoted g_P, by

$$G_P = (365/4)(g_P P_t)/N_t. \qquad (10.3.4)$$

Finally, the density of foxes, P_t, is determined by their numerical response to rabbits, r_P,

$$P_{t+1} = P_t \exp(r_P). \qquad (10.3.5)$$

The estimated numerical and functional responses of rabbits and foxes, as provided by Pech & Hood (1998), are shown in figure 10.3 .

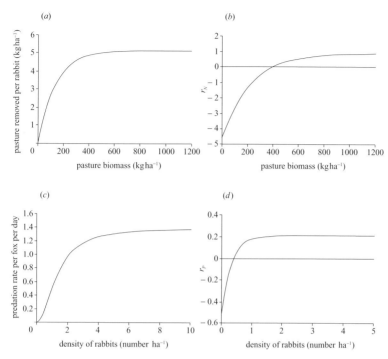

Figure 10.3. The four relationships that govern the interactions between
pasture and rabbits and between rabbits and foxes (from Pech & Hood (1998)).
(*a*) The quarterly consumption of pasture per rabbit (functional response),
(*b*) the quarterly numerical response of rabbits to pasture, (*c*) the functional
response of foxes to the density of rabbits, and (*d*) the quarterly numerical
response of foxes to rabbits. The pasture removed per rabbit is $C_t/N_t = v\,(365/4)$
$(w^{0.75})\,(1 - \exp(-(V_{t-1} + \Delta V)/f))$ where $w = 0.782$ kg, $f = 138$ kg ha^{-1} and
$v = 0.068$ kg kg$^{-0.75}$ day^{-1} (Short 1985). The quarterly numerical response of
rabbits to pasture is $r_N = -4.6 + 5.5\,(1 - e^{-0.0045V_{t-1}})$. The functional response of
foxes, given by Pech *et al.* (1992) as a predation rate per fox per day is $g_P =$
$(1096/782)\,N_{t-1}^2/(N_{t-1}^2 + 1.32^2)$ and the quarterly numerical response of foxes to
the density of rabbits is $r_P = -0.56 + 0.77\,(1 - e^{-3.2N_{t-1}})$.

The only source of environmental variability in the reduced model is
variability in rainfall. Pech & Hood (1998) found that when a cube-root
transformation was applied to historical records of summer, autumn,
winter and spring rainfall these distributions did not differ significantly
from normal. Therefore to consider the effect of environmental variability in quarterly rainfall on the dynamics of pasture, rabbits and
foxes, quarterly rainfall was simulated by generating values of R^3 where
R was taken from a normal distribution with mean μ_R and standard

deviation σ_R. The average value of R^3 is related to the mean and standard deviation of R by

$$E[R^3] = \mu_R^3 + 3\mu_R\sigma_R^2. \qquad (10.3.6)$$

In order to change the variability in quarterly rainfall while keeping the average value of R^3 constant, σ_R was varied and μ_R was determined by equation (10.3.6). In some parts of southern Australia rainfall is seasonal, either falling predominantly in winter or predominantly in summer. Only a single distribution for R was used so that this type of seasonal structure was not included.

The mechanistic model for foxes, rabbits and pasture was used to generate trajectories from which long-term mean values of abundance at the three trophic levels could be calculated. In order to avoid extremely low densities of foxes or rabbits (the model does not include demographic stochasticity, so it is not appropriate when population sizes are very small) the same minimum constraints as Pech & Hood (1998) were placed on N_t and P_t. In the case of the prey species this minimum density was 0.017 ha^{-1}, and for the predator 0.001 ha^{-1}. Pasture biomass was restricted to non-negative values. This means that the long-term mean values calculated from these trajectories are incorrect for the model given by equations (10.3.1)–(10.3.5), but correct for an amended model where $N_{t+1} = \max\{N_t \exp(r_N - G_P), 0.0017\}$, with similar changes for P_{t+1} and V_{t+1}.

Trajectories were produced for both the full model and the case without foxes. The latter case corresponds to the trophic interaction model proposed by Caughley (1987b) for red kangaroos and pasture. Figure 10.4 illustrates that in the absence of predation the average abundance of rabbits steadily decreases with variability in rainfall while average pasture biomass increases. Variability in the abundance of rabbits initially increases rapidly with increasing variability in rainfall but then declines for larger values of the coefficient of variation. These trends are reversed when foxes are present. Figure 10.5 shows that the average abundance of rabbits increases in an exponential fashion with increasing variability in rainfall. Average pasture biomass slowly declines, and the variability in rabbit abundance increases in much the same way as average abundance. The abundance of foxes declines in response to variability in rainfall. The variability in the abundance of foxes is similar to the variability in abundance of rabbits when foxes are absent (compare figure 10.5 with figure 10.4); at

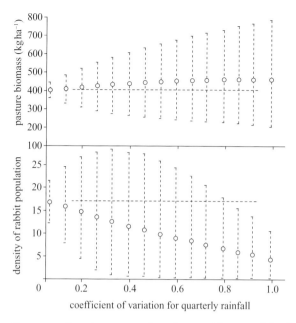

Figure 10.4. Long-term mean values of pasture biomass and the abundance of rabbits (ha^{-1}), in the absence of foxes, for increasing values of the coefficient of variation of quarterly rainfall. Means were estimated from model trajectories, each 10 000 years in length. Mean quarterly rainfall was always 95 mm. The upper and lower error bars represent the 90th and 10th percentiles. The horizontal dotted lines represent pasture biomass and abundance of rabbits when there is no variability.

first it increases with increasing variability and then declines for higher values of the coefficient of variation.

When quarterly rainfall is as unreliable as it is for much of southern Australia, the coefficient of variation for quarterly rainfall is typically greater than 0.5. For this amount of variability the model predicts population dynamics of rabbits that are characterized by occasional eruptions (figure 10.6). For most of the time foxes regulate rabbits, that is, in the presence of foxes the rabbit population rarely reaches densities in excess of 2 individuals ha^{-1}. Pech *et al.* (1992) estimated 2.5 ha^{-1} as the upper limit of the densities at which foxes can regulate rabbits. At this density both the numerical and functional responses of foxes are close to their maximum values (figure 10.3). However, during periods of high rainfall, rabbits can increase at a rate greater than the total predation rate of foxes such that an eruption occurs. In the model, foxes respond to high densities of rabbits

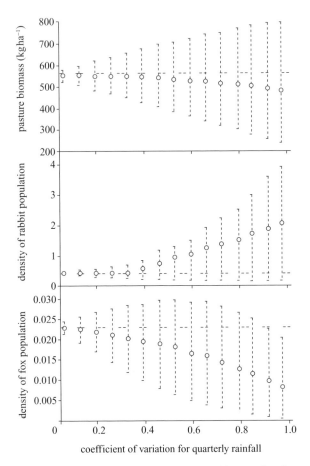

Figure 10.5. Long-term mean values of pasture biomass, abundance of rabbits (ha^{-1}), and abundance of foxes (ha^{-1}) for increasing values of the coefficient of variation of quarterly rainfall. Means were estimated from model trajectories, each 10 000 years in length. Mean quarterly rainfall was always 95 mm. The upper and lower error bars represent the 90th and 10th percentiles. The horizontal dotted lines represent pasture biomass and the abundance of rabbits and foxes when there is no variability.

by increasing in number but they do so at their relatively slow maximum rate of increase, $r_m = 0.21$. This is too slow a response to prevent the rabbits from increasing at a rate close to their maximum, which they will do while pasture is plentiful. Once rabbit densities realize values exceeding 2.5 individuals ha^{-1} they tend to continue to increase. Beyond this density the rabbits are no longer regulated by predation, but by their highly variable

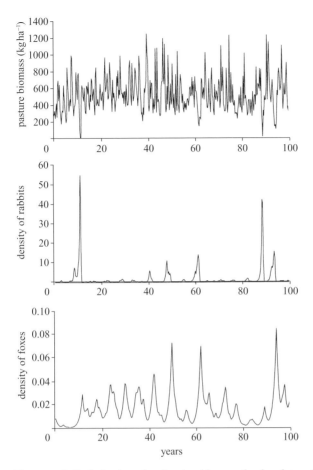

Figure 10.6. Typical trajectories of pasture biomass, the abundance of rabbits (ha^{-1}) and the abundance of foxes (ha^{-1}) over a 100 year time interval. The coefficient of variation of quarterly rainfall was 0.8 representing a highly variable environment.

food supply. Thus, when the period of high rainfall ends the rabbit population goes into a sharp decline that is due to insufficient pasture but is exacerbated by predation from an increasing number of foxes. On these occasions rabbits are capable of reducing pasture biomass to zero (see figure 10.6) and often crash to the minimum density of 0.17 ha^{-1}.

The eruptive dynamics of rabbits produced by the model are in broad agreement with the dynamics of rabbits that are observed in the field (Williams *et al.* 1995; Newsome *et al.* 1997; B. D. Cooke, unpublished data).

In the model, variability in rainfall is directly responsible for eruptions and the inevitable collapse to extremely low numbers. If there is no variability in rainfall then eruptions do not occur and pasture, rabbits and foxes settle on a constant value. As variability increases, the frequency of quarters in which rabbit densities are greater than 2.5 ha^{-1} (the approximate threshold for predator–regulation) increases monotonically (figure 10.5).

The model most clearly indicates that increasing variability in rainfall results in increasing variability at all trophic levels (figures 10.4 and 10.5). At high values of the coefficient of variation, pasture becomes increasingly unreliable, rabbits tend to display eruptive dynamics and fox populations, though buffered to some extent by being the highest trophic level, fluctuate over two orders of magnitude (figure 10.6). Rabbit densities appear to fluctuate the most dramatically, perhaps explaining why fox densities fluctuate as much as they do. In the absence of variability, rabbit densities settle at 0.4 individuals ha^{-1} but when the coefficient of variation for quarterly rainfall is 0.8, rabbit densities can exceed 55 ha^{-1} at the height of an eruption.

10.4. Discussion

In the absence of interspecific interactions, the effect of temporal variation in rainfall on a herbivore population having a density-independent convex numerical response to rainfall-driven pasture is to reduce the stochastic growth rate of the population. A negative stochastic growth rate implies that the population must decline in the long term. Thus environmental variability, in the form of unreliable rainfall, can determine whether or not populations persist. The inclusion of either density dependence in the numerical response, or a model for pasture that accounts for the plant material consumed by the herbivore, does not change the conclusion that in these single-species models populations do less well in variable environments than in predictable ones.

An immediate implication of the effect of environmental variability is that it may play a role in determining the range of a species if there is a geographical gradient from reliable to highly unreliable environments. For example, high variability in rainfall, combined with sensitivity to variability, is an obvious explanation for why local extinction (and recolonization) might be a feature of some herbivore species at the edge of their geographical range (e.g. Choquenot 1998; Cairns *et al.* 2000). A coarse test of the proposed effect of variability on the stochastic growth rate of populations is to compare the distribution maps for red kangaroos and feral pigs with each other, and with geographical patterns in the mean and variance of rainfall. Future work will include comparing the geographical distribution of kangaroos and pigs predicted from equation (10.2.6) with the actual distributions.

The questions of how variability in a species' primary resource affects its distribution and abundance may be applied to any species. Here, we have focused on a herbivore population responding to rainfall, but the theory applies to any animal population that has a convex numerical response to its primary resource, and where this resource is variable. For example, the response of foxes to increased variability in rainfall is very similar to the response of rabbits with predators removed. Just as variability in pasture decreases the average abundance of rabbits in the model for two trophic levels, the convex numerical response of the fox population means that variability in rabbit numbers decreases the average abundance of foxes in the model involving all three trophic levels. Perhaps more surprisingly this occurs even though the average abundance of rabbits increases in response to variability.

Variability in a species' primary resource may distort short- to medium-term population time-series data. During a time interval in which the resource is relatively constant a population would be expected to fare better than in time intervals in which the resource is highly variable. Differences in variability may therefore explain population trends within time-series data. Data on coyotes (*C. latrans*) in New Mexico (Windberg *et al.* 1997) may provide an interesting example of this. The authors relate the abundance of prey available to the coyotes to rainfall and suggest that a numerical response to abundance of prey is largely responsible for the population dynamics of the species. The authors draw attention to differences in variability in rainfall within the time-series data but do not propose this factor as an explanation for trends in coyote abundance.

Lewontin & Cohen (1969) demonstrated that for population growth in a random environment

$$E[r] = E[\ln \lambda] \cong \ln \bar{\lambda} - \frac{\sigma_\lambda^2}{2\bar{\lambda}^2}, \tag{10.4.1}$$

where λ is the finite rate of increase (population growth rate), $\bar{\lambda}$ is the mean of λ and σ_λ^2 is the variance of λ due to changing environmental conditions. The authors do not specify any interactions between the population and the environment, or between the population and other species, but suppose that λ is a random variable with no serial autocorrelation. Their approximation to the long-term growth rate is obtained by expanding ln λ around $\bar{\lambda}$, taking the expectation of the resulting expansion and dropping higher-order terms. The equation reflects the difference between the average value of lnλ and the logarithm of the average value of λ. It states that the greater the variability in λ the lower the long-term growth rate. Other authors have since proposed similar approximations for the effects of variability in λ on the long-term growth rate, but for more complex population models (Tuljapurkar 1997; Lande 1998).

At first appearance, the conclusions of Lewontin & Cohen (1969) and the conclusions presented here (equation (2.7)) appear similar – the effect of environmental variability is to reduce the long-term growth rate. However, this is not the case. The arguments used here, to show that the effect of variability in a species primary resource is to reduce the long-term growth rate, hold true only when the numerical response of the population is convex. Interestingly, if it is concave then the same arguments can be used to arrive at the opposite conclusion: the more variable the environment, the higher the long-term growth rate. It is unlikely that any population, herbivore or otherwise, has a concave response to their primary resource, but populations may have a concave response to a climatic variable. For example, the relationship between survival of calves in a population of elk (*Cervus elaphus*) and winter precipitation (R. A. Garrott, unpublished data) is concave. The apparent contradiction is reconciled by realizing that $\bar{\lambda}$ or σ_λ^2, or both of these, may depend on environmental variability, measured as the variance in the species' primary resource, precipitation or some other climatic factor, such that the net effect of increasing variability is to increase the long-term growth rate.

In the model for rabbits, foxes and pasture, the average abundance of the predator in the system declines with increasing variability, while the average abundance of the prey increases. The numerical response of rabbits to pasture is convex, so variability in pasture is expected to reduce the average abundance of rabbits. This is what is observed when foxes are removed, but not when foxes are present. The reason for this reversal is that the equilibrium abundance of rabbits when they are regulated by foxes (0.4) is so low in comparison with their average abundance when they are regulated by their food supply (5–15), that variability provides opportunities for the rabbits to escape predator regulation. The result is the plague dynamics shown in figure 10.6. The high densities reached during a plague have a large effect on the average abundance of rabbits, as is the case for any skewed distribution. More generally, the net effect of variability in predator–prey systems is likely to vary from one system to another, depending on the maximum rates of increase of predator and prey.

For the more complex model of rainfall-driven pasture, rabbits and foxes, it is not clear whether environmental variability in the form of unreliable rainfall either increases or decreases the likelihood of persistence of the herbivore in the system. The effect of unreliable rainfall is to profoundly change the dynamics at all trophic levels but whether this is to the advantage or disadvantage of the herbivore is difficult to judge, even though average abundance increases. In the presence of foxes, variability in rainfall increases the frequency of eruptions and eruptive dynamics can allow a species to colonize or recolonize unoccupied habitat. However, particularly in the case of rabbits, eruptions are followed by extremely low densities, providing a route to local extinction.

We thank Jim Hone, David Gordon, Russell Lande and Bernt-Erik Sæther for their helpful suggestions and encouragement. We also thank The Royal Society for their financial support that enabled S.D. to present the paper in London.

PETER J. HUDSON, ANDY P. DOBSON, ISABELLA M.
CATTADORI, DAVID NEWBORN, DAN T. HAYDON, DARREN J.
SHAW, TIM G. BENTON AND BRYAN T. GRENFELL

11

Trophic interactions and population growth rates: describing patterns and identifying mechanisms

11.1. Introduction

A long-term objective of population biology is to explain the spatio-temporal variations in abundance of organisms by understanding the factors that limit both distribution and changes in abundance. In general, theory predicts that a major determinant of distribution and dynamics is the instantaneous population growth rate, presented as r (where $r = \ln(\lambda) = \ln(N_{t+1}/N_t)$. Some models reveal the obvious, such that species will tend not to exist where their population growth is consistently negative and there is no immigration (see Chapter 2). But the models also expose intriguing dynamics; for example, simple single-species nonlinear models reveal that dynamics can vary from stability through oscillatory to chaotic behaviour simply by subtle changes in the population growth rate (May 1976). As such, it is not surprising that when we incorporate interspecific interactions, the stochastic vagaries of environmental conditions and dispersal, we reveal a Pandora's box of dynamical behaviours.

In the empirical literature, the population growth rate parameter does not enjoy the same importance as it does in the theoretical literature. Rarely do workers make an estimate of the intrinsic growth rate parameter (r) or its empirical equivalent, the maximum growth rate (r_{max}) which is simply the maximum rate of growth observed within a time-series. Changes in the observed growth rate at a specific time (r_t) may be recorded along with the factors associated with the reproductive output of individuals, but studies tend not to estimate the extent to which the growth rate is reduced by density dependent regulatory factors. This is essentially because estimating the population growth rate may not be simple.

Time-series data may not reveal the maximum growth rate simply because the population is close to equilibrium and therefore there is little variation in growth rate. Moreover, we can expect growth rate to fall with density, so by definition we can only obtain an estimate of r when the density is very low, although in some populations the growth rate may be subjected to Allee effects so that it is low or even negative. The growth rate will also be influenced by the age structuring when fecundity or survival varies with age. This population age structure will influence any estimate of growth rate, and a low density population is usually one that has suffered a massive decline in numbers; therefore the age structure is far from stable (Caswell 2001). Other means of estimating r exist, all of which incorporate certain errors. Nevertheless, the point is that if we are to challenge models with data then estimates of growth rates will be needed.

One of the first steps to bridge the gap between theory and empiricism is to examine if some basic predictions from models are supported by data. One approach is to make interspecific comparisons and predict from theory the expected differences in the dynamics between species in relation to variations in population growth rate. Such an approach is taken by Sæther & Engen (Chapter 5). An alternative approach is to undertake intraspecific comparisons and make predictions from models of how variations in growth rate will influence dynamics and equilibrium population size between populations of the same species. This is the approach we take in this paper with red grouse (*Lagopus lagopus scoticus*).

The red grouse is particularly suited to this type of study. First, the species is herbivorous with a diet that consists predominantly of the tips of heather (*Calluna vulgaris*), such that its habitat and distribution are limited by the extent of this shrub (Hudson 1992). The productivity of heather is determined by warm and wet conditions, and the quality of the heather is influenced by the underlying rock (Watson *et al.* 1984). Second, we have gathered hunting bag records from 352 individually managed upland estates where the numbers harvested have been carefully recorded for many years, frequently more than 100 years. These are distributed throughout Scotland, northern England and Wales and exhibit large variations in environmental and biotic conditions, such as the presence or absence of predators and tick-borne diseases (Hudson 1992). Third, preliminary evidence indicated that parasites might play an important role, so we have undertaken a series of replicated field experiments manipulating trophic interactions with parasites from the scale of the individual through to the population and community. We believe the combined approach of extensive estimates of population change coupled with experimental manipulations and modelling has allowed us to obtain a unique insight into how natural enemies influence the population growth rate and dynamics of red grouse. We focus our attention on the natural enemies, partly because we can obtain comparative data between populations and partly because the role of them in the natural regulation of red grouse has been contested (Moss & Watson 2000). We also believe that the majority of studies that have focused on growth rate have tended to apply single-species models, particularly when the population is age structured, so we wished to study trophic interactions and examine how different natural enemies can influence population growth rate.

In this paper, we take two contrasting approaches. The first is an examination of time-series data looking at the patterns theory predicts. We start by addressing the question 'what patterns in growth rate does theory predict we should see in time-series data?' We use models based on the Lotka–Volterra model that have been specifically adapted to understanding the interaction between parasites and their host, in particular the Anderson & May (1978) model and its derivative that has been tailored to the red grouse–*Trichostrongylus tenuis* system (Dobson & Hudson 1992). We use these models partly because we have good evidence to suppose that productivity of red grouse is determined by this macroparasite (Hudson 1986a; Hudson *et al.* 1992b) and partly because the tightly linked relationship between parasites and their hosts is one that lends itself to careful experimentation. However, this approach can only identify some ecological correlates that are associated with variations in maximum growth rates and patterns of red grouse population dynamics. Hence, in the second half of the paper we take an experimental approach and examine variations in the risk of infection between individuals and the consequence of this for population growth rate, population regulation and dynamics. Such an understanding then provides a foundation for us to explore community level questions about how other natural enemies such as predators, viruses and humans can interact to influence population dynamics and observed growth rates.

11.2.   What growth rate patterns does theory predict?

The fundamental model of trophic interactions is the Lotka–Volterra model with density dependence in the growth rate of the prey population (Hastings 1996). The model predicts stability in both prey and predator abundance, although the equilibrium is approached through damped oscillations and the predator population may fluctuate to very low levels before the equilibrium is reached. Increasing the growth rate of the prey population has no influence on the size of the equilibrium prey population and all populations stabilize to the equilibrium. This increase in the intrinsic growth rate of the prey population (r) simply feeds through to an increase in the size of the equilibrium predator population in what is sometimes referred to as the paradox of enrichment (Rosenzweig 1971). In other words, fundamental theory of trophic interactions predicts that changes in growth rate of prey populations will have no influence on prey equilibrium population size or dynamics, but may influence the predator equilibrium.

However, this pattern of no influence of population growth rate on dynamics or equilibrium population size breaks down once we start to consider nonlinearities in the density dependence, time-lags and specific heterogeneities associated with particular types of natural enemies (e.g. Crawley 1983). If we consider a specific derivative of the Lotka–Volterra model that examines the relationship between macroparasite and host, we see distinct and different patterns emerge. We use here a derivative of the Anderson & May (1978) model of parasite–host relationships developed by Dobson & Hudson (1992) that was specifically designed to describe the dynamics of the red grouse–*T. tenuis* system. The nematode is a directly transmitted macroparasite that is known to induce increased mortality and a reduction in the fecundity of grouse (Hudson *et al.* 1992*b*). The model examines changes in host, parasite and the population of free-living parasitic stages, and the consequences of changes in host growth rate on host and parasite dynamics. In this model, an increase in the host (or prey) intrinsic growth rate (r) does not lead to the same equilibrium but to some distinct changes in host dynamics, and in the predicted year to year growth rates $-r_t$ (figure 11.1*a*).

Several predictions emerge from this model with respect to the growth rate of the host population. When the growth rate in the model is low, then the host population comes to equilibrium with no variation in abundance and consequently no observed variation in growth rate. As growth rate is increased, so instability in the host population increases, variance in observed host growth rate (r_{var}) increases and the maximum growth rate observed (r_{max}) increases to asymptote at the point where $r_{max} = r$ such that we can assume r_{max} provides a reasonable estimate of the intrinsic growth rate. In real populations, this may not be so simple because the

(*a*)

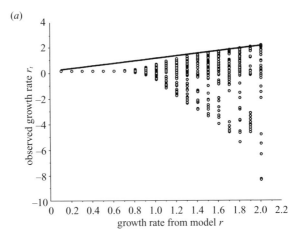

(*b*)

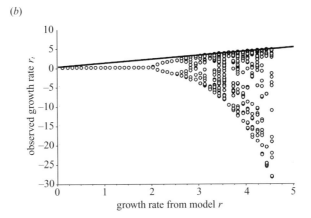

Figure 11.1. Model outputs predicting the relationship between observed growth rates per annum (r_t) and the intrinsic growth rates (r) included in the model: (*a*) from Dobson & Hudson (1992) host–macroparasite model showing that an increase in r leads to an increase in the variance of the observed growth rate, with r_{max} rising asymptotically to r and r_{min} decreasing nonlinearly; and (*b*) a similar result to (*a*) generated from the lagged logistic model.

host population may have to fall close to zero and below an extinction threshold before r_{max} approaches r. As the intrinsic growth rate increases, so the amplitude increases, leading to severe population crashes such that the minimum growth rate (r_{min}) decreases with r_{max}; it does so in a non-linear manner (figure 11.1*a*). However, this figure should be seen simply as a section through the broader parasite–host parameter space that corresponds specifically to the grouse–*T. tenuis* system. Variations in the final

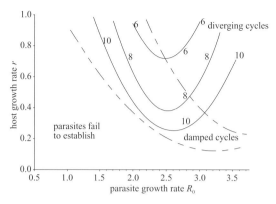

Figure 11.2. Dynamics of the Dobson & Hudson (1992) host–macroparasite model depend on the tension between the growth rate of the host population and the growth rate of the parasite population. Dashed lines indicate bifurcations from stable dynamics to damped oscillations and from damped to diverging oscillations, increasing instability resulting from an increase in the growth rate of the parasite. Solid lines indicate changes in cycle period of oscillations that decrease as host growth rate increases.

dynamics will depend on the biological features of the particular system and in particular the tensions between the growth rates of the host and the parasite (figure 11.2). Nevertheless, the dynamics with high growth rates are essentially oscillatory with cycle periods becoming shorter and amplitudes larger as host growth rate increases.

These predictions are made using the Dobson & Hudson (1992) macroparasite model as applied to the grouse–*T. tenuis* system, but it is gratifying to see that such general predictions hold true from a number of other fundamental models. For example, the single-species, discrete logistic model (Case 1999) provides a similar pattern of observed growth rate (r_t) against (r) (figure 11.1b). Essentially, general theory provides us with four predictions that should be seen when comparing time-series between species of animals, namely an increase in observed maximum growth rate r_{max} leads to the following:

(i)   an increase in the variance of the observed growth rates;
(ii)   a nonlinear decrease in the minimum growth rate r_{min};
(iii)   shorter cycle periods;
(iv)   cycles with greater amplitude.

Such predictions are not independent, as an increase in the variance of the growth rate is going to be associated with an increase in the minimum growth rate and cycles of lower amplitude.

11.3.   Growth rate patterns in grouse time-series

Red grouse inhabit the heather dominant moorland of the United Kingdom, much of which is managed by individually owned private estates to provide a surplus of birds to harvest each autumn (Hudson 1992). The majority of these estates maintain careful hunting records on the precise number harvested each year, providing a spatio-temporal dataset reflecting changes in abundance throughout the range of the subspecies. However, the value of any time-series of hunting records depends on the extent to which effort changes with density and how this may vary over time. The system of hunting on the majority of estates is based on driven grouse shooting where the effort per day varies little between days in that the number of hunters and the area harvested per day remains relatively constant, but the number of days harvested increases with the overall abundance of grouse. In addition, there is a minimum density below which harvesting is cancelled in order to preserve breeding stock (Hudson 1985, 1992; Hudson & Newborn 1995; Hudson & Dobson 2001; Hudson *et al.* 2002). This minimum tends to lead to an underestimate of abundance when numbers are low and consequently an overestimate of growth rate following recovery (Hudson *et al.* 2002). This minimum is reflected in the general relationship between the numbers harvested per square kilometre and numbers counted on 1 km² sample areas, but there is also a change in the variance of the bags with increasing counts (Hudson *et al.* 1999, 2002).

Interestingly, comparisons of counts within areas show a decrease in spatial variance with mean grouse density, supporting the observation that when abundance is low, grouse tend to be aggregated in small areas, and as density increases so grouse tend towards a more uniform spatial distribution (Hudson *et al.* 1998, 2002). Consequently, harvesting records probably provide a better reflection in abundance of what is happening at the population level than small scale counts. This is supported by a more recent rigorous analysis that examined the power law relationship between mean abundance and variance in both harvest and count data and shows that harvesting records provide a good reflection of the underlying population dynamics, and as such, a reasonable representation of the actual dynamics (Cattadori *et al.*, 2003). Even so, such records will incorporate biases: after all they are collected following breeding, so will tend to reflect changes in breeding production rather than simple changes in breeding density.

The time-series of hunting records were used to estimate growth rates. We obtained 352 harvesting time-series from independently managed estates throughout the United Kingdom. Zeros were not used and annual growth rate (r_t) was estimated as $\ln(N_{t+1}/N_t)$. From each time-series, three growth rate parameters were determined: the maximum growth rate observed (r_{max}), the minimum growth rate observed (r_{min}) and the variance in the growth rate (r_{var}).

(a)   What environmental factors are associated with maximum growth rate between populations?

Maximum growth varied greatly between populations and 92% of the cases had maximum growth rates between 1 and 4. Variations in growth rate were examined in relation to a series of habitat, environmental conditions and biotic factors. Previous studies on ungulates have found that maximum growth rates in herbivores are often associated with food quality (Chapter 8), and because grouse feed predominantly on *Calluna vulgaris* we examined growth rate in relation to the average climatic factors that would influence heather productivity, length of growing season and heather quality. Heather productivity, measured as the annual production of shoots (H) was estimated according to the relationship derived by Miller (1979) where heather productivity is a function of mean daily temperature between April and August $(T$ in $°C)$ and precipitation during the same period $(P$ in mm$)$:

$$H = 29.3T - 0.168P - 56.0.$$

A similar relationship was used by Albon & Clutton-Brock (1988) in a study of red deer in Scotland, and they proposed that this would reflect the quality of heather feed available to moorland herbivores in general. The growing season was estimated from the relationship between air temperature in degrees centigrade and corrected for altitude in metres as proposed by Smith (in Albon & Clutton-Brock 1988). Growing season will be correlated with heather productivity but was included as a separate variable from productivity. As with previous studies (Picozzi 1966), soil fertility and its presumed influence on heather quality were estimated from an index of base richness of underlying rocks using solid geology maps. Rock type was ranked from low (1) to high (5) base richness and an overall index estimated according to the area of each under each estate (see Hudson 1992). Other environmental conditions included altitude, average number days of snow cover and number of wet days. Predation pressure was estimated indirectly from the density of keepers (Hudson 1986*b*, 1992). The presence of ticks (*Ixodes ricinus*) and the tick-borne disease louping-ill were determined through a postal survey (Hudson 1992) and were also included as categorical variables. Data were standardized and where necessary $\log_{10}$ transformed.

A general linear model with maximum growth rate as the dependent variable and the independent variables base richness ($F_{1,144} = 6.00, p = 0.01$), ticks ($F_{1,142} = 7.98, p = 0.005$) and their interaction with heather productivity ($F_{1,138} = 5.85, p = 0.02$) explained 54% of the variance in growth rate. Not surprisingly for a herbivore, the maximum growth rate was associated with quality of the food plants available. The general finding that growth rate between populations was associated with variations in food quality is supported by a field experiment undertaken in northeast Scotland by Watson *et al.* (1984). They fertilized plots of heather moorland and counted the number of grouse and recorded their productivity over a period of years. They found that the grouse populations on the fertilized plots produced more young and that numbers increased faster to a higher density than on control plots, but this treatment did not prevent or reduce the extent of a cyclic decline in abundance. In this respect, food conditions seemed important for determining the maximum production rate and so maximum growth rate, but was not associated with the negative growth rates and the cause of the cycles. Returning to our general linear model, the only natural enemy that entered the model and reduced growth rate was the presence of the sheep tick that carries the viral infection louping-ill that causes significant mortality in grouse (Reid *et al.* 1978; Hudson *et al.* 1995). Again, this may reflect habitat conditions as ticks require wet damp conditions for survival and are frequently associated with poor quality vegetation with a thick mat layer dominated by bracken or other non-palatable vegetation (Hudson 1986*c*).

(b)   Do growth rate patterns reflect predictions from theory?

In §11.2, we made four predictions about the general patterns of growth rate in relation to variations between populations. In comparisons between populations of red grouse these predictions were, in general, supported. First an increase in the variance in growth rate was observed with the maximum growth rate (figure 11.3*a*; $F_{1,298} = 451.84, p < 0.001, r^2 = 0.603$), implying that populations became more unstable as population growth rate increases. On the one hand, such a relationship is not too surprising in that we may expect that any increase in the maximum growth

(a)

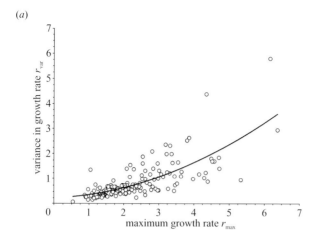

(b)

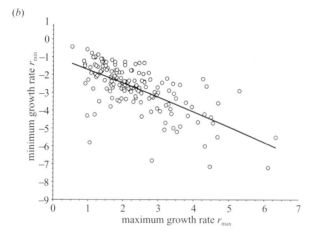

Figure 11.3. Variations in observed growth rate patterns from red grouse hunting records: (a) the relationship between the observed increase in the variance of annual growth rates with respect to maximum growth rate; and (b) the relationship between the annual observed decline in minimum growth rate and the observed maximum growth rate.

rate would lead to an increase in the variance, but on the other hand, if the populations were tightly regulated we may expect to see no increase in variance.

Second, minimum growth rate declined with maximum growth rate in a linear fashion (figure 11.3b; $F_{1,298} = 1109.25$, $p < 0.001$, $r^2 = 0.432$), whereas a nonlinear relationship was predicted from the models. One explanation is that year to year variation in the transmission parameter may change the shape of this relationship. We incorporated 10% stochastic variation in each of the parameters and found that in every instance the stochasticity reduced the size of the minimum growth rate (r_{min}), increased the observed maximum growth rate (r_{max}) and linearized the relationship with the intrinsic growth rate.

To examine the prediction that shorter cycle periods would be found in populations with large maximum growth rates, the tendency for each time-series to cycle and the cycle period were estimated using spectral analysis, full details of which are available in Haydon *et al.* (2002). Cycle period was only estimated for those time-series where there was a distinct and significant peak in the spectrogram. Overall, 57% of populations were cyclic. The tendency to cycle varied between regions such that populations in northeast Scotland were less likely to cycle than expected. Among the cyclic populations, cycle periods varied greatly between grouse populations from 3 to 14 years. Overall there was an increase in cycle period with latitude, although the relationship was weak (Haydon *et al.* 2002; $F_{1,163} = 11.962, p < 0.001, r^2 = 0.092$) with the northern regions of Scotland having significantly longer cycles than populations in northern England (figure 11.4). As predicted by theory, within the cyclic populations there was a significant decrease in cycle period with maximum growth rate $(F_{1,111} = 12.85, r^2 = 0.104, p < 0.001)$, although the percentage of variance explained was low (figure 11.5*b*). Examination of the figure tends to reveal that populations with a relatively high maximum growth rate tend to exhibit short cycle periods, whereas populations with a low maximum growth rate tend to show great variation in the cycle period observed. This may be because populations with low growth rates also have low amplitude and the signal for cycle period becomes difficult to discern (figure 11.5*b*).

We estimated the amplitude of each oscillation as the sum of positive growth rates and then expressed the mean amplitude for a series as the mean of these positive growth rates. As predicted from the general theory, amplitude increased with maximum growth rate $(F_{1,346} = 227.98, p < 0.001, r^2 = 0.397$; figure 11.5*a*).

When the intrinsic growth rates are high, populations of grouse tend to be unstable. These unstable populations, with their highly variable annual growth rates allow us to examine the mechanisms that influence these growth rates. Not surprisingly, for a herbivorous bird like the red grouse, our comparison of time-series data between populations indicated that part of the variation in the maximum growth rate could be

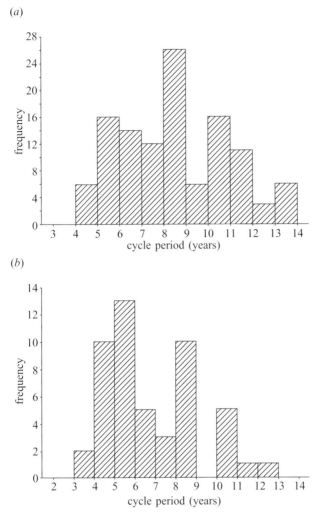

Figure 11.4. Frequency distributions of cycle periods determined from time-series of grouse harvesting data from (*a*) Scotland (*n* = 119) and (*b*) England (*n* = 50). Overall, English grouse populations have shorter cycle periods (mean ± s.e. = 6.73 ± 0.32) than Scottish populations (mean ± s.e. = 8.84 ± 0.36).

accounted for by the food quality and availability (see §11.3a). But such relationships are simply correlations and we have not started to identify the specific processes that lead to negative growth rates. Consequently, we must start to address questions such as: what mechanisms reduce the maximum growth rate?

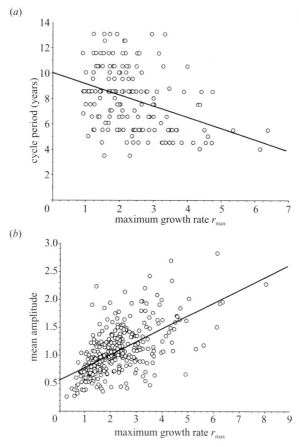

Figure 11.5. Relationships between grouse cycle period and amplitude with maximum growth rate, England and Scotland; (*a*) cycle periods of grouse time-series decreased with maximum growth rate although the variance explained was small, and (*b*) amplitude increased with maximum observed growth rate.

11.4. Individual level processes in influencing production

Population growth rate is in many respects a term that simply integrates the survival and productivity of all individuals within the population over a period of time. To understand how population growth is reduced, we need to know first how vital rates are influenced by specific factors, and also how the risk of exposure to these factors varies within the population. In this section, we start by describing experiments that examined the

impact of parasites on individual productivity and then the consequences
of this in the Dobson & Hudson (1992) macroparasite model. Much of this
section is based on previously published experiments, so these will be presented as a review illustrating how detailed experimental studies at the individual level, integrated with modelling, can provide an insight into the
mechanisms influencing growth rates.

(a)   Negative growth rates and productivity

Early studies on grouse reported that the negative growth rates of grouse
were often associated with heavy infections of the caecal nematode *T. tenuis*
(Cobbold 1873; Lovat 1911). More recent quantitative, longitudinal studies support this observation and show that within study areas, years with
negative growth rate were associated with heavy infections of *T. tenuis*
(figure 11.6*a*). A more careful inspection of the data demonstrates that
both over-winter loss (mortality plus net dispersal) and breeding mortality ($\log_{10}$ (maximum clutch size − brood size at 6 weeks of age)) were also
associated with the heavy burdens of *T. tenuis* (figure 11.6*b*), indicating that
parasites may have an important role in the year to year changes in population growth rate (Hudson *et al.* 1992*b*, 2002). Interestingly, earlier studies
(Jenkins *et al.* 1963) dismissed the relative importance of *T. tenuis* in causing
negative population growth rates because they focused solely on the possible role of parasites in influencing adult mortality and did not consider
the indirect influence parasites may have in increasing grouse breeding
mortality.

To determine if parasites reduced host reproduction, we undertook
replicated field experiments where we experimentally reduced the natural infections of grouse by treating some birds with an anthelmintic and
comparing their reproductive performance with an untreated, placebo
control group (Hudson 1986*a*; Hudson *et al.* 1992*b*). Both within and between year treatments demonstrated that parasites consistently reduced
the reproductive output of the grouse. Parasites also reduced host survival
but this was less severe relative to the impacts on host fecundity (Hudson
et al. 1992*b*).

(b)   Patterns of infection between individuals

Studies of age-structured populations have illustrated the importance of
considering variations in vital rates between age cohorts (Caswell 2001).
One advantage of such an approach is that age structured models can then
provide predictions of growth rates and future changes in growth rates

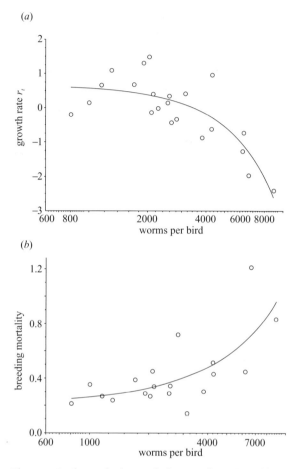

Figure 11.6. Changes in the population growth rate (*a*) and breeding mortality (*b*) of red grouse in relation to mean intensity of infection with the parasitic worm *Trichostrongylus tenuis*. Points represent results from different years during a longitudinal study of population dynamics on a Gunnerside estate, North Yorkshire, England (Hudson *et al.* 1992*b*).

according to changes in age structure (Caswell 2001). One limitation with such an approach, however, is that this may focus workers solely on age as the predictor of vital rates and not on the biological mechanisms that may influence variations in these vital rates between individuals, and hence their importance at the population level.

An important step in examining the influence of individual vital rates on population growth rate is to consider the frequency distribution of risk between individuals within the population. In terms of natural enemies, that is the risk of becoming infected with a virus, the risk of carrying a heavy parasite burden or the risk of being taken by a predator, and the consequences of these to survival and breeding production. In many species such risks are not evenly distributed through the population: there is frequently a small proportion of the population more susceptible to infection, predation or the impact of some other natural enemy. This is simply illustrated by comparing the frequency distribution of broods in different years where females have heavy and light infections of *T. tenuis* on one study area (figure 11.7a), or chicks have been exposed to heavy or light infections of the louping-ill virus on another study area (figure 11.7b). In both instances, mean brood size varied between the sampled years but there are still some individuals that do poorly in good years and some that do well in bad years. To understand how these natural enemies influence population growth rate we must first identify through field experiments that the natural enemies are indeed an important cause of these variations in productivity, and secondly, incorporate this frequency distribution of risk into the models. For both of these natural enemies we have undertaken field experiments that have demonstrated that these differences in breeding production are indeed caused by infection of *T. tenuis* (Hudson 1986a; Hudson *et al*. 1992b) or louping-ill (Laurenson *et al*. 1998).

Most macroparasites exhibit an aggregated pattern of distribution within their host population, with the majority of parasites aggregated in the minority of hosts (Anderson & May 1991; Shaw & Dobson 1995; Wilson *et al*. 2001). In the case of *T. tenuis* in red grouse, the degree of aggregation is relatively low and approaches a random Poisson distribution (Hudson *et al*. 1992b). By contrast, the distribution of the nymph ticks that transmit the louping-ill virus to red grouse chicks is highly aggregated between chicks and broods (Hudson 1992) and, as with many macroparasites, both frequency distributions are best described with the negative binomial distribution (Shaw *et al*. 1998). One consequence of this aggregation is that only a relatively small proportion of the grouse population will be exposed to the virus.

(c)   Modelling the impact of parasites on grouse

The fundamental model of macroparasite–host dynamics that incorporates the essential features of parasite induced reduction in fecundity and the pattern of parasite distribution between hosts is that of Anderson & May (1978). This was extended by Dobson & Hudson (1992) to include the

(*a*)

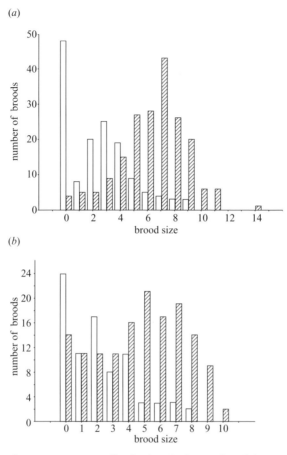

(*b*)

Figure 11.7. Frequency distribution of red grouse broods in years of high and low infection: (*a*) years of high (open bars) and low (hatched bars) infection of hen grouse with the nematode *Trichostrongylus tenuis* from Gunnerside, North Yorkshire, England; and (*b*) years of high (open bars) and low (hatched bars) seroprevalence in young grouse with the tick-borne louping-ill virus from Lochindorb, Morayshire, Scotland.

dynamics of the free-living stages and arrested development, and in so do-ing provided a suitable model of the grouse–*T. tenuis* system. Both studies identified that instability and variations in population growth rate will occur when the parasite-induced reduction in fecundity (δ) is large com-pared with the parasite-induced reduction in survival (α) with respect to the degree of aggregation as measured by the aggregation parameter k from the negative binomial distribution. More formally, this will occur

(a)

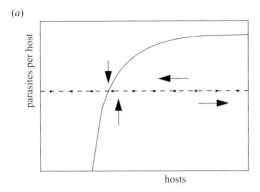

(b)

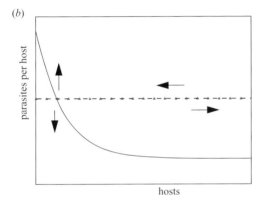

Figure 11.8. Phase plane analysis of the Anderson & May (1978) macroparasite model with the host (H) isocline $\delta H/\delta t = 0$ represented by the dashed line and the parasites per host (M) isocline $\delta M/\delta t = 0$ represented by the solid line: (a) parasite induced mortality (α) large relative to parasite induced reduction in host fecundity (δ) ($\alpha/k > \delta$) and (b) parasite induced mortality (α) small relative to parasite induced reduction in host fecundity (δ) ($\alpha/k < \delta$). In (a) the isoclines generate stability with populations showing damped oscillations while in (b) the parasite isocline moves and generates unstable oscillations.

when the risk of parasite-induced mortality is less than the reduction in parasite-induced fecundity:

$$\alpha/k < \delta. \tag{11.4.1}$$

A powerful way to illustrate this is through a graphical representation of the phase-plane analysis of the basic Anderson & May (1978) model (figure 11.8). When the impact of parasites on survival is large relative to that on fecundity ($\alpha/k < \delta$), the phase plane isoclines reflect the pattern seen for the basic Lotka–Volterra predator–prey model with density dependence in the predator population. The vectors also indicate that the dynamics are essentially stable, and perturbations lead to damped cycles that return to the equilibrium represented by the point where the two isoclines intersect (figure 11.8a). By contrast, when the parasite-induced reduction in survival is low relative to the impact on fecundity ($\alpha/k < \delta$), then the parasite isocline changes and leads to unstable and oscillatory dynamics (figure 11.8b).

In summary, the parasite-induced reduction in fecundity is particularly important to the stability of the observed growth rates. When this is large relative to the impact on survival, instability and variations in growth rate will be observed. The specific cycle period, patterns of oscillations and amplitude will relate to the specific tension between the growth rate of the host and the parasite (figure 11.2).

11.5. Population level patterns

In the previous section, we found that when the parasite-induced effects and distribution were incorporated into the model we could predict that the negative growth rates observed at the population level system are essentially caused by the impact of the parasites on host fecundity. This finding is supported by the patterns observed from long-term population monitoring at the population level where negative growth rates and increased breeding mortality occur in years with high worm burdens (figure 11.6). A clearer illustration is to examine the relationship between observed host growth rates and parasite growth rate; this plot (figure 11.9) illustrates that host growth rate in this study population tended to be positive when parasite growth rates were negative, and host growth rates negative when parasite growth rates were positive, indicating that the parasites were a regulating factor.

Before we can take our understanding of this system and the interaction with other natural enemies further, we need to test the model predictions and specifically the population level prediction that the parasites are the cause of the negative growth rates. To achieve this, Hudson *et al.* (1998) undertook replicated population level experiments. Initially they used the Dobson & Hudson (1992) model to predict when the growth rate of the study populations would become negative and when treatment should be applied. They also used the model to predict that more than 20% of the

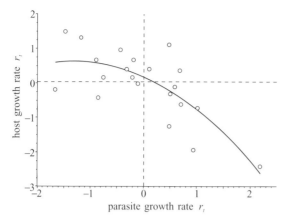

Figure 11.9. Observed host population growth rate in relation to observed parasite growth rate from a longitudinal study (the same study area as figure 11.6). Note that host growth rates tend to be positive when parasite growth rates are negative and host growth rates tend to be negative when parasite growth rates are positive.

population would need to be treated to provide sufficient power to identify a reduction in the size of the negative growth rate. Interestingly, the model predicted that it would be difficult to make growth rates positive and probably impossible to eradicate worms totally. This was because increased treatment would lead to increased survival and productivity, an increase in the density of grouse and a greater force of infection on the grouse population (Hudson *et al.* 1999). Returning to the experiment, a total of six study populations were used; the first two were left as untreated controls for the two population crashes observed. The next two populations were treated prior to the first period of negative growth rate such that the second period of negative growth rate provided a time control. The remaining two populations were treated during both periods of negative growth rate. In each of the six cases of treatment (in four populations), the treated populations exhibited negative growth rates lower than the controls and, overall, experimental treatment significantly reduced the variance in the growth rate of the populations, with two treatments almost completely removing the tendency to cycle (figure 11.10; Hudson *et al.* 1998), even though the population had been cyclic prior to the experiment. These results support the model predictions that negative growth rates are produced by the influence of parasites on the fecundity of grouse.

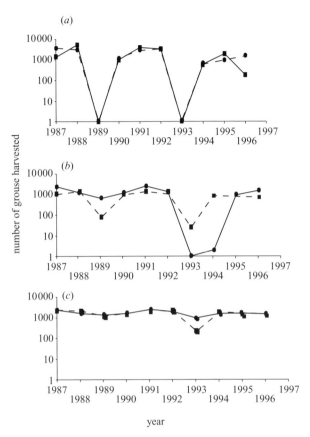

Figure 11.10. Results from a replicated population level experiment where grouse were treated with an anthelmintic to reduce parasite intensities prior to population growth rates becoming negative in 1989 and 1993 and hunting records recorded: (*a*) two control populations where no grouse were treated; (*b*) two experimental populations treated in 1989; (*c*) two experimental populations treated in both 1989 and 1993. Treatment reduced the negative population growth rates in both years (modified from Hudson *et al.* 1998).

11.6. Community level interactions with natural enemies

By integrating our findings from experiments, models and through long-term monitoring at the individual and population level, we have obtained an insight into the factors influencing the growth rate of red grouse populations. The initial comparison of growth rates identified that variations in the maximum growth rates between populations could be accounted for through variations in the factors likely to influence food quality (base richness) and the climatic conditions associated with food productivity, but that variations in annual growth rates between years were a consequence of the impact of parasites on the breeding production of individuals. Given that we now have a fairly good understanding of these factors, we can use the model and comparative data to start exploring the consequences at the community level and consider the range of other natural enemies. We start by examining how selective predators may influence patterns of growth rate and then consider the effects of the louping-ill virus and the influence of hunting.

(a)   Selective predation

Red grouse are prey to a wide range of mammalian and avian predators, notably red foxes, peregrines, hen harriers and eagles (Hudson *et al.* 1997). On all managed grouse moors there is legal control of most of the mammalian predators, notably the fox and stoat, but the killing of raptors such as peregrine, hen harrier and golden eagle along with the owls is illegal but does occur. In a study of grouse mortality, Hudson *et al.* (1992*a*) recorded the frequency distribution of parasites in red grouse taken from grouse killed by predators and compared them with a random selection of grouse shot. Grouse killed by predators carried greater intensities of worm infection than a random sample but lower infections than grouse that were found dead and presumably died from the infection. These findings, coupled with other data and experimental studies, indicated that predators were selectively removing the heavily infected individuals from the population, presumably because the individuals had been weakened by the parasites and the predators found it easier to either locate them or catch them (Hudson *et al.* 1992*b*).

If we incorporate predator selection into the Dobson & Hudson (1992) model then we discover that this selective predation does two things to the grouse population. First, the selection reduces the variance in growth rate and dampens the cyclic tendency of the population. This is essentially because the predation is increasing the parasite induced mortality and in so doing making $\alpha/k > \delta$ so the population no longer oscillates. In this respect the predation is reducing the variance in the growth rate of the host. A reduction in the tendency to cycle was identified in long-term intensive field studies at Langholm moor in southern Scotland where increased predation pressure from hen harriers prevented a predicted cyclic increase in grouse abundance (Thirgood *et al.* 2000). More intriguing is that the model predicts that the equilibrium population size increases

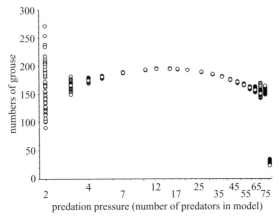

Figure 11.11. Bifurcation diagram of grouse numbers against predation pressure from the Hudson *et al.* (1992*a*) macroparasite–host model incorporating selective predation where predators selectively remove the heavily infected grouse. Note that as predation pressure increases so the oscillations dampen and the mean equilibrium level rises, although once predation pressure is high the population subsequently crashes.

(figure 11.11). At first this appears counterintuitive because we have increased mortality, but this has led to an increased equilibrium. However, the reason for this is quite simply that the selective removal of a few heavily infected individuals from the grouse population effectively removes more parasites than grouse and so reduces the regulatory role of the parasites that causes the instability and also leads to an increase in the equilibrium population size.

(b)   Louping-ill virus
Louping-ill is a tick-borne virus that causes 80% mortality in exposed grouse (Reid *et al.* 1978), although the evidence is that observed moderate levels of tick infestation have no effect on the grouse chicks (Hudson 1986*b*). Not all populations of grouse are exposed to ticks and not all of those populations with ticks necessarily have the louping-ill virus circulating in the tick population (Hudson *et al.* 1995). Moreover, the rate of exposure (as estimated indirectly from the proportion seropositive) varied between populations and can range from close to zero to up to 60% (Hudson 1992). We can incorporate this additional mortality into the model as an additive effect, and because we are considering just exposure rate to the virus we do not need to take account of the aggregated

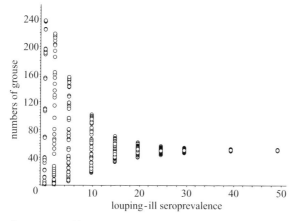

Figure 11.12. Bifurcation diagram of grouse numbers against seroprevalence of young grouse to louping-ill virus from the Dobson & Hudson (1992) model. Note that increased infection dampens oscillations and reduces mean equilibrium levels.

distribution of the ticks. The model predicts that increased exposure first leads to increased stability (reduction in the variance of the growth rate) and second to a decrease in equilibrium (figure 11.12). In contrast to the selective predation, the additional mortality induced by the virus is effectively reducing the population growth rate and through this process reducing infection with the worms, thus leading to stability. The stabilizing effect of louping-ill can also be seen in the time-series hunting record data where the second order partial correlation coefficient of the time-series (an estimate of the strength of the delayed density dependence) of populations with louping-ill was significantly lower than that from populations without louping-ill.

(c) Harvesting by humans

While mortality induced by both selective predation and the impact of louping-ill virus were stabilizing, we should note that the hunting records examined in the earlier sections of this paper were highly unstable. In other words, predation by humans does not appear to be stabilizing the population dynamics of the harvested grouse populations. Harvesting by humans regularly removes up to 50% or more of the population in the autumn, and yet the dynamics are still clearly unstable. This is somewhat surprising as this level of mortality when incorporated into the models should easily stabilize the populations.

One explanation for this observation is that the mechanisms that drive the oscillations occur before the onset of harvesting, such that harvesting reflects the cyclic changes in abundance rather than reducing abundance and infection and so dampening the oscillations. In this respect, it is interesting to note that much of the infection of grouse with *T. tenuis* takes place prior to harvesting (Hudson & Dobson 2001), and even then a large number of infective stages are on the ground and available for infection such that the 'die has been cast' before harvesting commences. If we accept that harvesting has not been highly stabilizing, then this means the important destabilizing effect must occur before harvesting and as such would imply that population changes caused by other mechanisms soon after harvesting, such as changes in spacing behaviour, would in general not influence the population cycles (contra Moss & Watson 2000).

11.7. Conclusion on studying population growth rate

The intrinsic growth rate describes the maximum growth rate of a population and, as such, is the key variable of population models which examine how biological mechanisms or demographic changes can alter observed growth rates. Our comparison of grouse time-series with models is gratifying in that the variation in growth rate can be explained by fundamental predator–prey theory. Furthermore, more than 60% of the variation in the maximum growth rate can be accounted for through food quality. Data analyses, modelling – and especially their synthesis in modern mechanistic time-series approaches – can lead to focused, testable hypotheses about underlying biological mechanisms. Ultimately though, hypotheses need to be tested through experimentation.

The observed population growth rate is the integral of the birth and death rate processes that occur at the individual level, and the distribution of these risks leads to these processes throughout the population. Risk here is the risk of an individual being killed by a predator, the risk of reduced fecundity through low food availability and the risk of being excluded from a breeding population by the behaviour of others. But the biological mechanisms are the interesting and important aspect of the growth rates that need investigating. This paper has attempted to show this by examining individual level variations in vital rates and how the risk of suffering from natural enemies was distributed through the population. Previous studies led us to suppose parasites may play a role, but as the variance in growth rate was high and parasite induced reductions in fecundity are known to be destabilizing (Anderson & May 1978), we thought it reasonable to suppose that the effects on breeding production would be an initial place to focus. An important component of such studies was to examine the patterns of parasite distribution between individuals and thus consider the overall impact of the parasites at the population level. Indeed, one of the great strengths of the deterministic Anderson & May (1978) model is that it incorporates both the impact of parasites on reduced fecundity and survival, and describes the frequency distribution of this risk within the population. The individual level experiments were incorporated into the model and used to predict population level patterns that could be tested with a replicated field experiment. Such an approach provided a sound framework from which we could start exploring the community level consequences of adding additional natural enemies including selective predation and viral induced mortality.

Clearly, the strength of the experimental approach is that we obtain a foundation of understanding of the processes important at the individual level, and when these findings are incorporated with the modelling this allows us to makes predictions at the population level. These predictions can then be tested at the population level: if they are confirmed, we can start to incorporate further complexities from the community level and make predictions that can be tested experimentally. Indeed, this close interaction between experiment, model and then prediction and experiment again provides a good working schedule for examining the important processes that influence growth rate. However, with many natural systems, field experiments on replicate populations may be impractical or unethical. In such cases, the experimental approach may still provide insight from 'model systems'. Such model systems could include free running animals kept in pens in the field (e.g. pheasants and macroparasites) through to laboratory cultures of invertebrates. Careful choice of the model may allow considerable insight into the population dynamics that can then be applied to field studies.

In conclusion, there are three points we wish to make. First, growth rates are an important summary parameter in population models, but considering them in isolation might neglect the many important biological and environmental factors that influence the birth and death process of individuals. These processes should be the focus of our attention and we need to know the processes influencing birth and death and how the risk of suffering from these is distributed within the population. Age structured models can provide a framework to explore these issues: they should seek to explain the reasons for variations in vital rates between and within age groups, and in so doing, integrate evolutionary life-history studies with demographic ecological studies. Second, time-series analysis, modelling and carefully monitored studies of individuals can provide important insights into the mechanisms influencing population growth rate, but without experiments they can never identify the true mechanisms involved. The ideal approach is simply a combination of all techniques, but they need to be based on a biological understanding that arises from experimental field manipulations. Third, we are keen that the overall purpose of this paper is not misunderstood. This is not a paper simply about parasites and grouse, aimed at showing that parasites have some role to play in host population dynamics. Host–parasite interactions are also particularly suitable for exploring general aspects of population dynamics, in particular the ways in which we can tease apart how trophic interactions operate in wild animal populations. Quantifying the risks is not always easy, but this study has illustrated that parasite–host trophic interactions are intimate and hence often quantifiable. Moreover, we can use the understanding we obtain at the individual level to make population dynamic predictions that can be tested and then developed to the community level.

Acknowledgments

We thank all our friends and colleagues who have been very helpful over the years working on the red grouse studies. Particular thanks go to Karen Laurenson, Lucy Gilbert, Simon Thirgood and David Howarth. Grouse moor owners and their keepers have been remarkably helpful with the collection of the bag record data and other extensive information. The Scottish Game Conservancy Trust and NERC supported much of this work. Mick Crawley was particularly helpful in pointing out some simple but important aspects of growth rates while Charley Krebs, Ken Norris and Jim Hone provided helpful comments. We also thank The Royal Society for support to attend the meeting. During the writing of this paper one of our greatest field workers, a German pointer named Fergie, died and this paper is dedicated to her memory.

1 2

Behavioural models of population growth rates: implications for conservation and prediction

12.1. Introduction

A common justification of ecological research is that it allows predictions of the consequences of environmental change. There is a considerable need to be able to make realistic and justifiable predictions. With many environmental issues, such as genetically modified (GM) crops, climate change, habitat loss and exploitation, there is an urgent need to be able to produce quantified predictions. Such quantified predictions are essential for policy-makers if they are to consider ecological consequences within their framework of social and economic costs and benefits (Sutherland & Watkinson 2001).

Conservation biologists regularly carry out analyses (usually referred to as population-viability analyses) to evaluate the likelihood of a population persisting in the presence of existing or novel conditions (for reviews see Beissinger & Westphal (1998); Norris & Stillman (2002)). These are then often used to determine the conservation measures required to maintain the population, such as the release of additional individuals, a reduction in exploitation levels or the expansion of the available habitat (see Beissinger (1995); Green *et al.* (1996); Hiraldo *et al.* (1996); Root (1998) for a range of avian examples that illustrate these applications).

The basic elements of all population models are the population growth rate in the absence of interspecific competition, the extent of density dependence and the level of stochasticity (Burgman *et al.* 1992). However, in practice, population-viability analyses very rarely use measured parameters for density dependence, particularly when models are applied to the management of endangered species due to the paucity of data (see Green & Hirons (1991)). We shall also argue that the frequently used methods for estimating population growth have serious problems associated with them.

In this paper we shall review the importance of density dependence in answering conservation questions, describe the problems associated with conventional methods of studying density dependence and then review the potential of using behaviour-based models of population ecology to answer ecological and conservation questions.

12.2. Density-dependent population growth – its measurement and use in conservation

(a)   Importance of density dependence

Density dependence is obviously central to any understanding of population ecology. It thus follows that it is essential for predicting the consequences of environmental change (Sibly *et al.* 2000b). For a population showing no consistent increase or decrease over time, if density dependence is absent then any reduction in the population growth rate results in eventual extinction while any increase leads to an infinite population.

The importance of density dependence is most straightforward when considering the consequences of a density-independent change in vital rates, such as a change in the levels of exploitation, mortality due to pollution, or breeding output resulting from predators or natural disasters. It is then conceptually straightforward to consider the change in population size resulting from a change in the vital rate assuming that the density-dependent responses stay constant.

The population response to habitat loss (Sutherland 1996a), habitat deterioration (Sutherland 1998) or human disturbance (Gill *et al.* 1996) depends critically upon the density dependence. In the absence of any density dependence, individuals would just move out of the lost/deteriorated/disturbed patches and occur at higher densities within the remaining/unaffected/undisturbed patches. There will only be population-level effects if the density dependence operates and the higher population densities in the unaffected patches result in increased mortality or reduced reproductive output. Thus Sutherland (1996a) showed how the extent to which habitat loss for migratory birds resulted in population declines (as usually stated by conservation lobbying groups objecting to the change in land use) or birds simply being accommodated elsewhere (as usually stated by those favouring some alternative use of the disputed land) depends upon the relative strengths of density dependence in the breeding and wintering grounds. By estimating these strengths it is then possible to estimate the relative importance of these two processes and thus show how quantitative predictions can replace verbal arguments.

Density dependence is similarly central to the understanding of exploitation. In the absence of any density dependence any increase in mortality due to increased exploitation inexorably leads to extinction. Thus, the essence of exploitation is that it reduces population size and as a result of density dependence the population grows and it is this growth that can be exploited.

(b)   Problems in measuring density dependence

There are three mains ways of studying how population growth rate changes with density, each of which has associated problems.

The first way is the most common and analyses a time-series of population estimates. As has been realized for some time, a serious problem that undermines such techniques results from the fact that populations are usually estimated with some level of measurement error (Bulmer 1975; Royama 1981). As a year with an atypically high level of measurement error is likely to be followed by a year with a more typical measurement error, then the analysis of rate of population change (N_{t+1}/N_t) plotted against population size (N_t) can result in inflated measures of density dependence. Although this phenomenon has been appreciated and there are a range of suggested solutions, it is still extremely difficult to overcome this problem if the measurement error is unknown (Shenk *et al.* 1998).

A second way of studying density dependence is to correlate some aspect of fitness (such as mortality or fecundity) against population size (e.g. Paradis *et al.* 2002). This has the considerable advantage of overcoming the statistical problems associated with measurement error of time-series as described above. A problem is that vertebrate populations usually show little variation in abundance over the study period, unlike many insect populations. Even if abundance does vary over time, the range of variation may be insufficient to reveal the underlying density-dependent function. One problem, especially with relatively short datasets, is that the population is often either increasing or decreasing over time and the correlation with a similar change in some component of the demography may be an artefact and not causal. Another problem is that the population changes are driven by changes in demography, which may be due to changes in the weather, food supply, predation level and other unknown factors. Thus periods of, say high food abundance, may result in both a high survival rate and high population size. Correlating the survival rate and population size will then provide underestimates of the strength of density dependence.

The third way of studying density dependence is to carry out field experiments. This is an excellent means, but it is rarely feasible to carry out well designed replicated experiments on vertebrate populations (but see Smith *et al.* (2000)).

(c)   Problems in measuring population growth

The intrinsic rate of population increase r_{max} (or R_{max} in the finite form) is an important measure. It determines the ability of a population to survive stochastic events or respond to environmental change or exploitation.

In the absence of an assessment of the level of density dependence, it is often impossible to assess the intrinsic growth rate. A common method is to measure life-history parameters, such as age-specific mortality and fecundity, and use these with a population matrix to determine the population growth rate (e.g. Lande 1988; Hiraldo *et al.* 1996). For populations near an equilibrium level, the value of population growth must inevitably be close to 1 – and for plant populations this is typically the case even though the actual values are likely to be considerably higher (R. Freckleton and A. R. Watkinson, personal communication).

As we shall describe in §12.3, a problem with applying the conventional means of population ecology to conservation issues is that even if it is possible to quantify the values of the intrinsic rate of population growth and of density dependence under existing conditions, these are likely to alter as a result of the conservation issue under consideration. Indeed, it is often these precise alterations that bring about the resulting changes in population size.

The problems of estimating the population growth rate are probably clearest in the determination of sustainable levels of exploitation. As a rough approximation, a proportion of individuals equivalent to the population growth rate can be removed each year in a sustainable manner. For a population at its natural equilibrium there will be no population growth and thus no exploitation is possible (Caughley 1977). However, it is common to calculate the population growth for an unexploited population at naturally occurring levels and then use this to calculate the response to exploitation. As the growth is negligible in unexploited populations at natural levels this inevitably leads to the conclusion that the population is sensitive to very small levels of exploitation, even though this may not be the case.

There will be maximum population growth at very low population levels (unless an Allee effect operates (Courchamp *et al*. 1999; Stephens & Sutherland 1999)), yet this value is sometimes recommended for calculating sustainable exploitation (Robinson & Redford 1991). An extreme version of this problem arises from measuring the population growth rate under optimal conditions (Robinson & Redford 1991). For example, measures of reproductive output from zoos, such as age of first breeding, age of last breeding, number of young produced per attempt, have been used to estimate the possible breeding success. Such estimates may well overestimate the breeding success of typical individuals in natural conditions. The measure that is required for sensible sustainable exploitation is the population growth rate at the exploited population size. It is often assumed that the highest sustainable growth will occur at *ca*. 60% of the unexploited population size (e.g. Caughley & Gunn 1995; Robinson & Redford 1991).

12.3. Even perfect knowledge of the demography may be insufficient

As described in §12.2, there are a range of problems associated with measuring population growth rate and density dependence. What if these problems could be overcome and there was a complete knowledge of population growth rate and density dependence? In this section we shall argue that even this rarely achievable ideal is insufficient.

In order to describe density-dependent processes adequately in wild populations, the ideal would be to follow populations that are changing in size over time from very low abundances to abundance levels at which there is evidence of regulation such that the population no longer grows. Ironically, conservation biology often offers such opportunities when populations are restored to areas from which they have been extirpated. For example, the Mauritius kestrel (*Falco punctatus*) population was reduced to only four known individuals in the wild in 1974 (Temple 1977) due to a combination of pollution and habitat loss. A recovery programme was initiated in the early 1970s, and in the late 1980s captive-reared individuals were released into the Bambous mountains in the east of Mauritius, a former part of the kestrels' range from which the last wild individual was recorded in 1955 (Jones *et al.* 2000). Since its initiation, this population has been intensively monitored using colour-marked individuals, and the complete life histories of virtually every bird that has existed in the population are known. A recent analysis indicated that very few (i.e. less than 10) nesting attempts in the history of the population have been missed (Groombridge *et al.* 2001), indicating that the demographic information available for this population is perhaps more complete than for any other vertebrate.

There is evidence that this kestrel population is now showing signs of regulation (figure 12.1). This means that it is possible to examine density dependence in particular vital rates as the population has grown in size. For example, the annual survival probability of wild-bred kestrels in their first year of life has declined dramatically as the population size has increased (figure 12.1*a*). Similar analyses could be repeated for other aspects of the birds' life history, and the relative importance of different density-dependent processes in regulating abundance assessed. This would permit us to model, in an extraordinary degree of detail, the dynamics of this population, at least in terms of regulatory processes. What insights might such a demographic model provide?

Conservation biologists use demographic models for two broad purposes – risk assessment and designing effective conservation management. The former aims to quantify the risk faced by a population as a result of a particular environmental change, whereas the latter concerns quantifying the potential benefits of planned changes to the environment. The question then becomes, how useful would our demographic model be for reliably predicting the impact of environmental change? The answer is that it depends on the extent to which environmental changes influence density-dependent processes. For example, one potential future threat to the persistence of the kestrel population at Bambous is an increase in the frequency or severity of cyclones. This is essentially a stochastic process, and one that could, at least in principle, act in a density-independent way. Heavy rain associated with cyclones can flood nest sites, destroying the nesting attempt and thereby reducing fledgling output. Our demographic model could then be used to estimate the rate at which population size is likely to recover following a cyclone, and examine how population persistence might be affected by the frequency and severity (in terms of nest mortality) of cyclonic events.

By contrast, it is also conceivable that cyclones could interact with density-dependent processes. For example, imagine that the negative

(a)

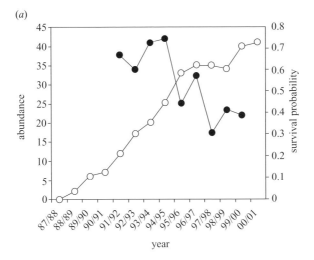

(b)

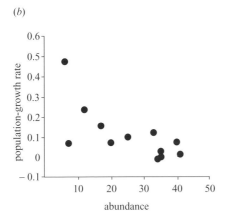

Figure 12.1. Time-series trends in the abundance and survival of a reintroduced population of Mauritius kestrels. (a) The population increase in the Bambous region of Mauritius and the juvenile survival probability rate (estimated using a mark-resighting analysis). White circles show abundance and black circles show juvenile survival rate. (b) The density-dependent population growth rate with $r = -0.694$ and $p = 0.012$. Note that censuses were complete, so there is no sampling error in the abundance estimates (M. Nicoll and C. G. Jones, unpublished data).

density dependence in juvenile survival (figure 12.1a) is driven by competition for food, such that vulnerable juveniles are more likely to starve when competing for food at relatively high population size. Further, imagine that during the heavy rain and wind associated with cyclones hunting efficiency is severely reduced, exacerbating the starvation risk of vulnerable juveniles, particularly at relatively high population sizes. This mechanism would mean that the form of density-dependent juvenile survival varied with respect to cyclonic conditions. Although a hypothetical example, data from long-term population studies are beginning to document such interactions between the environment and density-dependent processes (e.g. Soay sheep, *Ovis aries*; Coulson *et al.* 2001). The consequence of such a process in the kestrel population would be that the observed density dependence in juvenile survival would represent the interactive effects of both population density and cyclone frequency–severity on survival, the exact survival rates observed in the wild being determined by the frequency and severity of cyclones, and the form of density dependence under different cyclone conditions. If in the future cyclone frequency and severity increases, then the negative effects of population size on juvenile survival would become more severe, but in a way that is unlikely to be predictable from past population behaviour because the new combinations of environmental (cyclone frequency and severity) and population density experienced by the population are likely to be unique. Therefore, without detailed data that allow us to describe interactions between density-dependent processes and the environment, the predictive ability of any demographic model, even based on a very complete dataset describing the current population, would become compromised when applied to novel future conditions. This lack of data is the general rule for demographic information on endangered or threatened species, even in well-studied taxa such as birds (Green & Hirons 1991).

This is a very specific example, but illustrates two crucially important general principles. First, even a demographic model based on a complete set of data such as those available for the Bambous kestrel population would only provide reliable predictions about how the population (in terms of changes in abundance over time) would respond to future environmental change if the change influenced vital rates via a mechanism that acted completely independently of any density-dependent processes operating in the population. If the environmental change alters the form of density dependence in a way that is unpredictable from past population behaviour (and this is likely to be the general rule), then the model is no longer reliable. Second, it is clear from the hypothetical example that understanding how the environmental change is likely to impact on demography depends on developing an understanding of the mechanisms driving density dependence in particular vital rates. How can this be done in a way that provides insights for conservation?

12.4.   A role for behaviour-based models?

Behaviour-based models have been largely developed by behavioural ecologists interested in understanding decisions taken by animals competing for resources (e.g. Sutherland 1996b; Goss-Custard & Sutherland 1997; Pettifor *et al.* 2000a). These models examine the fitness consequences of various alternative decisions animals can take, and determine evolutionarily stable strategies (ESSs) that individuals within a population should adopt in order to maximize their fitness. When the 'optimal' decision for one individual is affected by decisions taken by others in the population, game theoretical approaches have been used to determine the ESSs. Perhaps the most influential and widely used theoretical frameworks employed to date are the ideal-free and ideal-despotic models originally formulated by Fretwell & Lucas (1970).

Behaviour-based models are linked to density dependence in demography because they describe the behavioural mechanisms underlying density-dependent processes. They can be used to derive density-dependent processes directly by including in a particular model a relationship that describes how fitness components such as survival or fecundity vary in relation to the resources acquired by individuals in the population. For example, simple prey depletion models of competition for food resources assume that individuals starve or emigrate if they fail to achieve a threshold rate of food intake (Sutherland & Anderson 1993; Pettifor *et al.* 2000a). In this way, density dependence in survival or fecundity can be derived from a behaviour-based model simply by re-running the model for variable numbers of competitors (e.g. Stillman *et al.* 2000).

Why is this approach potentially more useful for understanding the impact of environmental change than the more classical approach of measuring density dependence directly? We have already shown that the classical approach suffers from the problem that the form of density-dependent processes are likely to change in the future if environmental change affects habitat quality, and by implication resource availability. Behaviour-based models are robust in that models can be re-run for any plausible future environmental change scenarios because the principle that individuals attempt to maximize their fitness does not change. The environment provides the stage on which behavioural games designed to maximize an individual's fitness (by acquiring resources) are played out between individuals. If the stage changes, the individuals still 'play' by the same fitness maximization rule, although the demographic outcome of the game may be different. What insights might understanding such games provide to the conservation biologist? To answer this question we illustrate the approaches taken to address a range of specific conservation problems using behaviour-based models.

(a)   Mortality, sustainable population sizes and competition for food resources

Shorebirds overwinter in vast numbers on the coasts of northwest Europe. These habitats provide essential food resources that permit birds to survive the winter, and fuel their spring and autumn migrations to and from their high-latitude breeding grounds. The large numbers of birds means that many coastal sites are recognized as being of international importance for particular shorebird populations, and so are regarded as being habitats of high conservation value. Man also exploits these coastal habitats. Intertidal land is reclaimed to support development projects and so habitat is permanently lost. Habitat change also occurs because of a range of activities that include the dredging of sediment, the commercial exploitation of intertidal invertebrates and recreation.

As these habitats have high conservation value, conservationists often want to assess whether particular human activities are potentially damaging to particular bird populations. The ecological impact of habitat loss and habitat change occurs via a reduction in food availability. If food availability is limiting, then such changes might reduce the number of birds an area could support over a given time-period, or increase the mortality rate. Risk assessment, therefore, requires predictions about the magnitude of such changes in sustainable numbers or demography following habitat loss or change. We illustrate using two particular examples how behaviour-based models can be used to make such an assessment.

Oystercatchers (*Haematopus ostralegus*) spend the winter on predominantly sandy estuaries, where they are specialist consumers of bivalve molluscs, such as the edible cockle (*Cerastoderma edule*) and blue mussel (*Mytilus edulis*) (Goss-Custard 1996). The behavioural dynamics of competition between oystercatchers for food have been studied in detail in the Exe Estuary in southwestern England over the last 20 years. This work has produced a behaviour-based model that describes mechanisms of

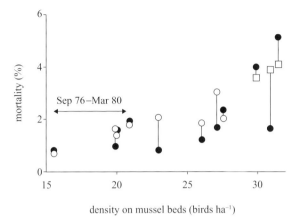

Figure 12.2. The predicted and observed density-dependent survival rates for oystercatchers on the Exe Estuary. Note that the model was created using data from the low densities occurring in 1976–80 but explains the subsequent response reasonably well. The black circles indicate observed rates and the white circles and squares indicate model predictions, but different estimates of food availability. (From Stillman *et al.* 2000.)

competition in this population, and derives estimates of mortality resulting from the impact competition has on the energy budgets of individual birds (Stillman *et al.* 2000). The model includes the effects of spatial and temporal variation in food availability, kleptoparasitism, feeding method, foraging efficiency, prey depletion and a range of other environmental factors on food intake, and has provided relatively accurate estimates of density-dependent mortality (figure 12.2). It is worth noting that the behaviour-based model was constructed using data collected prior to 1980, when the oystercatcher population size on the Exe Estuary was low and relatively stable (see figure 12.2). Subsequently, the number of birds wintering on the Exe increased, and the behaviour-based model predicted relatively accurately the resultant increase in mortality, even though the model was constructed with data prior to the population increase. A classical description of density-dependent mortality of the same initial period would have been inadequate to predict how mortality would have responded as population size increased. This highlights the potential importance of explicitly considering behavioural mechanisms in studying density-dependent processes.

The bivalve molluscs consumed by oystercatchers are also harvested commercially for human consumption by fishermen. Stillman *et al.* (2001)

used the behaviour-based model described in the previous paragraph to examine various aspects of fishery management on the mortality of oystercatchers exploiting the same prey population. Fishermen reduce food availability to the birds and so can exacerbate any density-dependent mortality by potentially increasing the risk of density-dependent starvation. Such interactions could be complex, and so would be virtually impossible to predict using previous observations of mortality rate in relation to population density and the activities of fishermen. Furthermore, risk assessment often involves asking how the population might respond to novel future conditions, such as an increase in fishing effort or a change in fishing methods. Such questions cannot be addressed using empirical data of population behaviour in the past. Figure 12.3a gives examples of some of the specific predictions that can be generated from the behaviour-based model when applied to fishery management issues such as fishing effort and fishing methods. Furthermore, the model can also be used to examine interactions between density-dependent mortality (due to starvation) and environmental conditions. For example, density-dependent mortality is likely to be more severe in oystercatchers in cold winters because competition for food resources at high population densities would lead to a greater number of birds failing to acquire sufficient resources and starving than would be the case in a mild winter. This has implications for assessing the impact of fishing because an increase in fishing effort, for example, would be more severe in a cold winter than in a mild one. Assessing such an impact using observations of past population behaviour is made difficult because the population may not have experienced the precise range of density, environmental and fishery management conditions that need to be assessed. However, the behaviour-based model can be used to generate predictions for a range of conditions, irrespective of whether these have been experienced by the population previously (e.g. figure 12.3b).

An alternative approach to risk assessment when changes in habitat availability affect food resources is to estimate the extent to which sustainable population sizes might decline if habitat is lost or degraded. Behaviour-based models that do this usually describe the process of scramble competition for resources, and to date have been primarily based on the spatial depletion model of Sutherland & Anderson (1993) (e.g. Sutherland & Allport 1994; Percival et al. 1996, 1998; Pettifor et al. 2000b; Gill et al. 2001a). This model is conceptually very simple. Consider an area of habitat, divided into discrete patches which vary in resource

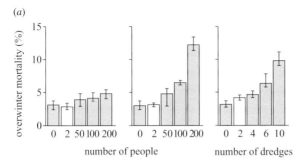

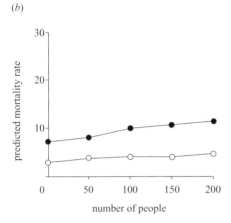

Figure 12.3. (a) The predicted overwinter survival rates of oystercatchers in relation to changes in (from left to right): the number of people hand-picking mussels; the number of people hand-raking mussels; and the number of dredges removing mussels. In each graph the light bar shows the current levels of fishing effort (from Stillman *et al.* 2001). (b) The predicted mortality rate of oystercatchers in relation to the number of people hand-picking mussels under different environmental conditions. The white circles show normal winter weather conditions and the black circles represent cold weather conditions (from Stillman *et al.* 2001).

availability. At the start of a season (e.g. winter), animals aggregate in the food patch(es) with the highest food availability. As food resources are consumed and prey in the best patches become depleted, food availability in the best patch(es) will eventually reach availabilities that are similar to other patches that were initially less rich in resources. At this point, individuals spread out and occupy these additional patches too. As the season progresses, resources get depleted and a greater range of patches become exploited. If one assumes that there is a threshold food availability below which a patch can no longer be used (e.g. birds would starve), then

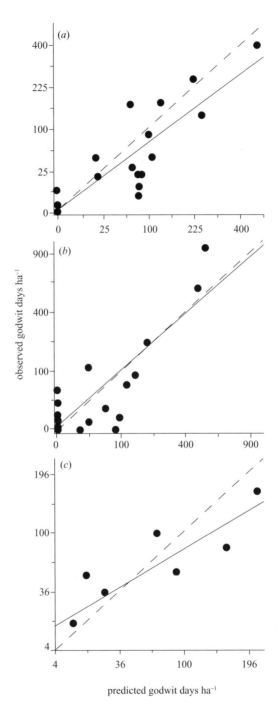

predicted godwit days ha^{-1}

this model can be used to calculate how many individuals the resources could support for a given period of time. Recent work by Gill *et al.* (2001*a*) on black-tailed godwits (*Limosa limosa islandica*) that feed on soft-shelled clams during winter has shown that a simple prey depletion model such as this can be used to predict population sizes at different levels of a spatial scale (figure 12.4). It is then a relatively simple process to incorporate habitat loss or degradation and re-run the model to predict the extent to which the maximum sustainable population size might decline at various scales.

(b)   Territoriality

Territorial behaviour is an obvious form of density dependence. At low population sizes individuals typically occupy large, high-quality territories. As the population density increases, territorial behaviour may change in three ways, each with clear implications for fitness. Territories may shrink as a result of the increased costs of territorial defence, poorer quality territories may be occupied or some individuals may decide not to breed (such individuals are usually known as 'floaters'). Thus, as the population increases the reproductive output declines.

Kokko & Sutherland (1998) analysed the question of when individuals should opt not to breed and become floaters. With a given distribution of habitat quality, perfect knowledge and no competition, individuals will obviously start by selecting the highest quality territory. Further individuals will then occupy the highest quality territory that is free. Assuming that searching for new territories and territory occupancy are mutually exclusive, there becomes a threshold territory quality at which it is better not to breed but to wait for a better territory to become available. The game theoretical solution can be derived for the optimal decision of individuals in deciding whether to breed in the best available territory or float. Once the demographic consequences of this decision are determined, it can be shown that the optimal game theoretical solution for deciding when to float is also the one that maximizes the numbers of floaters in the population. If the threshold is at a higher territory quality, then fewer individuals

Figure 12.4. The observed number of bird days plotted against the predicted number for a population of black-tailed godwits wintering on the east coast of England. The model incorporated data on the food abundance, searching efficiency and handling time. The fit seemed good whether for a comparison of different patches on mudflats (*a*), different mudflats on estuaries (*b*) or different estuaries (*c*). Details of the model and the fit to observed data are given in Gill *et al.* (2001*a*).

breed and so the total population is lower. If the threshold is at a lower territory quality, then the population is higher as more individuals breed and produce young, but the number of floaters is reduced as individuals that would have otherwise floated now breed in poor quality territories. Pen & Weising (2000) and Kokko *et al.* (2001) expand this result to consider a range of situations and its implications.

This theoretical abstraction seems to fit well with real world behaviour. Ens *et al.* (1992) described the breeding system of oystercatchers on Schriermonnikoog, The Netherlands, and showed that there were two main types of breeding behaviour (this behaviour is probably typical for many locations). Birds adopting one strategy, known as *residents*, obtain territories adjacent to the mudflats and then move onto the mudflats with their chicks to feed. The alternative strategy, known as *leapfrogs*, involves having a territory inland and flying over the residents to obtain food from the mudflats. As a result of the markedly reduced provisioning efficiency, the annual reproductive success rate of leapfrogs was only one-third of that of residents. Individuals usually cannot both occupy a leapfrog territory and wait for a resident territory to become vacant. Why then did some birds adopt the leapfrog strategy? After examining several explanations, including that the leapfrogs were making mistakes, leapfrogs were poorer quality or that leapfrogs lived longer, Ens *et al.* (1995) showed that the most probable explanation was that the two strategies are an ESS with similar average lifetime reproductive successes. There is considerable competition to become a resident that results in considerable delays in obtaining a territory, with many individuals dying while waiting for an occupancy. There is very little competition in becoming a leapfrog and so these can obtain a territory immediately. Leapfrogs thus, on average, breed for many more years but with a lower annual success rate.

Sutherland (1996a) extended the concepts and data of Ens *et al.* (1992, 1995) to produce a model of the consequences for the breeding output of a game theoretical model of whether individuals should take a leapfrog territory or join a queue to wait to occupy a resident territory. This model predicted the nature of the density-dependent breeding output. By incorporating data on the known survival rate between fledgings and returning to breed as an adult, it was then possible to model the density-dependent production of adult recruits. By adding the known annual mortality, the equilibrium population size was determined which seemed to fit well with the actual population size at Schiermonnikoog. This model can then be used to determine the consequences of changes in mortality on population size and was used to provide parameters for the model of habitat loss of migratory species (Sutherland 1996*a*).

Liley (1999) developed this approach to predict the consequences of human disturbance on populations of ringed plovers (*Charidrius hiaticula*). As with the model of the oystercatchers described in an earlier paragraph, there were empirically quantified differences in territory quality and the plovers showed preferences according to habitat type. From this it was possible to calculate the density-dependent reproductive success. Birds that were colour ringed when breeding at Snettisham were observed in winter across a wide area including France, southwestern Britain with some even heading north to winter in northern England. Thus, although there may be competition and density dependence acting upon ringed plovers in winter, when considering the Snettisham ringed plovers in isolation there is likely to be negligible density dependence – for example, doubling the population at Snettisham will, once the extra birds were spread across northwestern Europe, have a negligible effect on the total competition. Thus, by restricting the analysis to the Snettisham population, we can incorporate winter mortality as being density independent. It is then possible to calculate the equilibrium population size.

The main objective of Liley's study was to predict the consequences of human disturbance. Most papers on the disturbance of breeding birds tend to consider either the consequences on behaviour (e.g. time spent alert or off the nest) or the possible fitness consequences (e.g. number of nests trampled), but these do not determine the consequences for the total population. However, it is the consequences on populations that matter from a decision-making perspective. The model could easily be modified to consider the consequences of removing the quantified effects of humans trampling the eggs. The predicted increase in the population size would be 8%. In addition, as the ringed plovers avoided areas of high human densities, many of which had the characteristics of high-quality territories, the removal of the disturbance would be predicted to increase the ringed plover population by 71.3%. Removing both the effect of trampling and the avoided good territories would increase the population by 84.9%. Using a similar approach, the model was also used to predict the consequences of sea-level rise on the ringed plover population: a 25 cm rise is predicted to reduce the population by 4.3%.

(c)   Social behaviour

A recent development is to create demographic behaviour-based models incorporating social behaviour. These can be considered as a development of territorial models, but with the complication that individuals differing in rank will differ in reproductive success. Understanding the dispersal between groups, and the resultant mortality costs, is a key component.

Stephens *et al.* (2002) devised models of the alpine marmot *Marmota marmota* based on a 13 year field study within Berchtesgaden National Park (e.g. Arnold 1990a,b; Hackländer & Arnold 1999) in which about 95% of the study population was captured at least once a year. There were data on the major components of demography such as age of maturity, litter size, sex ratio, summer survival and winter survival.

Stephens *et al.* (2002) compared four models: (i) a population-based matrix model that incorporated environmental stochasticity and density-dependent fecundity; (ii) a group-based matrix model (i.e. similar to (i) but subdivided into groups with density dependence acting within groups); (iii) a spatially and temporally explicit individual-based model using field data to decide the probability of individual fates; and (iv) a behaviour-based model in which the optimization of an individual's residual fitness determined the individual fates. The main behavioural decision was when individuals dispersed, with individuals dispersing at the time that resulted in the highest lifetime fitness. The game theoretical solution thus depends upon factors such as the individual's dominance status and the mortality risk associated with dispersal. Model (iii) led to the most un-realistic results with the greatest deviation from the data (figure 12.5). As the decisions of individuals were based on fixed probabilities, individuals would often make irrational decisions, such as dispersing from a territory in which they would otherwise be just about to reproduce. This resulted in a lower growth rate than did the other models (figure 12.5).

The behavioural-based model was reasonably successful in predicting the dispersal behaviour, distribution of group sizes and rates of turnover of dominant animals (figure 12.6). An emergent property of this model was the Allee effect, the reduction in reproductive success at lower population sizes.

(d)   Buffer effects

A particularly important behavioural process underlying density de-pendence in survival and fecundity in animal populations is the buffer effect. This term is used to describe the process whereby at relatively low

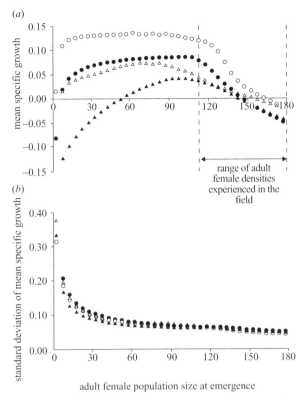

Figure 12.5. The mean (*a*) and standard deviation (*b*) of population growth of the alpine marmot *Marmota marmota* as predicted for different female emergence population sizes using four different models. Matrix model (white circle), matrix-group model (black circles), individual-based spatial model (black triangles) and behaviour-based model (white triangles) (from Stephens *et al.* 2002).

population sizes most animals in the population are able to occupy high-quality habitat, but as the population size increases an increasing fraction of the population is forced to occupy progressively poorer quality habitat. As this results in a reduction in fitness in animals occupying poor-quality habitat, the buffer effect is an important density-dependent mechanism. Its specific impact on demography is dependent on the stage of the life cycle in which it operates and the components of fitness influenced by habitat quality experienced during that life-cycle stage. It is obvious that the buffer effect can arise in breeding populations in which animals maintain exclusive breeding territories. However, there is also extensive evidence

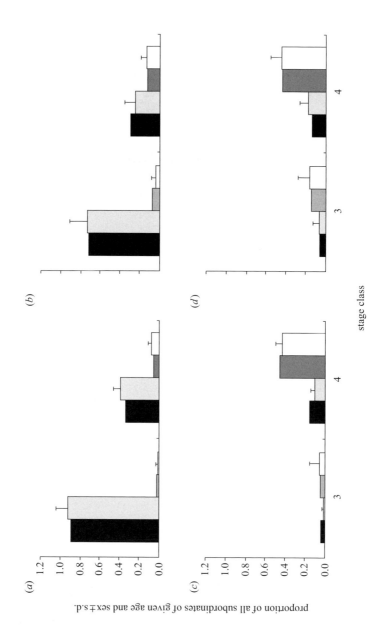

proportion of all subordinates of given age and sex ± s.d.

stage class

that the buffer effect can operate during the non-breeding period and affect fitness components other than fecundity. Recent work on shorebirds has also shown that the fitness consequences of the buffer effect can be cross-seasonal, in that fitness components outside the season in which the buffer effect is evident can be affected (Gill *et al.* 2001*b*).

Although the simple description of a buffer effect cannot necessarily elucidate the behavioural mechanisms driving it, the description has some qualitative value for conservationists. This is because it permits some ecologically relevant measure of habitat quality to be made by comparing abundance trends within habitat patches to overall changes in population sizes, and examining how such patterns relate to fitness. For example, recent studies on black-tailed godwits (*Limosa limosa*) revealed a buffer effect operating during the non-breeding period when birds occupy estuarine habitats on the coasts of south and eastern England (Gill *et al.* 2001*b*) (figure 12.7). The population increase is buffered on the south coast by individuals moving to the east coast. Those godwits wintering on the south coast had higher food intake rates in spring, higher survival rates and earlier arrival dates on their breeding grounds in Iceland, than birds wintering on the east coast. This buffer effect shows that both survival and arrival date are density dependent in this population. These population-level density-dependent processes would be affected by habitat loss, but in different ways depending on whether good (south coast) or poor (east coast) habitat was lost or degraded. If a south coast site was lost, birds would be forced to occupy poorer quality habitat and experience reduced survival and later arrival dates. By contrast, birds occupying poor-quality habitat already have reduced fitness, so the loss of a poor-quality site is likely to have a relatively less severe impact on survival and arrival date.

The important general issue highlighted by this example is that an understanding of the behavioural mechanisms driving density-dependent processes provides potentially much greater insights than simply

---

Figure 12.6. The fates of subordinate alpine marmots *Marmota marmota* as predicted by a behaviour-based model. (*a*) Males and (*b*) females that stayed in the territory. The shading shows those that remained subordinate (empirical, black), those that remained subordinate (model, light grey), those that became territorial (empirical, dark grey) and those that became territorial (model, white). (*c*) Males and (*d*) females that were dispersed from natal territory. The shading shows those that became territorial in the neighbourhood (empirical, black), those that became territorial in the neighbourhood (model, light grey), those that were dispersed beyond the neighbourhood (empirical, dark grey) and those that were dispersed beyond the neighbourhood (model, white) (from Stephens *et al.* 2002).

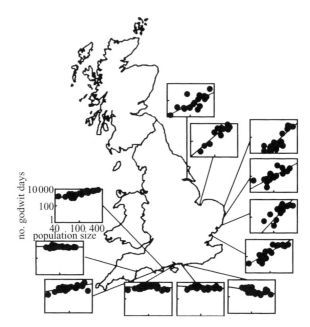

Figure 12.7. The change in the abundance of black-tailed godwits on different estuaries in Britain. Each graph shows the count (expressed as number of godwit days) on an individual estuary (y-axis) plotted against an index of overall population size in the UK (x-axis). Those sites with high rates of population increase as the total population has grown are associated with less food, lower intake rate, higher mortality and later arrival in the Icelandic breeding grounds (from Gill *et al.* 2001*b*).

describing the population-level processes directly. If survival rates and arrival dates of colour-marked birds had been estimated annually as the godwit population has grown in size over the past 20 years, we could have described density dependence in these life-history variables. However, we would have been unable to comment on whether the loss of a south or east coast site to development might have had equivalent or disproportionate effects on the population. A simple description of the buffer effect provides such insights; insights that are crucially important if coastal habitats are to be managed in a way that accommodates potentially conflicting (e.g. conservation and development) 'uses'.

12.5. Future challenges

The models we have described tend to describe the situation under standard conditions. In reality there are likely to be considerable differences between years. Thus, much of the population regulation may take place under occasional years that are particularly cold, wet, dry or with food shortages. That is, there is likely to be an interaction between environmental conditions and density-dependent processes. Understanding such processes is obviously very important in predicting how populations might respond to future environmental change, given that the change itself is likely to vary in space or time. Certain behaviour-based models deal with the issue explicitly. For example, the model of Stillman *et al.* (2000) of wintering oystercatcher populations includes density-dependent starvation by calculating an energy budget for each individual. Those that lack sufficient resources at any point starve. The resources required to avoid starvation obviously vary temporally depending on weather conditions, so the model includes an interaction between weather and density-dependent starvation. This is important because understanding such an interaction is crucial to risk assessment applications of the model (see Stillman *et al.* 2001). It is also worth reiterating that the model is capable of considering the impact on survival of any combination of weather conditions and population density, irrespective of whether the population has experienced similar conditions in the past, due to its mechanistic nature. A wider application of the behaviour-based approach to a range of systems and problems in the future will tell us how well these models cope with similar interactive effects.

12.6. Conclusions

We have argued that current demographic models used for conservation purposes are often inadequate due to the lack of data, particularly with respect to the description of density-dependent processes. Even if extremely detailed data exist on the life histories of individuals within a population that has experienced a wide range of population sizes, demographic models based on such data are unlikely to be reliable in the face of future environmental changes that modify density-dependent processes. This is because population responses to such changes are unlikely to be predictable from past population behaviour. Instead, we suggest that studies of the behavioural mechanisms that underlie density-dependent processes are likely to be of more predictive value to conservationists interested in understanding how populations might respond to future environmental change. This is because such studies require more precise thinking into exactly how environmental change impacts on populations, and behaviour-based models provide tools that can be used to make quantitative predictions about impacts. We have illustrated how this approach might be used for both risk assessment and the design of conservation management for endangered species. This latter application of behaviour-based models is in its infancy but the technique is potentially very valuable because population processes are difficult to describe in small, poorly studied populations, but behavioural mechanisms involved in competition for resources are a much more tractable ecological research problem.

How does a behaviour-based approach relate to the paradigms in population ecology described by Sibly & Hone (Chapter 2)? There are obvious parallels with the mechanistic paradigm in that behavioural mechanisms can explain how demography is linked to environmental drivers such as food availability, temperature, etc. However, it is distinct in that all of the existing paradigms can be used to describe how a population has behaved in the past, but only a more mechanistic approach that incorporates decision making by individual animals can provide reliable insights into how populations might behave in the face of future environmental change that modifies density-dependent processes. This is an important progression for population ecologists to make in the sense that it means the subject is moving from being a descriptive science to a more predictive one. This is also of potentially immense practical importance, in that understanding how populations respond to future environmental changes will be pivotal to their effective conservation.

The authors thank R. Freckleton, J. Gill, J. Hone, J. Ridley, P. Stephens, T. Sinclair and A. Watkinson for useful discussions, J. Hone and R. Sibly for inviting them to this meeting and to The Royal Society for funding this conference.

13
_____

Comparative ungulate dynamics: the devil is in the detail

13.1. Introduction

Recent increases in the number of time-series long enough to provide an adequate description of population fluctuations clearly show that population stability varies widely among animals with similar longevities and rates of reproduction, as well as between species with contrasting life histories (Caughley & Krebs 1983; Gaillard *et al.* 2000). For example, among grazing ungulates, populations may either show little variation in size across years, irregular oscillations, semi-regular oscillations resembling the stable limit cycles found in some smaller mammals or dramatic oscillations occasionally leading to extinction (Peterson *et al.* 1984; Fowler 1987b; Coulson *et al.* 2000). While many ecological differences probably contribute to these differences (including predation, disease and human interference), the fact that stability varies widely among naturally regulated ungulate populations living in environments where human intervention is minimal and predators are absent (Boyd 1981*a,b*; Boussès *et al.* 1991; Clutton-Brock *et al.* 1997*a*), suggests that variation in population dynamics may often be caused by interactions between populations and their food supplies.

Theoreticians have explored the possibility that contrasts in population dynamics may be consistently related to differences in life histories or in the temporal or spatial distribution of resources (e.g. Peterson *et al.* 1984; Sinclair 1989; Sæther 1997; Illius & Gordon 2000; Owen-Smith 2002). While it is likely that both these differences may contribute to variation in dynamics, attempts to explain observed variation mostly assume that the causes of contrasts are sufficiently simple to be explained by general models derived from first principles (Caughley 1977).

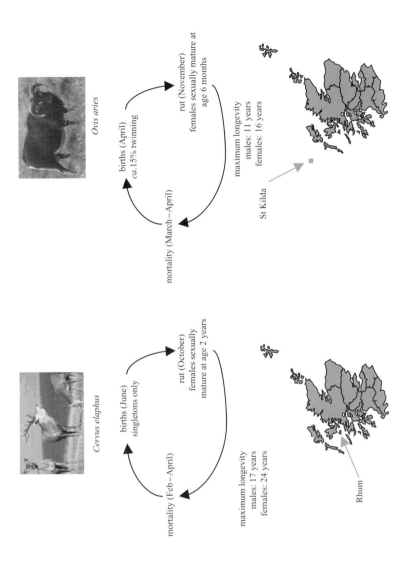

Figure 13.1. Summaries of the life cycles of red deer and Soay sheep.

Cervus elaphus

births (June)
singletons only

rut (October)
females sexually
mature at age 2 years

mortality (Feb–April)

maximum longevity
males: 17 years
females: 24 years

Rhum

Ovis aries

births (April)
ca. 15% twinning

rut (November)
females sexually mature at
age 6 months

mortality (March–April)

maximum longevity
males: 11 years
females: 16 years

St Kilda

Another possibility is that contrasts in population dynamics are a consequence of detailed differences in the demographic processes affecting dynamics and are driven by specific interactions between breeding systems
and life-history parameters and the distribution of resources. If so, current attempts to predict variation in population dynamics using general
models may meet with little success until we have a better understanding
of the specific causes of contrasts in dynamics (Sutherland 1996b).

In this paper, we compare the dynamics and demography of two populations of food-limited ungulates (red deer *Cervus elaphus* L. and Soay
sheep *Ovis aries*) on different Hebridean islands over the same years
(figure 13.1). We show that a detailed knowledge of demographic processes and population structure is necessary to predict changes in population size successfully and to explain the contrasts in population dynamics
between the two populations.

Research on the red deer population of the North Block of Rum
has continued since 1972, when the annual 14% cull of the population
of around 200 deer was terminated (Clutton-Brock *et al.* 1985b, 1997a).
After 1972, numbers rose rapidly to around 300, stabilizing by 1980 although the adult sex ratio continued to change in favour of females
(see figure 13.2a). Demographic processes varied between the initial period

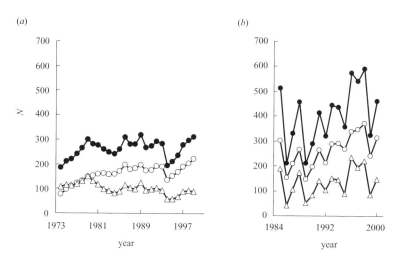

Figure 13.2. Time-series for (a) deer and (b) sheep numbers. Filled circles, total;
open circles, total females; open triangles, total males. The ranges on the y-axes
for sheep and deer are identical to allow comparison of the relative size of
fluctuations in population size.

of population growth and the subsequent years when deer numbers had reached ecological carrying capacity (Albon *et al.* 2000); so, to maximize comparability with the sheep (see below), we have restricted our analysis of dynamics in the deer to the period between 1985 and 2001.

Since 1985, we have also monitored the dynamics of the Soay sheep on Hirta, the largest island of the St Kilda archipelago, *ca.* 120 km to the northwest (figure 13.2*b*). The sheep population was originally introduced from the neighbouring island of Soay and has been naturally regulated since 1932 (Grubb 1974*a–c*; Grubb & Jewell 1974; Clutton-Brock *et al.* 1991; Grenfell *et al.* 1992). Soay sheep are derived from domestic stock that were probably introduced to the Hebrides over 2000 years ago and have re-mained on Soay since then (J. Clutton-Brock 1981). Compared with most other time-series for large mammals, both of our datasets are unusual in that virtually all individuals in both populations are recognizable as in-dividuals and their life histories have been monitored from within a few days of birth, when most lambs and calves are caught, weighed, sexed and skin-sampled for genetic analysis (Clutton-Brock *et al.* 1982*b*, 1991; Pemberton *et al.* 1996). As a result, we are able to identify the contributions of specific demographic changes to variation in population size with un-usual accuracy.

The habitats occupied by the two populations at the two sites are broadly similar, with areas of herb-rich or *Agrostis*-dominated grassland at sea level grading into heather-dominated communities interspersed with flushes on the slopes of the surrounding hills (Jewell & Grubb 1974; Jewell *et al.* 1974). Densities of sheep reach higher levels than those of the deer, rising to 25 km^{-2} compared with *ca.* 15 km^{-2} in years of high density. Compared with deer, the sheep show relatively high annual population growth rates (figure 13.3*a*), partly because many females conceive for the first time at 7–8 months instead of at 2–3 years (Clutton-Brock *et al.* 1997*a*; figure 13.3*b*), partly because most females over a year old conceive each year (figure 13.3*c*) and partly because, on average, *ca.* 15% of females produce twins (figure 13.3*d*).

13.2. Comparative dynamics

Between 1985 and 2001, neither the deer population on Rum nor sheep numbers on Hirta have shown a consistent temporal trend. However, while deer numbers have been relatively stable, sheep numbers have fluc-tuated widely between successive years. For example, while deer numbers have never declined by more than 17% in a single winter (figure 13.2*a*), over

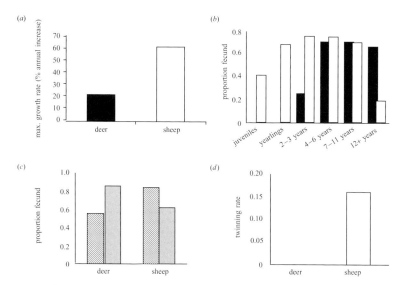

Figure 13.3. Fecundity in deer (black bars) and sheep (white bars). (*a*) Maximum annual population growth measured as the maximum percentage annual increase in population size. (*b*) Proportion of animals conceiving offspring at different ages. (*c*) Proportion of milk (hatched bars) and yeld (grey bars) females giving birth. 'Milk' are those that reared an offspring successfully until (at least) the onset of the winter in the previous year; 'yeld' are those that failed to do so either because they did not give birth or because their calf died during the summer. (*d*) Proportion of individuals bearing twins – note that red deer never twin on Rum.

60% of the sheep in autumn can die in the course of 2 months in late winter (see figure 13.2*b*). When high mortality occurs in the sheep, this not only removes the increment in population size that has occurred in the course of the last year but, on average, reduces the population to less than 65% of the maximum number that has been known to survive the winter. By contrast, in the deer, winter mortality never reduces spring numbers much below 90% of observed maximum winter numbers. These contrasts in winter mortality are associated with differences in growth rate between the two populations. In the deer, numbers rarely increase by more than 10% per year; for example, it took the population 7 years to rise by 50% following the termination of culling in 1972 (see figure 13.2*a*). Sheep numbers, on the other hand, can increase by over 50% in the course of a single season and commonly double in the course of 2 years (figure 13.2*b*).

In both species, high winter mortality affects some sex and age categories more than others. To permit comparison, we have plotted mortality for different categories of animals of each species against mortality

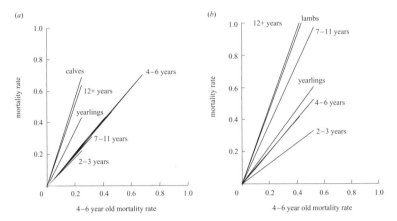

Figure 13.4. Mortality rates for different age categories of (*a*) female deer and (*b*) female sheep plotted against the mortality of females in their prime. Probability of mortality was the proportion of individually identified animals entering that age category that died before leaving it. All lines are forced through the origin and age categories printed next to the lines.

levels among mature females, defined here as 4–6 years to allow comparison between species (figure 13.4*a,b*). Relative to females of 4–6 years, juveniles, yearlings and old females show relatively high levels of mortality in both species. Yearlings show higher mortality than 4–6-year-old females in the deer, probably because they are still growing (Clutton-Brock & Albon 1989), 2–3-year-olds show relatively low mortality in the sheep compared with the deer, perhaps reflecting the fact that growth in the sheep has largely ceased by the end of the third year of life (Jewell *et al.* 1974) while 7–11-year-old sheep show relatively high mortality as a consequence of earlier ageing.

In both species, variation in winter mortality is affected by population size (Clutton-Brock *et al.* 1982*b*; Coulson *et al.* 1997). In the deer, winter mortality increased with population density only in calves, though there is a tendency for mortality to increase in yearlings (Clutton-Brock *et al.* 1997*a*) as well as in older adults (figure 13.5*a*). In the sheep, only lambs and older adults were affected (figure 13.5*b*).

Winter weather conditions are also important (Albon *et al.* 1987; Benton *et al.* 1995). Winters in the North Atlantic region can either be wet and windy or drier and colder; these contrasts are associated with large-scale atmospheric fluctuations over the North Atlantic, called the NAO (Rogers 1984). When pressure is low over Iceland and high over the Azores

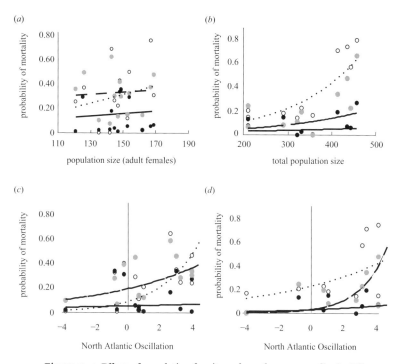

Figure 13.5. Effects of population density and weather on mortality in different age classes: (*a*) effects of density on deer; (*b*) effects of density on sheep; (*c*) effects of variation in the NAO on deer; (*d*) effects of NAO on sheep. Winters with high NAO values are relatively wet and windy; those with low NAO values are drier and colder. Open circles and dotted lines represent juveniles, black circles and solid lines represent prime aged adults and grey circles and dashed lines represent older adults.

(high NAO), strong westerly winds bring warm, wet weather north into Europe, and gales are common. By contrast, when pressure is high over Iceland and low over the Azores (low NAO), cold, dry weather spreads west from Siberia, and winters in northern Europe are calmer and colder. Fluctuations in NAO explain much of the variation in winter weather conditions; for example, in the Outer Hebrides, annual changes in the NAO winter index account for 61%, 56% and 23% of the variance in winter temperature, winter rainfall and number of winter days with gales, respectively (Forchhammer *et al.* 2001).

In both sheep and deer, winters of high NAO are associated with increased mortality in juveniles and older adults (figure 13.5*c,d*). Winter weather condition interacts with population density to produce high mortality in years when high population density is associated with adverse weather conditions (Coulson *et al.* 2001; Forchhammer *et al.* 2001; Milner *et al.* 1999; T. H. Clutton-Brock, unpublished data). In the sheep, such years were sometimes associated with the death of over 60% of the animals entering the winter, though lambs and males of all ages were always more strongly affected than mature females (Clutton-Brock *et al.* 1991, 1997*a*).

13.3. Variation in population structure and its consequences

Because relative survival differs between age categories (figure 13.4), the age structure of both populations varies widely between years (figure 13.6*a,b*) and these changes are not closely correlated with population density. For example, the proportion of the population made up of mature females varied from 24% to 41% between years in the deer and from 16% to 46% in the sheep (figure 13.6*a,b*); in neither species are these changes in the relative proportion of different age categories consistently correlated with density. Changes in age structure affect the number of animals that are likely to die in a particular year and so introduce an additional factor affecting fluctuations in population size. Fluctuations in population structure combined with variation in the relative mortality of different age categories contribute substantially to the magnitude of changes in population size.

Since the number of animals in different age categories varies, differences in survival do not necessarily reflect the extent to which particular categories of animals contribute to changes in total population size (Brown & Alexander 1991). For example, while the survival of red deer calves fluctuates more than that of mature animals, variation in the survival of mature females contributes more to fluctuations in population size (see figure 13.7*a*) (Albon *et al.* 2000). Similarly, despite the greater variability of changes in lamb and yearling numbers in the sheep, variation in adult numbers contributes more to variation in population size than changes in lamb and yearling numbers (see figure 13.7*b*) (Coulson *et al.* 2001). Comparing the two populations, it is clear that lambs and yearlings contribute more to changes in population size in the sheep than the deer, although both show higher mortality relative to mature females in the deer (see figure 13.4*a,b*).

As a result of both these effects, models which assume a constant age structure and ignore the contrasting effects of density and climate on survival and reproduction in different age categories, fail to predict changes

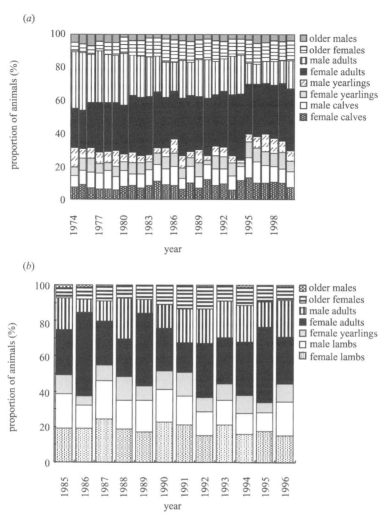

(a)

Figure 13.6. Age structure in different years for *(a)* deer and *(b)* sheep. The figure shows the proportion of animals of different ages in different years. Population structure varies more between years in the sheep.

in population size accurately (see figure 13.8). For example, stochastic, un-structured models of variation in sheep numbers account for only 21% of variation in population size (figure 13.8*c*) (Grenfell *et al.* 1992, 1998), while the inclusion of variation in age structure and in the responses of different age and sex categories in age-structured Markov models raises the proportion of the variation accounted for to nearly 90% (figure 13.8*d*) (Coulson

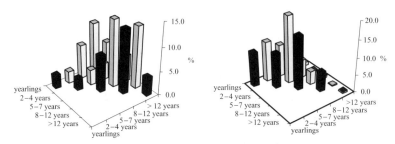

Figure 13.7. Contributions of different age classes to the relative change in population size in (a) deer and (b) sheep. Columns show the extent to which particular age classes (and the covariation between them) contribute to relative changes in population size between successive years. Black columns on the centre diagonal represent the percentage contribution of each class to the relative change in population size over the study period (see Coulson *et al.* unpublished; Albon *et al.* 2000). The off-diagonals (pale bars) represent the percentage contribution of the covariation between different age classes to the relative change in population size.

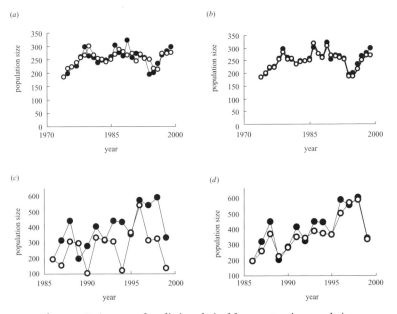

Figure 13.8. Accuracy of predictions derived from contrasting population models for (a,b) deer and (c,d) sheep. Black circles represent the observed population size and open circles predicted population sizes derived from (a,c) time-series models fitted to count data and winter weather (see Grenfell *et al.* (1998) for sheep; Coulson *et al.* (2000) for deer) and (b,d) age-structured Markov models that incorporate the effects of variation in age structure (see Coulson *et al.* (2001) for sheep; Coulson *et al.* (unpublished) for deer).

et al. 2001). The age structure of the deer population does not vary so widely between years so that the contrast in accuracy between structured and unstructured models is less pronounced in the deer (see figure 13.8*a*,*b*).

13.4. Comparative effects of density and climate on recruitment

In both species, the rate of recruitment to the population (which is a function of the number of females breeding and the proportion of calves surviving to weaning) is related to density as well as to density independent factors. In the deer, the fecundity of adult females ($\geq$3 years) decreases with increasing population density (figure 13.9*a*). During the early years, when population size was still increasing, changes in age at first breeding

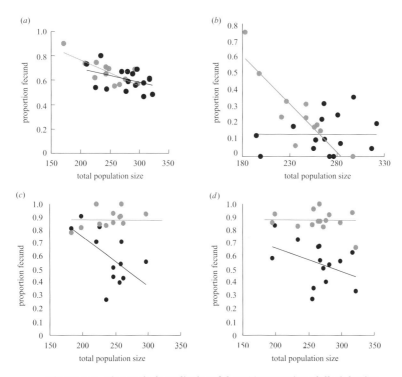

Figure 13.9. Changes in fecundity in red deer: (*a*) proportion of all adults that produced calves; (*b*) proportion of 3-year-old females that produced calves in early years (grey) and later years (black); (*c*,*d*) proportion of multiparous females that produced calves in (*c*) the early years (1973–1984) and (*d*) later years (1985–2001); black, females that had reared a calf to 6 months the previous season; grey, those that had failed to do so.

contributed extensively to this trend while, in later years, the proportion of animals breeding for the first time at 3 years was consistently low and did not vary with population size (figure 13.9b). In both periods, female deer that had successfully reared calves (milk hinds) showed reduced fecundity when numbers were high, while the fecundity of those that had failed to do so was unaffected by density (figure 13.9c,d). This contrast between milk and yeld hinds is caused by differences in body mass that are known to affect fertility (Mitchell & Brown 1974; Albon *et al.* 1983b, 1986); female deer that have reared calves continue to suckle them until the early winter and, compared with yeld hinds, show lower body masses and fecundity in the autumn rut (Mitchell *et al.* 1976). Variation in the birth rate was also affected by winter weather – relative to population density, fewer milk hinds bred after wet, windy winters, probably partly because a higher proportion of animals failed to conceive and partly because more embryos were aborted or resorbed in the course of the winter (Albon *et al.* 1986; Kruuk *et al.* 1999).

In the sheep, there is a weaker relationship between density and fecundity overall (figure 13.10a) though, as in the deer, the proportion of animals that produced offspring in their first year after reaching breeding age showed a more marked decline (figure 13.10b). In contrast to the deer, females that had reared lambs successfully the previous season were *more* likely than 'yeld' females to give birth again (figure 13.10c), presumably reflecting a difference in phenotypic quality. A similar pattern has been demonstrated in bighorn sheep (Festa-Bianchet 1998).

The contrasting effects of previous reproduction on fecundity in the two species probably reflect differences in the timing of breeding and the duration of parental investment (Clutton-Brock *et al.* 1997a). Female deer that have reared calves through the summer months (milk hinds) are of substantially lower mass than those that have not done so. By contrast, female sheep that have reared one or two lambs through the summer regain lost condition after June, when lactation virtually ceases, and enter the winter at similar masses to those that have raised young.

Variation in recruitment rate also has contrasting effects on the population dynamics of the two populations. In the deer, density dependent changes in age at first breeding and in the fecundity of milk hinds begin to depress recruitment at relatively low population densities (see figure 13.9). As a result, density dependent changes in recruitment slow population growth, contributing to population stability (Clutton-Brock *et al.* 1982b, 1997a). By contrast, density dependent changes in fecundity have relatively little effect on population growth rates in the sheep

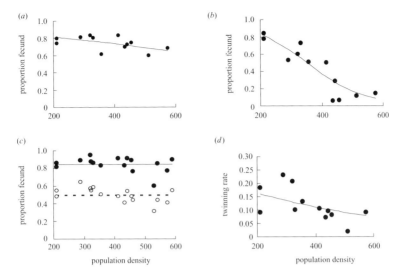

Figure 13.10. Changes in fecundity in sheep: (*a*) proportion of females of breeding age (12 months and older) that produced lambs (all potential breeders); (*b*) proportion of juveniles (12 months) that produced lambs (first time breeders); (*c*) proportion of mature females (>12 months) that had raised at least one lamb the previous season (all adults excluding first time breeders) (black circles and solid line, 'milk ewes') and those that had failed to do so (open circles and dashed line, 'yeld ewes'); (*d*) All adults.

(figure 13.10). Although birth rate and neonatal survival are depressed at high densities, these effects are relatively slight (Clutton-Brock *et al.* 1997*a*). Moreover, changes in fecundity usually occur in the year *after* a winter of high mortality (Clutton-Brock *et al.* 1997*a*) and consequently do little to slow the rate of population growth, though they may delay recovery of the population (Clutton-Brock *et al.* 1997*a*).

13.5. Comparative effects of density and climate on development

Contrasts in the timing of reproduction also affect the impact of density and climate on early development and neonatal mortality in the two species. In both the deer and the sheep, environmental factors affecting foetal growth generate differences in birth mass between cohorts, which exert an important influence on subsequent growth, survival and breeding success (Albon *et al.* 1983*a,b*; Clutton-Brock & Albon 1989; Forchhammer *et al.* 2001). In the deer, birth mass is unaffected by population density (figure 13.11*a*) but varies with temperature in April and May, the last 2 months of gestation (figure 13.11*b*) (Albon *et al.* 1983*a,b*).

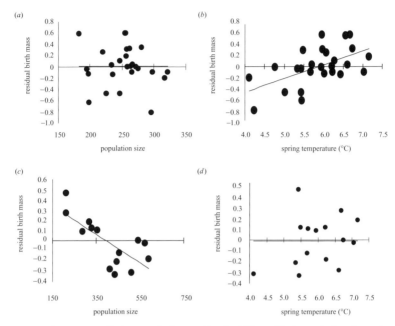

Figure 13.11. Birth mass of calves and lambs. Effects of (*a*,*c*) population size and (*b*,*d*) spring temperature on birth mass in (*a*,*b*) deer and (*c*,*d*) sheep. Spring temperature was measured as the mean average daily temperature in April and May. The *y*-axis (residual birth mass) shows the difference between the mean birth mass of each cohort and the population mean for the study period of both populations. Solid lines derived from linear regression models.

By contrast, in the sheep, high population density depresses birth mass (figure 13.11*c*) while variation in climatic conditions in spring has no consistent effect (figure 13.11*d*). Like differences in the effects of environmental variation on fecundity, these contrasts are probably a result of differences in reproductive timing. Because of their relatively long gestation period (32 weeks), red deer bear their young in late May or June, so that the last 2 months of gestation occur in April and May, when grass growth is variable and can be strongly influenced by seasonal fluctuations in temperatures (Clutton-Brock & Albon 1989). As a result, spring temperatures are the dominant factor affecting birth mass and neonatal mortality in the deer. By contrast, the shorter gestation period of the sheep (21.5 weeks) and their earlier breeding season means that the last 2 months of gestation (February, March) precede the principal onset of spring growth, so that food availability and maternal condition are influenced to a greater extent by the effects of population density and climatic factors on food availability and energy expenditure during the winter months.

The effects of density and winter weather on the two species generate contrasting differences in survival and breeding performance between cohorts. In the deer, neonatal mortality is density independent and is affected principally by temperature and rainfall before and immediately after birth (Clutton-Brock & Albon 1989). Density independent variation in birth mass generates density independent differences in growth and breeding success between cohorts that persist throughout the lifespans of their members; females born below average mass remain below average mass as adults and produce small calves throughout the whole of their lives (Albon *et al.* 1987; Clutton-Brock *et al.* 1988). These differences are substantial; the average mass of calves produced over their lifetime by members of cohorts born between 1970 and 1979 varied by 40%, ranging from 5.0 kg to more than 7.0 kg (Clutton-Brock & Albon 1989). Cohorts that produce small calves show consistently high levels of calf loss, and offspring survival during the first 2 years of life varied between cohorts of mothers from less than 10% to over 60% (Clutton-Brock & Albon 1989). In the sheep, where birth mass is depressed by population density as well as by wet winter weather, cohorts show density dependent differences in neonatal mortality, growth and fecundity (Forchhammer *et al.* 2001). However, the demographic effects of these differences are reduced, for individuals born after a winter when population density and mortality were high join a cohort with access to superabundant food supplies, which can offset their initial disadvantages.

While variation in early development exerts strong effects on the subsequent life histories of individuals in both populations, as in other vertebrates, these have limited impact on population dynamics in both species. In the deer, this is because cohort variation is density independent, so that, although the occasional sequence of 'good' or 'bad' cohorts may generate variation in population size (Albon & Clutton-Brock 1988), the demographic impact of cohort variation is limited. In the sheep, where birth mass and cohort performance vary with population density, cohort variation might be expected to have larger demographic effects. However, partly because cohorts conceived at high density commonly experience low density during their first year of life and show compensatory growth, and partly because years of high mortality are relatively frequent (and so interrupt runs of 'good' and 'bad' cohorts – see above), the demographic effects of variation in development between cohorts are slight. This will not always be the case and cohort effects may have important consequences for population dynamics in species where early development varies with density and population size is comparatively stable.

13.6. Sex differences in survival

So far, we have focused exclusively on the dynamics and demography of females. However, in both populations, high population density and adverse winter conditions affect the growth and survival of males more than females (Clutton-Brock *et al.* 1985*b*, 1997*a*). In the deer, where males are *ca.* 8% heavier than females at birth, sex differences in survival occur during gestation. As population density increases, the proportion of males at birth declines among offspring born to yeld hinds and, when these effects are allowed for, birth sex ratios decline after wet winters (Kruuk *et al.* 1999). Since the effects of density and weather occur after conception, they are presumably a consequence of variation in rates of abortion or resorption rather than of variation in the sex ratio at conception. By contrast, birth sex ratios in the sheep do not vary consistently either with population density or with winter climate (Lindström *et al.* 2002), perhaps because sex differences in birth mass are smaller (*ca.* 4%).

Neonatal mortality does not vary between the sexes in either species (Clutton-Brock *et al.* 1985*a*, 1992) while mortality is higher in males than females both during the first 2 years of life and in prime-aged animals (Clutton-Brock *et al.* 1985*a*, 1991, 1992) (figure 13.12*a,b*). Sex differences in mortality tend to be more pronounced in the sheep and, after years when mortality is high there can be eight times as many mature females as males among the survivors (figure 13.12*c,d*). In younger animals, this is probably partly because male lambs and yearlings are actively involved in the rut (Coltman *et al.* 1999*a,c*) and consequently enter the winter in comparatively poor condition, often with relatively high parasite loads (Wilson *et al.* 2003). By contrast, differences in relative mortality of mature males between the two species are related to the timing of the rut; the relatively early (October) rut of the deer allows males to regain some of the mass lost during the mating season before midwinter, while male sheep (which rut in November and December) have little opportunity to regain condition before the onset of the regular period of starvation in late winter.

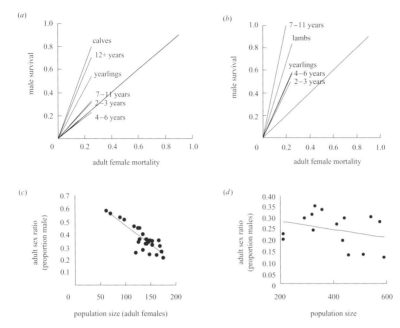

Figure 13.12. Mortality of males: (*a*) mortality of different age categories of male deer plotted on the mortality of 4–6-year-old female deer; (*b*) mortality of different age categories of male sheep plotted on the mortality of 4–6-year-old female sheep; (*c,d*) adult sex ratios in spring plotted on population size the previous autumn for deer and sheep. In both cases, population size was the number of individuals entering the winter while the sex ratio was estimated the following spring.

13.7.   Sex differences in dispersal

The sexes also show contrasting patterns of dispersal in the two species. While most female deer remain in the area where they are born throughout their lives (Clutton-Brock *et al.* 1982*a*; Coulson *et al.* 1997), males commonly disperse between the ages of 2 and 4 years, and better grown adolescents are more likely to do so than smaller animals (Clutton-Brock *et al.* 2002). High population density increases the proportion of males that disperse and discourages immigrants, contributing to female biases in high density populations. By contrast, both male and female sheep remain in their natal area (heft) though neither associate closely with their relatives (see §13.1).

Density dependent increases in mortality and emigration among males relative to females generate consistent relationships between population density and the adult sex ratio that affect the intensity of male competition and the distribution of mating success (Rose *et al.* 1998; Coltman *et al.* 1999*c*). In the deer, high population density is associated with a progressive bias towards females in the population sex ratio (figure 13.12*c*). This leads to increases in the proportion of males (especially of younger males) that hold harems (Clutton-Brock *et al.* 1997*b*) and to a reduction in the period for which individual males hold harems. In the sheep, males conceived at high population density and born immediately after a year when mortality is high enter a population that is heavily skewed toward females (figure 13.12*d*) and show relatively high mating success in their first and second years of life (Coltman *et al.* 1999*a–c*).

13.8. Discussion

Our comparison of the population dynamics and demography of the deer and sheep emphasizes four main points. Changes in population density as well as in climatic variation commonly affect young and old animals more than those in their prime, and males more than females (see figure 13.5). Similar differences in the effect of environmental variation on different age and sex categories have been observed in a wide range of other ungulates (Caughley 1977; Fowler 1987b; Gaillard *et al.* 2000). As our analysis shows, variation in the numbers of different categories of animals and in their reproductive potential mean that differences in survival do not necessarily reflect the extent to which different categories of animals contribute to changes in population size (see figure 13.7). In both our study populations, variation in the survival of mature animals contributed more to changes in population size than juvenile survival.

Our work emphasizes the extent to which demographic processes change during different stages of population and the need to control for these effects in comparisons between populations (see figures 13.5 and 13.9). During the period of population growth in the deer, changes in density had a larger impact on variation in recruitment than winter weather conditions, while the situation was reversed after the population had ceased to increase (Albon *et al.* 2000). Variation in birth rate also accounted for a larger proportion of the observed changes in recruitment during the period of population growth than in the subsequent period while the relative contribution of adult survival rose in later years.

Our results show that differences in the potential rate of reproduction underlie the contrasts in stability between the two populations. The higher rate of recruitment in the sheep, combined with the lack of density dependent changes in the early stages of population growth, permits sheep populations close to ecological carrying capacity to increase by over 50% in the following summer (when resources are not limiting) and to enter the following winter at levels that cannot be sustained by winter food supplies, generating over-compensatory mortality that reduces population size to a level below ecological carrying capacity (Clutton-Brock *et al.* 1992, 1997*a*; Grenfell *et al.* 1992). As might be expected, the extent of winter mortality is not controlled by population density alone and is affected both by climatic factors in winter and by the age structure of the population (Grenfell *et al.* 1998; Coulson *et al.* 2001). The potential for over-compensation to occur is greater in the sheep than the deer, where population size rarely increases by more than 15% in a single season so that the population's capacity to exceed winter carrying capacity is constrained. Consequently, both population size and over-winter mortality do not vary as widely as in the sheep. As might be expected, maximum values of over-winter mortality appear to be larger in ungulates characterized by high potential rates of increase than in species showing lower values of *R*. Recent studies have recorded substantial fluctuations in population size in several other ungulates with high potential rates of increase (Boyd 1981*a*; Boussès *et al.* 1991; Milner-Gulland 1994; Coulson *et al.* 2000).

Finally, our analyses emphasize that the timing of reproductive events, relative to seasonal changes in food abundance and energetic needs, exert important effects on the demographic impact of variation in population density and climate. For example, it is only because lambs are weaned so early in the summer that female sheep are able to recoup the energetic costs of lactation before the onset of winter, with the result that fecundity escapes from the effects of rising density (figure 13.10), generating high rates of population increase at relatively high population densities (figure 13.2*b*) and leading, eventually, to over-compensatory mortality (Clutton-Brock *et al.* 1997*a*). Differences in the timing of reproductive events between the two species are probably responsible for contrasts in the effects of reproductive success on subsequent breeding performance (see figures 13.9 and 13.10) as well as in the contrasting effects of density and winter on differences in birth mass, neonatal survival and breeding success between cohorts (see figure 13.11). In addition, specific aspects of the ecology of the two populations affect the impact of demographic changes on population dynamics. In both populations, there are substantial differences in early development between cohorts but, for different reasons, this variation has few consequences for population dynamics. In the deer, this is because cohort differences are density independent so that, although they may produce short-term changes in population performance and size, these have few protracted effects. In the sheep, cohort differences are density dependent but their impact on survival and growth usually occurs immediately *after* years of high mortality and is also quickly eroded by subsequent fluctuations in density. In species that combine the early birth date of the sheep with the protracted lactation of the deer, cohort effects might both be density dependent and have an important impact on population dynamics.

General models of population dynamics have played an important heuristic role in predicting the effects of demographic change on dynamics. However, our results suggest that predictions of change in population size accurate enough to be used for management purposes will need to be based on more specific models that incorporate the effects of variation in age structure and in the responses of different age and sex categories to changes in population density and climate. General attempts to account for variation in population dynamics between populations and species will need to recognize the effects of relatively detailed differences in environmental seasonality, development, reproductive timing and population substructure on demographic processes as well as on their impact on population size. At the moment, a detailed understanding of the effects of variation in demography on dynamics is available for very few mammals (Fowler 1987b; Gaillard *et al.* 2000) and the extent of these effects is unclear. The immediate need is to explore the demography and dynamics of a wider range of species.

We are grateful to Scottish Natural Heritage for permission to work on Rhum, and to their staff on Rum, in Inverness and in Edinburgh for logistical support; to Josephine Pemberton, Angela Alexander, Colleen Covey, Ailsa Curnow, Ali Donald, Derek Thomson, Jill Pilkington and many others who have helped to maintain long-term research on reproduction and survival on Rhum; to Steve Albon, Bryan Grenfell, Josephine Pemberton and Richard Sibly for comments on the manuscript; and to the Natural Environment Research Council for supporting the work.

1 4

Population growth rate as a basis for ecological risk assessment of toxic chemicals

14.1. Introduction

Ecological risk assessment tries to predict the likely impacts of human activities on ecological systems (USEPA 1992). In the case of toxic chemicals, the raw materials for ecological risk assessment involve exposure assessment based on predictions or measurements of environmental concentrations of toxic chemicals and an assessment of hazards, i.e. the potential of those chemicals to cause ecological harm. Hazard assessment is generally based upon observations on survival, growth or reproduction in a few individuals in a few species. We shall refer to these responses as individual-level variables. Variability in responses among species is expressed only in terms of differences in these traits as measured under standard laboratory conditions and hence only reflects physiological variability in sensitivity to chemicals. It is presumed that these kinds of observations are relevant for protecting populations and ecosystems. However, this raises at least three different questions, as follows.

(i) To what extent do individual-level variables underestimate or overestimate population-level responses?
(ii) How do toxicant-caused changes in individual-level variables translate into changes in population dynamics for species with different life cycles?
(iii) To what extent are these relationships complicated by population-density effects?

We have addressed these questions, which go to the heart of the ecological relevance of ecotoxicology, using the population growth rate as an integrating concept. We have limited our attention to modelling the links between development, fecundity and survival to population growth rate. Other models go beyond this to relate physiological processes to growth, fecundity and survival (e.g. Gurney *et al.* 1990; Kooijman 1993), but these are demanding in their requirement of detailed data.

For each question, we shall present a short review of the work done to date followed by one or two examples based on experimental data to illustrate how population growth rate can be used in a practical way to provide a more sound basis for ecological risk assessment. Our aim in the first instance is to develop an approach that can be applied to generalized ecological risk assessments of toxic chemicals, as is often required in the regulatory arena. This means that we are using population growth rate analyses to compare the effects of toxic chemicals on the same species among exposure concentrations (i.e. to derive concentration–response relationships), on the same species among toxic chemicals (i.e. to rank chemicals in terms of their relative ecological hazard) and on the same chemical among species (i.e. so that the effects of chemicals on test species can be extrapolated to the effects on other, untested, species). Our analyses, therefore, have to be limited in detail and cannot take account of immigration–emigration effects or interactions with other species, the importance of which will vary from one habitat to another. In principle, when it comes to considering detailed impacts of specific chemical(s) on specific populations in particular habitats, it is possible to develop more detailed models. However, such models have had very limited applications in the regulatory arena to date.

14.2. Within a life-cycle type, to what extent are responses in individual variables more or less sensitive to increasing toxicant concentration than population growth rate?

(a)   Review
In practice, ecotoxicological tests focus on individual-level variables (i.e. survival in response to high chemical concentrations over short periods of time; survival, growth or fecundity in response to low chemical concentrations over long periods of time). However, most often, the targets of ecological risk assessments are not individuals but entire populations as, within limits, individuals can be removed from populations without any adverse effects on population persistence. Thus, for such individual-level responses to be useful endpoints, it is necessary that they adequately and consistently reflect the impacts of chemicals on populations. This

means that there are two main issues of concern. The first is whether individual responses measured in terms of survival, fecundity or growth–development are more or less sensitive to chemical impacts than effects measured in terms of population growth rate. The second is whether there is consistency in the relationship between changes in the individual traits and changes in population growth rate, such that it is possible to identify those traits that are the best predictors of effects on population growth rate.

There is a number of reasons why differences in sensitivity between individual-level and population-level responses to chemicals may occur. These may arise from the nonlinear relationship between population growth rate and the demographic variables contributing to it; from the relative size of the demographic variables with respect to each other (i.e. life-cycle type); from species- and chemical-specific differences in the relative sensitivity of the demographic variables to chemical exposure; and from the demographic state from which the population starts (i.e. growing, stable, declining; see Forbes & Calow 1999). Clearly, whether population growth rate is expressed as the intrinsic rate of increase (r) or population multiplication rate (λ, where $\lambda = e^r$) is also important, particularly if proportional changes are used as a measure of relative sensitivity.

In a review of 41 studies, which included a total of 28 species and 44 toxicants, Forbes & Calow (1999) found that, out of the 99 cases considered, there were only five where chemical effects on population growth rate (most of which were expressed as r) were detected at lower exposure concentrations than those resulting in statistically detectable effects on any of the individual demographic variables. In 81.5% of the cases considered (out of a total of 81), the percentage change in population growth rate (expressed as r) was less than the percentage change in the most sensitive of the individual demographic traits: 2.5% where the percentage change in population growth rate was equal to that of the most sensitive trait and 16% where the percentage change in population growth rate was greater than the percentage change in the most sensitive demographic trait. Despite the fact that any proportional changes in population growth rate were significantly correlated with the proportional changes in fecundity and with time to first reproduction, these correlations were rather weak and trend analysis indicated that these relationships were nonlinear. Surprisingly, the correlation between the proportional reduction in survival (i.e. the most frequently measured trait in ecotoxicological studies) and the proportional reduction in population growth rate was not statistically significant. Overall, there was no consistency in which of the measured individual-level traits was the most sensitive to toxicant exposure, and none of them, considered individually, could be said to be very precise predictors of toxicant effects on population growth rate.

Another way of approaching this question is to consider the sensitivity of population growth rate to changes in the life-history traits contributing to it. This can be formalized by assessing the percentage change in λ that arises from small percentage changes in individual-level variables. This quantity is referred to as the elasticity of λ with respect to the individual-level variables (De Kroon *et al.* 2000; Caswell 2000). It should be noted that these elasticities do not strictly represent contributions to λ and therefore do not necessarily sum to 1 (i.e. in this context there is no reason to expect that λ is a homogenous function of the individual variables (Caswell 2000, p. 232)). Using a simplified two-stage life-cycle model, Forbes *et al.* (2001*a*) were able to show that when λ is close to 1 and the generation time is greater than or equal to 1, the elasticities with respect to the individual life-cycle variables were less than or equal to 1. In other words, in the neighbourhood of the population steady-state, a small percentage change in the individual life-history variables, for example brought about by a toxicant, would result in at most the same percentage change in λ. However, if λ is allowed to increase above 1, the situation becomes more complex and in certain circumstances it is possible that a given proportional change in some of the individual-level traits leads to a proportionally greater change in λ.

(b)   Example 1

The conclusion from the review is that, in general, the most sensitive individual life-cycle traits will be at least as sensitive as population growth rate to any increases in toxicant concentration. However, this assumes that the most sensitive trait(s) will always be measured in ecotoxicological assays, but this need not be the case. The following example, based on a life-table response experiment using the polychaete *Capitella* species I exposed to nonylphenol (Hansen *et al.* 1999*a*) illustrates this point. Table 14.1 shows the percentage reductions in life-cycle traits and population growth rate at the nonylphenol concentration at which the most sensitive trait was significantly impaired relative to the control. At this concentration there were wide differences in percentage effects on individual traits and population growth rate, with the reproduction effects being the most severe. There were no effects on either juvenile or adult survival over the entire

Table 14.1. *Percentage changes in individual demographic variables and* λ *for the polychaete* Capitella *species I exposed to 174* μg *nonylphenol* g^{-1} *dry wt sediment.*
(The value of λ *in the control was about 2.5 week^{-1}. Data are from Hansen et al. (1999a).)*

Trait	Change relative to control (%)
juvenile survival	0
adult survival	0
time to first reproduction	+17
time between broods	+25
total number of broods per individual	−44
total number of offspring per individual	−78
population growth rate (λ)	−24

concentration range used. So if the analysis had initially focused on just survival, a risk assessment based on this individual-level variable would have concluded that no effects of this chemical occurred at a concentration at which a 24% reduction in λ was calculated. Fecundity was the most sensitive trait to nonylphenol, with brood number and total offspring being reduced by 44 and 78%, respectively. Although the time to first reproduction was delayed by only 17%, a decomposition analysis of these data (Hansen *et al.* 1999a) showed that this trait contributed more to the effect on λ than did fecundity.

This example clearly demonstrates why life-table studies should take all effects into account and the value of population growth rate as an integrating variable. In addition, it highlights the necessity of considering both the toxicological sensitivity of individual-level variables and the demographic sensitivity of λ to changes in these variables for understanding the mechanisms of toxicant effects on population dynamics.

14.3. How do toxicant-caused changes in individual-level variables translate into changes in population dynamics for species with different life cycles?

(a)   **Review**

In ecotoxicology, concentration–response relationships are obtained for individual-level variables across a number of standard test species. It has been implicitly assumed that these responses have the same meaning, independent of the species involved; for example, 50% mortality is presumed to have the same effect, in terms of population dynamics, for all species. Alternatively, variability in individual-level traits across species has been used to construct species-sensitivity distributions that are now used as a basis for ecological risk assessment (Van Leeuwen & Hermens 1995). The implicit assumption here is that the variability in individual-level variables is directly related to variability in population responses. However, this ignores the possibility of complications from life-cycle differences across species. The analysis of demographic models very clearly shows that, indeed, different life-cycle types respond differently to changes in their corresponding demographic input variables (e.g. Stearns 1992; Caswell 2000).

Calow *et al.* (1997) considered a series of simplified but plausible scenarios to illustrate how information from ecotoxicological tests can be used to explore the effects on population dynamics for different life-cycle types. A number of general conclusions arose out of this analysis. (i) As expected, the effect on population growth rate of a toxicant that reduces juvenile survivorship or fecundity will be greater for semelparous species (i.e. species that reproduce once) as compared with iteroparous species (i.e. species that reproduce more than once), and the reverse will be the case for the effects of toxicants on adult survival. (ii) Iteroparous species with life cycles in which the time to first reproduction is shorter than the time between broods will be more susceptible to toxicant impacts on survival or fecundity than will species in which time to first reproduction is longer than the time between broods. (iii) Anything that shortens time to first reproduction relative to the time between broods (e.g. increased temperature, increased food availability) is expected to increase the population-level impact of toxicant-caused impairments in survival or fecundity. (iv) Lengthening of the time to first reproduction should lessen the population impact of toxicant-caused impairments in survival or fecundity. An additional important outcome from this analysis was a clear demonstration of the importance of a population's demographic starting point for relating toxicant-caused impairments on demographic traits to consequences at the population level. This is particularly so for time to first reproduction, t_j. Whereas, in growing populations, toxicant-caused delays in t_j have a negative effect on population growth rate, in shrinking populations such delays can have an ameliorating effect in that they slow the rate at which the population approaches extinction.

Extrapolating the effects of toxic chemicals from individual-level responses from a few species to the effects on entire communities is increasingly performed by fitting available ecotoxicological test data to a statistical distribution and using this to estimate the chemical concentration that is unlikely to impair most of the species in the distribution. However, the distributions of sensitivities based on individual-level variables are likely to differ from distributions based on population growth rate as discussed in the previous paragraph. Moreover, the species used to provide the input data for these distributions rarely, if ever, reflect the actual distribution of the life-cycle types in natural communities. Many of the species used routinely in ecotoxicological tests are chosen because they have life-cycle features that are amenable to laboratory work, and therefore cannot be considered to represent a random sample from nature. Forbes *et al.* (2001*a*) took an initial step towards exploring the importance of these considerations in ecological risk assessment by comparing sensitivity distributions that were based on the response of juvenile survival to a chemical with sensitivity distributions based on λ for communities of varying life-cycle distributions. Nine scenarios were simulated in which the sensitivity of the life-cycle types and their proportions in the community were varied. For all of these cases, the sensitivity distributions based on juvenile survival gave lower effect concentrations (i.e. the concentration at which a defined percentage of the species was negatively affected) than the distributions based on λ and would therefore tend to provide an added measure of protection if used for risk assessment. However, the proportions of life-cycle types in a community must vary across communities and may have an important influence on sensitivity distributions. They therefore need to be given further consideration in ecological risk assessment.

(b)   Example 2

Here, we provide two examples as follows: Example 2a illustrates how the same response of an individual demographic trait to a toxicant can have markedly different consequences on population growth rate as a result of life-cycle differences; Example 2b illustrates how very different responses of individual demographic traits to a toxicant can result in similar consequences for population growth rate as a result of life-cycle differences.

(i)   Example 2a

Forbes *et al.* (2001*a*) considered the life cycles of the most widely used ecotoxicological test species (i.e. a green algal species, an iteroparous fish, a

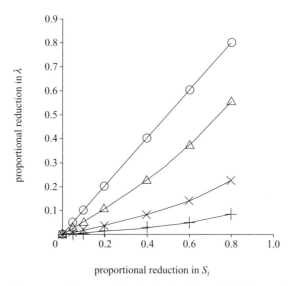

Figure 14.1. The proportional reduction in λ resulting from a given proportional reduction in juvenile survival (S_j) for different life-cycle types. Circles, benthic macroinvertebrate life cycle; crosses, fish life cycle; pluses, daphnid life cycle; triangles, algal life cycle (after Forbes *et al.* 2001*a*).

daphnid and a semelparous benthic invertebrate) and, using a simple two-stage demographic model, estimated the proportional decline in population growth rate (expressed as λ) resulting from a proportional decline in juvenile survival. Figure 14.1 summarizes the results of these analyses and shows that, for stable populations, the same toxicant-caused reduction in survival would have very different effects on population growth rate, dependent on the life cycle. Daphnid population dynamics would be the least sensitive to impairments in juvenile survival, followed by the iteroparous fish and the alga, with the semelparous benthic invertebrate dynamics being the most sensitive to any impairments in juvenile survival. For all life cycles it could be shown that, for starting values of λ close to 1, toxicant-caused impairments in survival, fecundity or timing would result in equivalent (benthic invertebrate) or smaller (the fish, daphnid and algal life cycles) impacts on the population growth rate. Figure 14.1 shows that a 10% reduction in juvenile survival (hereafter referred to as LC_{10}) would result in a 10% reduction in λ for a semelparous benthic invertebrate life cycle, a 5% reduction in λ for a green alga life cycle, a 2% reduction in λ for an iteroparous fish life cycle and only a 0.6% reduction

Table 14.2. *Life-history traits of three sibling species of* C. capitata *grown under the same conditions in the laboratory.*
(Conditions: <63 μm sediment with 6.6% organic matter, 30% seawater, 18 °C, constant darkness. For details on experimental design and sibling species, see Linke-Gamenick et al. *(2000).)*

Trait	Species S	Species M	Species I
juvenile survival (proportion)	0.20	0.77	0.79
adult survival (proportion)	0	0.61	0.79
age at first reproduction (days)	115 ± 20	58 ± 5	76 ± 10
time between broods (days)	–	11 ± 4	19 ± 9
number of offspring per brood per reproductive individual	10.3	15.6	13.4
type of larval development	direct	lecithotrophic	lecithotrophic
population growth rate (λ, d^{-1})	1.05	1.42	1.30

in λ for a daphnid life cycle. Clearly, a chemical having similar effects on juvenile mortality would be expected to have vastly different population-level consequences for these life cycles. In a similar manner, although the benthic invertebrate might have a higher LC_{10} value than the daphnid for a given chemical, its population dynamics could nevertheless be more sensitive. Our analysis indicated that a 5% reduction in juvenile survival of the benthic invertebrate life cycle would have the same effect on λ as an 80% reduction in juvenile survival of the daphnid life cycle.

The degree to which the responses of the individual demographic traits overestimated the impacts on population growth rate varied widely among the life cycles, and this may have important practical implications for risk assessment. An extension of the analysis showed that for starting values of λ that were much greater than 1, the effects on the individual demographic traits could, for at least some of the life-cycle types, result in proportionally greater effects on population growth rate.

(ii)   Example 2b
Linke-Gamenick *et al.* (2000) examined the effects of the polycyclic aromatic hydrocarbon, fluoranthene, on survival, reproduction and the development time of three *Capitella capitata* sibling species (I, M and S) and employed a two-stage demographic model to assess the consequences of the measured effects on population growth rate. In the absence of fluoranthene, the three species differed markedly in life-cycle traits and in λ (table 14.2).

Table 14.3. *Percentage changes in individual-level traits and* λ
between the control populations of three Capitella *sibling species and*
populations exposed to 50 μg g^{-1} *of fluoranthene.*
(*The values of* λ *under control conditions are given in table 14.2. Data*
are from Linke-Gamenick et al. (*2000*).)

Trait	Species I	Species M	Species S
juvenile survival	−15.2	−29.9	50.0
adult survival	−20.3	23.0	0
time to first reproduction	15.7	15.4	0
time between broods	−47.2	45.9	0
number of offspring per brood per reproductive individual	−9.7	−32.1	34.0
λ	−4.6	−10.5	4.3

The percentage changes in the individual-level variables and λ following exposure to 50 μg g^{-1} of fluoranthene during a period of 25 weeks are
summarized in table 14.3 . Without going into too much detail, it is clear
that for all species some of the percentage changes in individual-level variables were greater than the percentage changes in λ. By comparing among
the species studied, table 14.3 shows that there were appreciable differences between the responses in the individual-level variables that were
not matched by changes in λ. This can be explained in terms of (i) within
species, some changes in individual-level variables acted to increase λ despite impairments in other variables (e.g. in species I, the time between
broods was shortened in the exposed populations; in species M, the adult
survival rate was increased in the exposed populations), or (ii) λ was relatively insensitive to changes in those variables that were impacted by
toxicant exposure.

   It should be noted that we have chosen the concentration range for
fluoranthene to illustrate the point that changes in individual-level variables do not necessarily translate into changes in population growth
rate. However, at concentrations beyond this range, juvenile survival was
markedly reduced and reproduction was completely inhibited in species
S, so that its population growth rate was reduced to zero. By contrast, there
was little further effect either on individual-level variables or on population growth rate in the other two species up to a concentration of 95 μg
fluoranthene g^{-1} dry weight of the sediment.

   As a result of life-cycle differences among species (table 14.2), λ is not
equally sensitive to changes in each of the individual-level traits. Elasticity

Table 14.4. *Elasticities of three sibling species of* Capitella *estimated from the slopes of relationships between each individual trait and λ (both on a ln scale), while holding all other traits constant (Caswell 2000, p. 226) and using a simple two-stage life-cycle model (Calow* et al. *1997). (The values were rescaled so that the elasticities for each species summed to 1. n.a., not applicable.)*

Elasticities	Species I	Species M	Species S
juvenile survival	0.199	0.210	0.391
adult survival	0.048	0.040	n.a.
time to first reproduction	−0.465	−0.477	−0.218
time between broods	−0.089	−0.065	n.a.
number of offspring per brood per reproductive individual	0.199	0.208	0.391

analysis can be used to identify particularly sensitive life-cycle types, i.e. life cycles whose population dynamics are very responsive to small changes in the individual survival, reproduction or timing, and to identify, for different life-cycle types, those demographic variables that have the greatest influence on the population dynamics. By combining traditional ecotoxicological measures of chemical effects on survival, reproduction and growth with demographic elasticity analysis, it is possible to tease apart the relative contributions of physiology and life cycle in determining the susceptibility of different species to toxicant exposure. This can be illustrated by considering the elasticities of the three sibling species of *Capitella* that have been estimated from the slopes of ln λ plotted against ln trait (Caswell 2000, p. 226) and summarized in table 14.4. From table 14.2, it is clear that species S has the lowest juvenile survival and fecundity of the three species under unexposed conditions. The analysis summarized in table 14.4 indicates further that the life history of this species is such that its population growth rate is approximately twice as sensitive to changes in these traits compared with the other two species. In addition, species S was physiologically more sensitive than the other species in that its fecundity was reduced to zero at the highest exposure concentration. The greater sensitivity of species S to toxicants compared with the other two species results from a combination of physiological and life-cycle differences. In addition, it is known that species I is more widely distributed, particularly in heavily polluted sediments (Linke-Gamenick *et al.* 2000), than species S, and this can be explained in the same way.

14.4. To what extent does density complicate any conclusions drawn from observations on the response of individuals to toxicants under nonlimiting densities?

(a)   Review

When, as is rarely the case, population growth rate is measured in ecotoxicological tests, it is generally done under conditions in which food and space are unlikely to limit population growth. There is concern, therefore, that conclusions drawn from such tests may have little relevance for field situations when populations are regulated by density dependence. Whether the combined effects of density and chemical exposure on individual survival, growth and reproduction interact to produce additive, more-than-additive or less-than-additive effects on population growth rate depends on (i) how, in combination, density and chemical exposure affect individual performance; (ii) the type of density-dependence operating (i.e. scramble or contest); and (iii) the life-cycle type of the species in question. Although the interactions of chemical exposure and density on population growth rate are theoretically predictable, the number of factors influencing the outcome is large and therefore simple, general, *a priori* predictions are not feasible (Forbes *et al.* 2001*b*). A few simulation studies have been performed, and these indicate that density is likely to ameliorate the effects of chemicals on population growth rate (Grant 1998; Hansen *et al.* 1999*b*). Experimental studies have produced mixed results, with some showing additive interactions between density and chemical effects on population growth rate (Winner *et al.* 1977; Klüttgen & Ratte 1994), whereas others have found less-than-additive effects (Marshall 1978) or more-than-additive effects (Chandini 1988). There is even some indication that the form of the interaction may vary across a chemical concentration gradient, with effects shifting from less-than-additive at low-toxicant concentrations to more-than-additive at higher toxicant concentrations (Linke-Gamenick *et al.* 1999). As recent work on the effects of pulsed pesticide exposures on freshwater trichopteran populations has demonstrated, compensatory interactions between population density and toxicants on population dynamics can persist throughout a cohort's life cycle even if the period of toxicant exposure is very brief (Liess 2002). In addition, it appears that the kind of interaction that may be observed is, to an important extent, constrained by the kind of experimental design employed (Forbes *et al.* 2001*b*). The design most likely to approximate natural conditions is the so-called 'bucket test' (Sibly 1999) in which populations are initiated

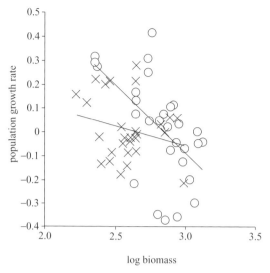

Figure 14.2. Population growth rate (r, week^{-1}) as a function of (Log transformed) population biomass in control populations (circles) and populations exposed to 50 μg fluoranthene g^{-1} dry wt sediment (crosses). The lines are linear regressions through the data: control: population growth rate = 1.63–0.57 (Log biomass), $n = 30$, $r = 0.57$, $p = 0.001$; 50 μg fluoranthene g^{-1} dry wt sediment: population growth rate = 0.44–0.17 (Log biomass), $n = 30$, $r = 0.24$, $p = 0.20$ (after Forbes *et al.* 2003).

with different combinations of food and chemical exposure and assayed over time. We illustrate in the example below how this might be put into practice.

(b)   Example 3

Forbes *et al.* (2003) used a bucket-test design to explore the interaction between population density and toxicant effects in the polychaete, *Capitella* species I. Populations of worms were initiated with a stable age distribution and with different combinations of food availability and exposure to fluoranthene (0, 50 and 150 μg g^{-1} dry weight sediment). The experiment was conducted over a period of 28 weeks. Further details of the experimental design can be found in Forbes *et al.* (2003). Figure 14.2 plots population growth rate against log density for control populations and those exposed to 50 μg g^{-1} fluoranthene (populations exposed to 150 μg g^{-1} fluoranthene became extinct by the eighth week of exposure). The results indicate that increasing population density ameliorates the effect of fluoranthene on population growth rate, and similar though weaker amelioration by high-population density of the effects of a toxicant were obtained for *Tisbe battagliai* exposed to pentachlorophenol (Sibly *et al.* 2000b). Thus, in these particular cases, tests carried out under non-density-limiting situations would tend to overestimate the effects of the toxicants at high density, which might be closer to natural conditions. However, as the review indicated, this conclusion should not be taken to be a general one, and it underlines the point that density effects need to be taken more seriously in the design of ecotoxicological tests.

14.5. Conclusions

We have shown that individual-level variables are equally or more sensitive to increasing concentrations of toxic chemicals than is population growth rate. Hence, the concern that small effects on individual survival, growth or reproduction are magnified into large effects on populations is not supported by the available data. This is an important message for environmental protection given the large number of chemicals that have to be considered. However, the validity of relying on individual-level endpoints depends on the most sensitive variables always being measured. Due to the fact that these vary across species and chemicals, it is not feasible to identify which variables will be generally the most sensitive to toxicants, or the best general predictors of population growth rate *a priori*. Moreover, differences in life cycles across species mean that similar effects of chemicals on individual-level variables in different species can have vastly different consequences for population growth rate. Alternatively, very different effects on individual-level variables can sometimes translate into similar or no differences in effects on λ among species. Finally, the effects of toxicants in situations involving density limitations can differ from the effects recorded under low-density circumstances (as is usually the case in ecotoxicological tests) and it is not straightforward to predict actual outcomes *a priori*. More attention, therefore, needs to be given to the inclusion of appropriate and realistic density conditions in test scenarios.

Clearly, if we want to know more about the effects of toxic chemicals on population dynamics then we need to carry out more work on the relationships between individual-level responses and population growth rate under increasing toxicant concentrations. Moreover, if for particular chemicals we want to develop more ecologically relevant risk assessments then this should be done in terms of population growth rate rather than just observations on individual-level responses. Population growth rate analysis can, in principle, also indicate which species within communities will be the most or least susceptible to chemical pollution and which, after a pollution event, will be the most or least likely to recover and at what relative rates. For example, work with the polychaete *Capitella* would indicate that there are large differences among sibling species in their ability to persist in and recolonize polluted habitats and that these differences are at least partly due to life-cycle differences among species (Linke-Gamenick *et al.* 2000). For the species that are not amenable to laboratory testing, modelling the effects of toxicants using population growth rate analyses could be used to characterize their relative susceptibility to different toxicants.

These kinds of analyses have more general implications for understanding the ways that individual-level variables contribute to population dynamics. For example, in pest control, understanding how population growth rate responds to manipulations of different parts of the life cycle can enable the development of more effective control programmes (e.g. McEvoy & Coombs 1999). A similar case can be made for the development of effective conservation strategies (see Chapter 12).

In conclusion, with appropriate time and resources, population growth rate analysis should form the basis of ecological risk assessment. It is desirable to incorporate more detail into models with regard to individual-level variables and the ecological context in which the species exist and pollution occurs. However, it will rarely be possible to conduct population growth rate analysis for all species and chemicals to a sufficient level of detail. One possible solution to this is to use population growth rate analysis to identify the most vulnerable species and to focus our assessments on them. Not only could this approach be useful for carrying out ecological risk assessments, but it could also contribute to the development of conservation strategies.

Financial and logistical support from The Royal Society and the Novartis Foundation are gratefully acknowledged.

15

Population growth rates: issues and an application

15.1. Introduction

This paper consists of two parts: the first is a general discussion of population growth rates, the factors that determine them and their role in studies of population regulation; the second is a specific example of the calculation of a population growth rate and its use in an evolutionary ecological study. The first part is based on the summary talk given in The Royal Society Discussion Meeting from which the collected papers are assembled in this issue, but was also much influenced by an informal discussion meeting hosted by the Novartis Foundation that immediately followed the meeting at The Royal Society, and which was attended by most of the speakers as well as other population biologists (see the Acknowledgements). But though the strongly expressed views of many of the participants have often affected or determined what we have written, this is not an attempt at a consensus, and the blame for any muddle-headedness rests with us alone.

15.2. Population growth rate and population regulation

(a) Population growth rate
Is it justified to give the population growth rate such a central position in population biology? Is it not just one of a spectrum of useful concepts that permeate population dynamics, a particularly important one perhaps, but not deserving of so pivotal a role? We see several strong arguments supporting the importance of population growth rate, but with a number of caveats.

First, population growth rate is the central tool in population projection, the exercise in which data from contemporary studies are used to say what the potential growth rate of an organism's population is, assuming that present schedules of birth and death (and possibly immigration and emigration) remain unchanged. The father of population projection is surely Malthus (1798) who in his 'Essay on the principle of population' recognized the essential geometric nature of population growth, and that this inescapably meant that either what we might now call extrinsic and intrinsic population factors must come into play to limit population growth. In the simplest case of an unstructured population, population growth rate is easily calculated as the number of births minus deaths per unit time. When the population is structured by age, size or another variable, calculating population growth rate is more complex as normally it can only be expressed implicitly as the root of the Euler–Lotka or equivalent equation (Chapter 3), but today with the ready availability of powerful personal computing this is no longer an obstacle to its easy use. There is presently a huge armoury of techniques that can be used to explore how population growth rate is influenced by age- or size-specific entries in the organism's life table, or by variables that determine jointly multiple entries (Caswell 2001), and these are used routinely in applied areas of ecology such as conservation (e.g. Seamans *et al.* 1999), pest management (e.g. Shea & Kelly 1998) and ecotoxicology (e.g. Chapter 14). Although the theory was originally developed for deterministic populations, an equivalent stochastic body of theory is now available (Tuljapurkar 1990), though obviously its use is more hungry of data.

Second, population growth rate is often the most natural response variable for the statistical analysis of the factors influencing a species' population dynamics. In the density-independent case, Caswell (2001) and others have developed statistical tools to partition observed variability in population growth rate into components associated with different life table entries, a process called retrospective analysis, conceptually distinct though related to population projection. Most modern techniques used to search for density dependence are based on analysing how population growth rate varies with density (e.g. Chapter 4), as are attempts to search for periodicity in time-series (Kendall *et al.* 1999) and more exotic dynamic behaviour such as chaos (Sugihara & May 1990; Hastings *et al.* 1993; Ellner & Turchin 1995).

Finally, population growth rate intimately links population dynamics and evolutionary biology. Exactly how this link operates depends upon which model of evolutionary change is assumed. It is clearest when evolution is viewed as the successive invasion of new traits into a resident population (Metz *et al.* 1992; Rand *et al.* 1994). The course of evolution depends upon whether a trait increases or decreases in frequency, which involves determining whether the population growth rate of the clan of individuals with the invading trait is positive or negative (here assuming a log scale) in an environment some of whose properties, particularly the nature of the density-dependent feedback, are set by the resident trait. If density dependence in the population operates through age-independent mortality then evolution will maximize population growth rate, but this is not a general result and depending on exactly how density dependence operates, a variety of other criteria will be maximized (Mylius & Diekmann 1995). This theory can be extended to take into account stochastically varying environments (Ellner 1985).

In the quantitative genetic model of evolutionary change, additive genetic variation in different traits is assumed always to be present and the progress of evolution is determined by the variance–covariance matrix representing this variation and a vector of selection gradients describing how different traits contribute to the organism's fitness (Lande 1982). In the simple case of frequency-independent selection, the element in the selection gradient vector corresponding to trait x can be interpreted as the change in logarithmic population growth rate (log λ) brought about by a small change in the trait ($\partial \log \lambda / \partial x$) which is related but not identical to sensitivities ($\partial \lambda / \partial x$) and elasticities ($\partial \log \lambda / \partial \log x$) from classical population projection theory (Lande 1982; van Tienderen 2000). Here, population growth rate provides the mean fitness of a population. Again, it is possible to extend the theory to more complicated scenarios such as stochastic environments, and traits with frequency-dependent fitness (Lande 1982).

But study of population growth rate is not synonymous with population dynamics and there are circumstances where it is not the central focus of enquiry. Consider, for example, conservation biology where species viability plans and programmes are frequently used to predict extinction risks and as management tools to help recovery programmes. If the population growth rate of a current population is measured, and no account is taken of how density dependence may alter population growth rate as density changes, then simple population projections may be quite seriously misleading (e.g. Grant & Benton 1996; Chapter 12). A very rare species with a positive growth rate may be in far greater peril than expected if density dependence acts at low population densities, while a similar species with a negative growth rate may be relatively easily conserved if above a population density threshold it begins to increase quite rapidly (as with an Allee effect). A population growth rate of zero may say little about the fate of a population if the species is at an equilibrium set by limited habitat extent. Mills *et al.* (1996) and Brook *et al.* (1997) provide examples from grizzly bears and the Lord Howe Island woodhen where the incorporation of density dependence has a major effect on population viability analysis (see also Chapter 12). Similar arguments can be made from other areas of applied ecology, and these indicate that while population growth rate is a critical component of population dynamics, it should not be divorced from other aspects of population change.

(b)   The role of models

Population biology has always advanced on two fronts, theory and experimentation, but the relationship between the two aspects of the subject has not always been cordial. A common criticism by some experimental ecologists has been that many models are overly simplistic, are hard to apply to real situations, and ignore the temporal and spatial heterogeneity inescapably present outside the laboratory. Much of this criticism has arisen when strategic models have been misapplied or misinterpreted as tactical models. By strategic models we mean those that are designed to explore the ramifications of general questions in ecology, questions such as whether an interaction between a predator and a prey with discrete generations and a random search by the predator give rise to persistent population cycles. Such questions can be answered only by using the tools of mathematics, and provide general hypotheses that can then be tested experimentally. Tactical models however, are tailor made for particular systems and are designed to predict their fate. They are used extensively, for example, in pest management to predict exactly when a crop should be sprayed, in epidemiology to determine the proportion of the population that needs to be culled or vaccinated (Anderson & May 1979; Anderson *et al.* 1996) and in fisheries to determine the size of the harvest that maintains stocks (Clark 1985). Some models of this type are purely statistical using explanatory variables such as current population densities and weather to predict future population densities or other parameters using methods that incorporate no explicit biological processes. Others include detailed biological processes, often in the form of large simulation models that use abiotic inputs as driving variables (e.g. Gutierrez *et al.* 1993).

We maintain that both types of model, in their right place, have served the subject well. Study of very simple single-species models first showed the potential dynamic complexity of populations, and contributed materially to the discovery of chaos in dynamical systems. Strategic models have shown us that two identical species cannot coexist in a temporally and spatially constant environment, and the resulting studies of niche segregation, competition/colonization trade-offs, and coexistence in environments that vary in time both through biotic and abiotic agencies have enormously enriched community ecology. Simple models have also shown that parasites and parasitoids may, but do not necessarily, regulate the densities of their host populations, and have stimulated a large experimental programme to discover whether real host–natural enemy systems possess these properties. Strategic models on the one hand are often poor at incorporating the biology of real animals and plants, and poor at making quantitative predictions that can be used to test their applicability in the field (Chapter 7). Tactical models on the other hand, while often successfully predicting changes in density, frequently do so at the expense of generality or an understanding of underlying processes. Those based purely on statistical prediction make no pretence at this, while large simulation models are often too complicated to disentangle the factors driving dynamics.

The distinction between strategic and tactical models has never been absolute, and especially in fields such as epidemiology there have always been some models that bridge the two classes (Anderson & May 1991). Recently, this blurring has been accelerating, giving a real prospect of analytical theory that better serves field biologists. The two factors that we suspect are most responsible for these changes are the increasing use of modern statistical methods in population dynamics (e.g. Bjørnstad et al. 1998, 2001; Lande et al. 1998; Yao et al. 2000; Fromentin et al. 2001; Lingjaerde et al. 2001) and the readiness of ecologists to construct relatively complicated models appropriate to particular systems, but from which the classic models of population ecology can be derived as limiting cases. The latter greatly assists in understanding the dynamics of the full model. This evolution has been occurring in all areas of population biology, but we choose two examples from our own fields to illustrate these trends.

Parasitoids are insects that lay their eggs in the bodies of other insects, and their hosts are invariably killed as the larval parasitoid achieves maturity (Godfray 1994). They are abundant members of virtually all terrestrial ecosystems and can be a source of major mortality for their hosts. Entomologists have successfully controlled outbreak pests by introducing parasitoids into an area, thus demonstrating the potential for parasitoids to control their hosts, though the extent to which they are the prime factor regulating hosts in natural ecosystems is still not clear.

Traditionally, host–parasitoid dynamics have been modelled in two main ways. First through computer-intensive simulations, typically of hosts that are major agricultural pests, and often with the explicit aim of improving pest management rather than understanding the underlying biology (e.g. Gutierrez & Baumgartner 1984). The second way is based on the discrete-generation form of the Lotka–Volterra model and was initiated by Nicholson & Bailey (1935) and particularly developed by Hassell, May and colleagues in the 1970s and 1980s (reviewed by Hassell (2000)). The basic Nicholson and Bailey model, which assumes among other things random search by the parasitoid, shows divergent oscillations and extinction of one or both insects. This is caused by the parasitoid overexploiting the host which leads to the density of first the host and then the parasitoid crashing; the host recovers, and then so does the parasitoid, but with a lag that causes the next crash to be even more catastrophic, and so on until one party goes extinct. Strategic models have shown that for the interaction to persist there must be some refuge for the host during periods of high parasitoid densities, and that this refuge may be an actual physical refuge, a fraction of the population that is genetically resistant to the parasitoid, or a fraction that escapes parasitism in time or in space. Examples of all these types of refuge have been identified in field systems, but these simple population models have not proved effective in describing real interactions in the field.

Murdoch et al. (1987) introduced a way of modelling host–parasitoid systems that facilitates the incorporation of considerable biological detail. Technically these are age-structured models, which assume demographic rates are constant within predefined stage classes, the assumption allowing the models to be phrased as delay-differential equations (alternative formulations have been explored by Hastings & Costantino (1987)). They have been called 'models of intermediate complexity' (Godfray & Waage 1991) as they lie between the simple strategic and the strategic complex models described above, but the critical point is that although they are frequently too complex to study fully analytically, the Lotka–Volterra, Nicholson–Bailey or other strategic models can be derived as limiting cases. The original application by Murdoch et al. (1987) was to understand the persistence of the interaction between the red scale pest of orange trees and its parasitoids. The unusual biology of the host meant that it was poorly described by classic host–parasitoid models that either assumed discrete generations or omitted developmental time-lags, and Murdoch *et al.* (1987) were able to show that host–parasitoid persistence was possible if certain life-history stages were both long-lived and resistant to the natural enemies.

These models also proved valuable in explaining why some hosts and their parasitoids in relatively aseasonal tropical environments show population cycles with a period approximately equal to one host generation (Godfray & Hassell 1989). Traditionally, the cycles had been explained by temporary synchronization of the host age structure by rare events such as droughts or hurricanes. However, age-structured host–parasitoid models suggested that synchronization would decay quite quickly, but also that persistent generation cycles could be maintained if the parasitoid generation time was approximately half that of the host. In this case the system has two time-delays, which interfere with each other causing the cycle (similar phenomena had been observed previously in electronics). More biologically, the persistence of the cycle can be explained by the fact that the majority of parasitoid eggs are laid at the peak of the host generation cycle giving rise to a burst of parasitoids when host densities are at a trough in density, so reinforcing the cycle.

The models just described are still relatively simple, but the approach can be extended to more complex systems where this is justified by our biological understanding of the interaction. For insect–natural enemy interactions, most of this work has involved laboratory systems (though see Rochat & Gutierrez (2001)). For example, Begon, Thompson, Sait and colleagues have conducted a long series of experiments with the Indian meal moth, *Plodia interpunctella*, a small insect whose larvae feed on bran and other stored products (Sait *et al.* 1994*a*,*b*, 1997; Begon *et al.* 1995, 1996*a*,*b*). The moth is attacked by a variety of natural enemies, including several parasitoids and a granulosis virus (a type of baculovirus). The interaction is particularly complex as the moth itself in single-species culture shows generation cycles which are caused by asymmetric and highly age-specific interactions between different age classes (Nisbet & Gurney 1983; Briggs *et al.* 2000). The parasitoid only attacks certain age classes, while depending on which stage the virus attacks it may cause death or the establishment of a sublethal infection that alters development time and adult fecundity. In experiments with the host, a parasitoid and a virus, Sait *et al.* (2000) found that the dynamics exhibited by the three species system depended on the order in which it was assembled, indicating alternative dynamic attractors. A stage-structured model constructed on the basis of a knowledge of the natural history of the interaction predicted dynamics similar to those observed in the laboratory, although analysis indicated not only a single stable dynamic attractor but also a dynamic repellor about which population trajectories might orbit for long periods of time as transients (Sait *et al.* 2000). A criticism of this approach is that the model construction and the identification of the important time-delays in the system are to a degree *ad hoc*. It is thus interesting that a statistical analysis of the data using modern time-series analysis (see below) identified the same number and magnitude of time-delays as had been assumed based on the natural history (Bjørnstad *et al.* 2001). The two analyses had been conducted separately, but increasingly we see model formulation and statistical analysis proceeding hand-in-hand, and a good example of this is the work on ungulate populations described in this issue (Grenfell *et al.* 1998; Coulson *et al.* 2001; Chapter 13). Other examples of the application of models of intermediate complexity to insect systems include those of Hastings & Costantino (1987) and Costantino *et al.* (1997).

Our second example comes from studies of plant population dynamics. Because higher plants are rooted to one spot, the classical Lotka–Volterra models of ecological competition provide only a limited understanding of how diverse plant communities may persist (Grubb 1977). There have been a number of different approaches to developing a more realistic theory of plant competition (e.g. Tilman 1994; Pacala *et al.* 1996). Here, we describe an approach that involves building spatially-explicit models parameterized from field data. These models are complex and can only be solved numerically, but then simpler 'models of models' can be constructed to understand the underlying processes that determine their results.

Consider a community of annual plants on a sand dune or similar habitat. What maintains their diversity in such a simple habitat? Obviously one explanation is niche differentiation, but other processes such as a competition–coexistence trade-off might also be involved. The system can be studied by modelling how the growth of spatially localized plant individuals is influenced by the species composition of other plants growing in the immediate neighbourhood, and then by modelling how growth translates into changes in population growth rate. Using this approach on a large spatially structured dataset, Rees *et al.* (1996) were able to show that density-dependent processes within different species' populations lead to stable dynamics and that interactions between species were weak. They attribute these weak interspecific interactions to niche partitioning, spatial segregation and the rarity of competitive dominants (Rees *et al.* 1996). Detailed experimental studies, which manipulate the local species composition using seed sowing experiments, support this interpretation (Turnbull *et al.* 1999), as do the results of recent modelling studies (Levine & Rees 2002).

This detailed mechanistic approach has been taken by Pacala *et al.* 1996) to study succession and coexistence in temperate mixed forests. Here, the statistical parameterization of the model involved studying the growth and survival of trees in different neighbourhoods. The purpose of the model is to extrapolate from measurable fine-scale and short-term interactions among individual trees to large-scale and long-term dynamics of forest communities. The model demonstrates the importance of initial abundances, which continue to affect community composition well into succession (for more than 300 years for some species), and provides a remarkably accurate description of successional dynamics in temperate mixed forests.

We finish this section by pointing to a few other current trends in population modelling that we think are also narrowing the gap between theory and experimental work. The explosion of interest in behavioural ecology in the 1970s and 1980s prompted the question of whether an explicit consideration of how natural selection acted on animal behaviour might help the parameterization of population models. The issue here is not whether the demographic parameters used by population biologists are influenced by natural selection – of course they are – but whether evolutionary ecology can give insights into how they might respond to changing conditions in a way that would be difficult to go out and measure in the field. Too often, a behavioural ecological study is glossed by the comment that it will aid understanding population dynamics when in fact the variation in fecundity or survival under discussion is so small it will be swamped by the typical variation in the parameter that is observed in the field (we do not exempt ourselves from this criticism!). This is especially true for those animals that experience large fluctuations in population densities.

Recent studies of bird populations have, however, shown how behavioural ecology can be a valuable tool for population dynamicists. A good example comes from habitat use in shore birds such as oystercatchers (Stillman *et al.* 2000, 2001; Durell *et al.* 2001) and godwits (Chapter 12). Here, the problem is to estimate quantities such as winter survival and condition at the beginning of the breeding season from how birds distribute themselves across estuaries and other habitats that consist of mosaics of patches that differ in quality for the birds. Behavioural ecological theory predicts that in the simplest case birds should distribute themselves according to the IFD (Smith & Fretwell 1974). At the IFD, all birds have equal fitness and no bird can improve its fitness by changing its location; in practice this means that good patches have many birds feeding together, and poorer patches fewer competitors. The theory can be made more sophisticated by assuming heterogeneity in competitive ability, both through exploitation and interference competition, and by assuming that food depletion by birds has a negative feedback on habitat quality. Given a knowledge of the availability of potential habitats and the density of different types of bird, IFD theory can be used to predict how birds distribute themselves across the habitat and how their food intake influences their risk of mortality and their ultimate condition at the start of the breeding season. Behavioural ecological arguments have also proved valuable in understanding the dynamics of territorial birds and mammals in which optimum territory size, or the existence of territories at all, is influenced by population density (e.g. Lewis & Murray 1993).

A second recent trend is the increasing attention being paid to stochastic effects in population dynamics. Including stochastic effects in population models is of course not new and some of the most important results are quite old (Bartlett 1957; May 1974), but the last few years have seen a growth in interest in how demographic and parametric stochasticity can combine to influence the fate of populations. The need for better models in conservation biology has partly been responsible for this, while interestingly there has been a fruitful application of techniques from population genetics to ecological problems (Lande 1993, 1998; Lande *et al.* 1998). Many field workers are distrustful of deterministic models that predict rigid, stable equilibria – 'figments of modellers' imagination' as one speaker at the meeting described them – and while it is true that strict deterministic equilibria are hardly if ever observed in the wild, the stochastic analogue of these equilibria are much more realistic. Turchin (2001) describes an equilibrium as a probability distribution of densities towards which a population tends to return if disturbed, and this idea implicitly underlies many of the models of Australian vertebrate populations described in this book by Bayliss & Choquenot (Chapter 9) and Davis *et al.* (Chapter 10). The latter study in particular explored how observed variation in rainfall might affect mammal densities in two or three trophic level systems, while Bayliss & Choquenot (Chapter 9) and Sinclair & Krebs (Chapter 8) argued that many mammal populations exhibited alternative stochastic equilibria. But many organisms do exhibit dynamics that never approach an equilibrium (a number that is probably increasing as global change accelerates) and these must be studied using non-equilibrium approaches.

Finally, we mention briefly the huge growth in spatial modelling that has occurred in the last decade. Partly fuelled by the enormous expansion in desktop computing power, a large array of new techniques have been developed to study spatiotemporal population processes (excellently summarized by Dieckmann *et al.* (2000)). At present, theory far outstrips data, as is natural in a new field (though see Lande *et al.* (1999) for an example of a complex model that gives simple testable predictions). Perhaps ironically, the greatest progress in parameterizing spatial models has come in plant ecology (Pacala & Silander 1990; Cain *et al.* 1995; Pacala *et al.* 1996). Early ecological theory was seldom applied to plants because of their sedentariness, but this same immobility both makes spatial effects more important and makes it easier to parameterize spatially explicit models. In animal ecology, the most important body of spatial theory to emerge concerns metapopulations, the dynamics of ensembles of populations loosely coupled by dispersal (Hanski 1999). A number of metapopulations are now emerging as model systems, in particular the butterfly *Melitaea cinxia* in the Åland islands of Finland. Using a combination of modelling and experimentation, Hanski's group has shown, among other things, the existence of threshold conditions for the persistence of the metapopulation, as well as how inbreeding depression may be involved in population extinction (reviewed in Hanski 1999).

(c)   Observation and experimentation

Since the days of Charles Elton's analysis of cycles in the numbers of lynxes trapped for their fur, population ecologists have studied time-series. Their two main goals have been the discovery of cycles and more complex dynamic behaviour, and the identification and measurement of negative feedback on population growth (density dependence). Traditional time-series techniques such as spectral analysis and autocorrelation are valuable for identifying significant cycles and their periodicity, but are limited in the extent to which they can be used to detect chaotic dynamics. Schaffer (1985) spearheaded a move to use model-free techniques to reconstruct dynamic attractors from time-series data, and then to look directly for the signature of chaos in the shape of the attractor, although it is now clear that these methods require too much data for most biological applications (Hastings *et al.* 1993). The first attempts to look for data in time-series involved fitting simple population models to data and then examining the model's dynamics (Hassell *et al.* 1976). Today, a direct descendent of this approach is most often used to search for chaos, but now a very flexible model in time-lagged population densities is fitted to population growth rates using response surface, splines or neural net techniques (Ellner & Turchin 1995).

Population ecologists in the 1950s debated vigorously whether populations were regulated by density-dependent or density-independent processes, and although the debate has subsided (and much of the disagreement was largely due to semantic confusion), there is still considerable interest in trying to measure the effects of direct and delayed density dependence on animal and plant populations. A series of statistical techniques of increasing complexity have been developed, although no single method is superior in all situations. Detecting delayed density dependence is particularly difficult, and Lande *et al.* (Chapter 4) show how in age-structured populations it is important to disentangle how delayed density dependence operates through the organism's life history to determine current population growth rates.

Krebs (Chapter 7) characterizes this approach to studying populations as the density paradigm which he contrasts with the mechanistic paradigm which seeks to identify the effects influencing population densities. The mechanistic approach dates back to key-factor analysis pioneered by ecological entomologists (Varley *et al.* 1973). The relationship between the two approaches is complex, but frequently the former is used in preliminary analysis when no other data are available, and if successful, for example in identifying significant population feedbacks, it may be very valuable in determining on which potential mechanisms to concentrate.

Probably all population ecologists would follow the mechanistic paradigm if they could, but one of the problems has been to integrate potential explanatory variables with the organism's life history into analyses of time-series. Traditional techniques have problems incorporating both nonlinear terms and complex error structures. Recently, a series of new techniques, which incorporate mechanistic details of the organism's population biology, have been introduced to population ecology (e.g. Ellner *et al.* 1997, 1998; Turchin & Hanski 1997; Bjørnstad *et al.* 1998, 2001; Rees *et al.* 2002). These techniques allow stronger inferences to be made because they incorporate more information, and often outperform classical time-series methods. This approach when combined with recent developments in the statistical analysis of partially specified models (Wood 2001), where constraints on the properties of the functions fitted are imposed rather than assuming *a priori* functional forms, holds considerable promise. However, in the presence of temporally correlated (coloured) noise there may be real limits to the degree to which population mechanisms can be diagnosed from time-series data (Lundberg *et al.* 2000; Ranta *et al.* 2000; Kaitala & Ranta 2001; Jonzén *et al.* 2002). A different approach to introducing mechanism more explicitly into theoretical population dynamics is to produce models based on explicitly physiological processes. These models may be quite complex but offer the possibility of introducing new insights into such issues as how body size and other parameters influence population dynamics (Nisbet *et al.* 2000).

One goal of observational studies of time-series and other population data is to discover cross-species patterns in population processes. Silvertown *et al.* (1996) were one of the first groups to do this, comparing woody and herbaceous perennials and which elements of a population projection matrix had the greatest effect on population growth rates. Sæther and colleagues (Sæther & Bakke 2000; Chapter 4) have used bird demographic data to show how fecundity and survival influence population growth rate and how these effects are correlated with other aspects of the bird's life history. In analyses of this kind it is important to consider statistical non-independence among species arising through phylogeny, and to control for this using modern comparative methods (Harvey & Pagel 1991). Population dynamics has always been important in life-history theory; for example, the theory of r and K selection, in its original sense, is a hypothesis about how a population dynamic constraint will influence life-history theory. But whereas large datasets on life-history characters are relatively easy to construct, data on population dynamics, with the possible exception of mean population size, are much harder to collect. Charnov (1993) was able to identify a series of life-history invariants – dimensionless combinations of life-history parameters – in a study in which he essentially assumed populations were perfectly regulated. As more comparative population data become available, it will be interesting to see if deviations from the patterns identified by Charnov are correlated with patterns in the dynamics.

While observational studies are invaluable in generating hypotheses about population processes and may be the only means of investigating processes that involve evolutionary time-scales, experimental studies, particularly in the field, are still the gold standard for testing specific hypotheses about population processes. For example, red grouse populations show cyclic behaviour on British moorlands. Because they are an economically important game bird, long time-series of bag sizes are available and there is considerable information about their biology. As Hudson *et al.* (Chapter 11) describe in more detail, there are two hypotheses to explain the cycles, (i) that it is due to the interaction between the grouse and its main parasite, a nematode worm, and (ii) that it is due to intrinsic variation in grouse behaviour and quality over the course of the cycle. To test the first hypothesis, Hudson *et al.* (1998) carried out a large experiment in which they applied anti-helminthic drugs to four populations of birds just before an expected crash in population densities. Treated populations either did not decline in density, or the drop was much less than controls, strongly implicating parasites as a cause of the population cycle. But other field experiments also indicate that intrinsic factors may be involved. It is possible to manipulate grouse behaviour using testosterone implants and this influences territory size and survival (Moss *et al.* 1994). These are individual-level rather than the population-level experiments, but indicate how field experiments may be used to distinguish between different hypotheses (Fox & Hudson 2001).

This example illustrates the value of large-scale experimental studies but also the huge investment in time and resources needed to carry them out properly, and these are by no means the largest field experiments in population ecology. Tilman (1993, 1994, 1996, 1997) has set up a large series of experiments at Cedar Creek in Minnesota designed to test a wide range of hypotheses about the mechanisms of plant coexistence and the relationship between community and ecosystem processes. These experiments involve manipulating and maintaining community composition in thousands of experimental plots, and the results obtained have become clearer and easier to interpret the longer the experiment has run, illustrating the value of long-term studies. The Kluane experiment in northern Canada designed by Krebs *et al.* (2001c) as a factorial experiment to test how predation and food supply influence snowshoe hare dynamics was so ambitious that the relevant research council had to invent new procedures to deal with it.

Population ecology is by no means unique in requiring large experiments, and in fact our most grandiose schemes are paltry compared with the costs of a new particle accelerator. But the particular problem faced by ecologists is the diffuseness of the subject and the difficulty the community has in agreeing on model systems that can attract cross-community support for major investment. There is a trade-off between concentrating large amounts of resources to get a more accurate answer to one particular question, and dispersing resources more widely to get more approximate answers to a wider range of questions. Deciding the optimum resource allocation is difficult without knowing how the results from one system generalize to all systems, an issue about which there is little consensus in the field. A related question is the degree to which systems of applied relevance – invasive species, fisheries and timber trees – for which funding for long-term research is easier to obtain, give us answers to general problems in the subject. Where ecology does differ from most other experimental sciences is in its need for long-term experiments, their duration often set by the lifespan of the study organism that may be of the same order of magnitude as the lifespan of the experimenter. Thus, a further argument for establishing model systems is to break the tie between the length of a study and the length of an ecological career.

An issue on which it is perhaps easier to agree is the need to improve accessibility to long-term population data, both observational and experimental. Some schemes do already exist, for example the GPDD (http://cpbnts1.bio.ic.ac.uk/gpdd/) administered by the NERC Centre for Population Biology in the UK which contains 5000 time-series of greater than 10 generations in length. Except where the organizations providing the data imposed conditions, these data are freely available to the community for study. However, the data in the GPDD are relatively simple, in the sense that they consist of raw time-series data with little annotation, with none of the ancillary material generated by detailed field studies and experimentation. Creating a true data repository of the types now available in molecular biology would require the development of means of making data portable, and accessible across different platforms. Recent advances in metadata formats that are beginning to be applied to ecological data may help to provide a solution to this (http://knb.ecoinformatics.org/). A further issue is the degree to which individual workers should be asked to make data available to the community. Much population data are so hard to obtain that it seems only fair that the investigator has a 'fair crack of the whip' before it is made generally available, but there is need for the community to agree on equitable rules on data sharing.

Our discussion of experiments has focused on field studies, but what is the role of laboratory experiments in studies of population, of growth rate and population regulation? One view is that they are too divorced from reality to say anything cogent about processes operating in the field, an opinion perhaps most often put forward by people working on larger organisms whose size and complex behaviour make them poor laboratory models. But we tend to agree more with Kareiva (1989) who argued that 'ecologists gave up bottle experiments too soon'. We mentioned above a series of experiments using insects pests of stored products and their natural enemies, and we argue that these provided valuable insight into how age-structured interactions may lead to different types of population cycling. Using a similar system, Bonsall & Hassell (1997) were able to show how one species may drive another to extinction when their only contact is a shared natural enemy, something long predicted by theory but which has proved difficult to study in the field. Much more complex laboratory communities can be assembled using micro-organisms. For example, using communities of protists, Kaunzinger & Morin (1998) found that more enriched habitats can support longer food chains, as predicted by classical theory. However, both higher enrichment and longer food chains are thought to be destabilizing, indicating dynamic constraints on food chain length. Protist microcosm experiments also first demonstrated the destabilizing effects of enrichment (Luckinbill 1973) and increasing food chain length (Lawler & Morin 1993). Theory predicts that metapopulation dynamics might stabilize otherwise-unstable predator–prey interactions, a prediction first confirmed in replicated experiments by Holyoak & Lawler (1996). At their worst, laboratory experiments are so artificial that they are merely analogue computers solving the problem set by the investigator, but at their best they can test between novel hypotheses and aid both further theory and the design of experiments on more natural systems in the field.

Finally, we ask to what degree individual species or groups of closely interacting species can be treated in isolation. Do the majority of species in natural systems interact with so many other species that considering them in isolation is doomed to failure? Does this underlie some of the problems with the repeatability of studies highlighted by Krebs (Chapter 7)? These questions are easier to pose than to answer, or to think about how experiments could be designed to address them. However, this is an area where plant ecologists are probably leading the way in developing a quantitative community ecology. We have already discussed how spatial models of plant communities have been constructed and parameterized, and then used to investigate questions of community succession and coexistence. Another large plant community project is the network of 50 ha plots set up by the Centre for Tropical Forest Science at the Smithsonian Tropical Research Institute in which all trees above 1 cm diameter are censused at regular intervals. The original 50 ha plot was set up on Barro Colorado Island in Panama, but there is now a network of six or so large plots and others that are slightly smaller in all the major regions of tropical forest of the world (http://www.ctfs.si.edu/). This project in population ecology perhaps comes nearest to the mega-projects of particle physicists and astronomers, and has already been used to explore global patterns in rainforest diversity, the role of Janzen-Connell effects in maintaining diversity, and to study the comparative demography of trees (Condit *et al.* 2000). It has also stimulated radical new theory (Hubbell 2001). Though not experimental, we suspect that such projects may be our best hope of understanding the dynamics of complex, long-lived communities.

15.3. Integral projection models

In this second part of the paper we switch from a general discussion of issues in population dynamics to a much more specific application of population growth rate to a problem in plant population and evolutionary ecology. This application does, though, illustrate some of the themes of the preceding section: the value of population growth rate and population projection, the constant flow of new techniques into the field, and the intimate link between population dynamics and evolution.

Population projection has been widely used in plant ecology and has largely been based upon matrix models in which individuals are either classified by age or size (Caswell 2001). This works well when the classes are discrete and easily recognized, but can lead to problems when what is essentially a continuous distribution in a variable such as size is cut up into discrete stages. Algorithms exist to help choose the optimal partition of the size distribution (e.g. Vandermeer 1978), but these cannot eliminate the essentially arbitrary nature of the division, and the problem that the number of classes chosen may influence the apparent relative importance of different biological processes (Enright *et al.* 1995).

To help solve these problems, Easterling *et al.* (2000) introduced integral projection models that retain many of the analytical advantages of traditional matrix models but which avoid the need to specify arbitrary distinct stage classes. As an example, they apply the technique to a herbaceous perennial, *Aconitum noveboracense* (northern monkshood), which reproduces by a mixture of sexual and vegetative means, calculating population growth rate and the elasticities and sensitivities to different size classes. Here, we briefly describe integral projection models and how they might be applied to a plant with a different life history, the monocarpic large-flowered evening primrose *Oenothera glazioviana*, and used to answer questions in evolutionary as well as population ecology. For further details of integral projection techniques see Easterling *et al.* (2000) and for its application to *Oenothera* see Rees & Rose (2002).

(a)   The technique

In matrix projection, a vector n_t representing the numbers or densities of individuals in different stage classes at time t is acted upon by a transition matrix K to predict the numbers in the different stage classes at the next time-step, n_{t+1}. In symbols

$$n_{t+1} = K n_t. \tag{15.3.1}$$

The transition matrix contains terms representing stage-specific fecundity and survival. The structure of an integral projection model is very similar except now the vector is replaced by a function representing the numbers or densities of individuals in stage y at time t, $n(y, t)$. For narrative simplicity we will henceforth view y as plant size. As with matrix models, time is treated as discrete and we can calculate the distribution of individuals in different size classes at the next census point using

$$n(y, t+1) = \int k(y, x) n(x, t) \, dx. \tag{15.3.2}$$

Here $k(y, x)$, the kernel function, describes the number of transitions between an individual in size class x at time t and size class y at time $t + 1$. A transition might occur through growth or through sexual or vegetation reproduction. Biologically, it is useful to distinguish between these two processes,

$$k(y, x) = p(y, x) + f(y, x), \tag{15.3.3}$$

where $p(y, x)$ is the survival–growth function, the probability that an individual size x survives and grows (or conceivably shrinks) to size y (a plant that remained the same size would contribute to a self-transition or $p(x, x)$ term); and $f(y, x)$ is the fecundity function, the number of offspring that are sized y at the next census that are produced by a plant currently of size x.

For the matrix case (equation (15.3.1)), classical mathematical results prove that if K is a time-invariant matrix, the population will come to grow (or decline) at a rate λ that is independent of the initial values of the population vector (Caswell 2001). The population growth rate, λ, is the dominant eigenvalue associated with K, while the right and left eigenvectors represent respectively the stable stage distribution and the vector of the Fisherian reproductive values associated with each stage class. Exactly equivalent quantities can be calculated for the integral projection matrix; λ as before is a simple scalar while the eigenvectors are replaced by functions that describe the distribution of individuals of different size $(w(y))$ and how reproductive value changes with size $(v(y))$. Much of the complex machinery described by Caswell (2001) for matrix models translates to the integral projection case, but here we highlight just one example, elasticity. As mentioned earlier, the elasticity $(e(y,x))$ is the proportional change in population growth rate consequent on a proportional change in one element (y, x) of the transition matrix or, in this case, of the kernel

$$e(y, x) = \frac{\partial \log \lambda}{\partial \log k(y, z)} = \frac{k(y, z)}{\lambda} \frac{\partial \lambda}{\partial k(y, z)} = \frac{k(y, z)}{\lambda} \frac{v(y)w(x)}{\langle v, w \rangle}$$

(15.3.4)

where $\langle v, w \rangle = \int v(x)w(x)\,dx$ is a normalizing constant. The elasticity can be understood intuitively as proportional to the product of the size of the size class *from* which the transition occurs $(w(x))$ and the importance of the size class *to* which the transition occurs $(v(y))$.

(b)   The dynamics of *Oenothera glazioviana*

The kernel can be decomposed further beyond $k(y, x) = p(y, x) + f(y, x)$ as the biology of the system dictates. With *O. glazioviana* we have extensive knowledge of its population ecology thanks to the work of Kachi (1983) and Kachi & Hirose (1985). In their sand dune study site, *O. glazioviana* exists as rosettes of varying size which survive from year to year (t is measured in years) until they flower which is always followed by the plant's death. Based on their work, fecundity can be expressed as a product of three functions

$$f(y, x) = p_f(x)f_n(x)f_d(y, x),$$

(15.3.5)

each of which can be statistically estimated from data (Kachi & Hirose 1985). $p_f(x)$ is the probability that a plant of size x flowers in the current year and was estimated as a logistic function of x. $f_n(x)$ is the fecundity of plants of different size and it was found that the logarithm of seed number was linear in x. Finally, $f_d(y, x)$ is the size distribution of recruits. Data were not available on the size of recruits derived from seeds of plants of different size, but evidence from other systems indicates a low maternal effect on recruit size (Weiner *et al.* 1997), and this function was estimated as normal on a logarithmic scale, independent of x.

The growth and survival function was also treated as the product of three functions

$$p(y, x) = (1 - p_f(x))s(x)g(y, x). \qquad (15.3.6)$$

As flowering always results in the plant's death, growth and survival are contingent on *not* flowering, which is given by the first term. The second term, $s(x)$, is the size-specific probability of flowering which was well estimated by a simple linear term. Finally $g(y, x)$ is the probability of growing from size x to size y which was estimated as a normal distribution with fixed variance and the mean a linear function of rosette size.

Putting all this together we can estimate the shape of the kernel, $k(y, x)$ (figure 15.1) and based on this we can calculate $\lambda = 1.041$. Unfortunately, we do not have field estimates of λ with which to make a direct test, but we can compare this result and the predicted stable size class distribution with the output of a detailed IBM described in Kachi & Hirose (1985). The IBM predicted $\lambda = 1.04$ and the stable stage distribution was accurately described by the integral projection model (figure 15.2). We conclude that the integral projection model captures the same detail of the biology as the IBM, but in much more concise and analysable form.

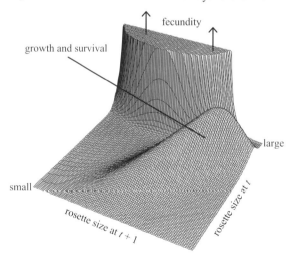

Figure 15.1. The shape of the kernel function $k(y, x)$ for *Oenothera glazioviana*: the number of transitions between an individual in size class x at time t and size class y at time $t + 1$.

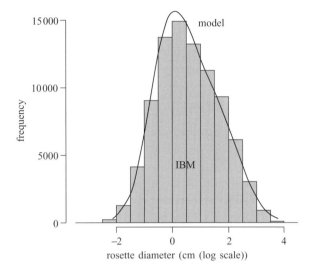

Figure 15.2. A comparison between the stable size distribution predicted by the integral projection model (curve) and a detailed IBM developed by Kachi & Hirose (1985).

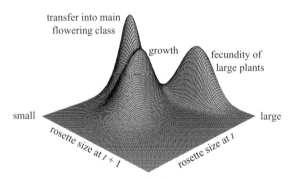

Figure 15.3. The elasticity of population growth rate to variation in the transition kernel. The biological interpretation of the different peaks is indicated.

Finally, in this section, we calculate the elasticity surface (figure 15.3). As indicated in the figure, the most important determinants of population growth rate correspond to, in decreasing order, the fecundity of large plants, the growth of plants into the chief flowering category and the growth of younger plants.

(c)   The evolution of size at flowering

One of the most critical life-history decisions of a monocarpic plant such as *O. glazioviana* is the size at which to flower. We can extend the integral project model to analyse this question, but first we have to be explicit about exactly where density dependence operates. The available evidence indicates that density dependence acts purely on seedling establishment with no later intraspecific competition (Kachi & Hirose 1985; Rees *et al.* 2000), in which case evolution maximizes R_0, total lifetime reproduction (Mylius & Diekmann 1995). If the population was at demographic equilibrium with flowering strategy $p_f(x)$ we could then explore whether any alternative flowering strategy, $\hat{p}_f(x)$, was able to invade. In fact the population is slightly increasing ($\lambda = 1.041$) and so we reduced the probability of seedling establishment slightly so that $\lambda = 1$ for the resident strategy.

We modelled the probability of flowering as a logistic function of plant size

$$\text{logit}\, p_f(x) = a + bx, \tag{15.3.7}$$

which indicates that flowering size is determined by the two parameters a and b. This formulation, while statistically well-grounded, is biologically rather difficult to interpret as the mean and variance of the size at flowering are both functions of a and b. In Appendix A we show how these can be disentangled and in figure 15.4 show the fitness (population growth rate) of flowering strategies that differ from the resident in mean size and variance in the size at flowering. The results show clearly that the current mean size at flowering is an ESS, though it does indicate that natural selection should act to reduce the variance in size at flowering. This selection for a threshold size at flowering is expected from general models of evolution in a constant environment (Sasaki & Ellner 1995).

There are at least two possible explanations for the variance in flowering time not being an ESS. First, the variance may be a constraint, either genetic or phenotypic, that prevents the plant from evolving more of a threshold response. However, studies of the monocarpic species, *Cynoglossum officinale*, have demonstrated sharp threshold flowering strategies are possible (Wesselingh *et al.* 1993). A similar analysis of another monocarpic plant, *Carlina vulgaris* suggests an alternative explanation (Rose *et al.* 2002). In this study an IBM model was used to study the invasion of alternative flowering strategies. When the equivalent of a time-invariant matrix or kernel was assumed, the observed mean size at flowering was larger than the predicted ESS and no variance was predicted. However, in that study,

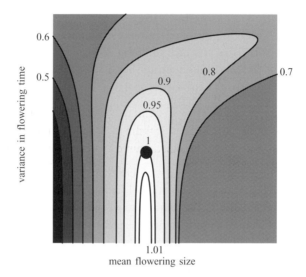

Figure 15.4. The fitness of alternative flowering strategies when invading a population with the observed phenotype in the field (shown by a black disk). As explained further in the text, the flowering strategy can be characterized by the mean size of flowering (on the *x*-axis) and the variance around this mean. No strategy with a different mean can invade, but the analysis indicates that strategies with lower variance have higher fitness.

data were also available on the temporal variance in life-history parameters and when these were included in the IBM non-zero variance was predicted (Rose *et al.* 2002); in a variable environment, some degree of spreading of risk is selected (Sasaki & Ellner 1995). We conjecture that were data available on temporal variance in the parameters of *Oenothera* life history, some degree of variance in size at flowering would be selected. Extending integral projection techniques to encompass temporal variance in parameters, as has been done for projection matrices (Caswell 2001), would be a useful future step.

We are grateful to many people at the two meetings for valuable discussion including Peter Bayliss, Tim Benton, Mike Bonsall, Peter Calow, David Choquenot, Tim Clutton-Brock, Tim Coulson, Stephen Davis, Patrick Doncaster, Valery Forbes, Jeremy Fox, Alastair Grant, John Harwood, Jim Hone, Peter Hudson, Charles Krebs, Xavier Lambin, Russ Lande, Ken Norris, Bernt-Erik Sæther, Richard Sibly, Tony Sinclair, Rob Smith, Nils Chr. Stenseth, Bill Sutherland, Michelle Walter and Ian Woiwod.

Appendix A

In this appendix we show how the fitted parameters of a logistic regression can be interpreted in terms of an underlying, unobserved distribution of threshold sizes for reproduction. Assume there is some distribution of threshold sizes for flowering within a population; then for a particular size, x, if 50% of plants flower, this implies that 50% of the population have a threshold less than x (Wesselingh & de Jong 1995). In this way we may interpret the fitted logistic curve as a cumulative distribution function; hence, differentiating equation (15.3.7) gives the probability density function of threshold sizes for reproduction within a population. Calculating $\partial p_f(x) / \partial x$ we obtain

$$f(x) = \frac{b \exp(a + bx)}{(1 + \exp(a + bx))^2},\qquad(15.A.1)$$

which defines a general logistic distribution with scale parameter $1/b$ and location parameter $-a/b$. This distribution has mean and variance given by

$$\mu = -a / b,\qquad(15.A.2)$$

and

$$\sigma^2 = \frac{\pi^2}{3b^2}.\qquad(15.A.3)$$

In this way, we may express the distribution of threshold sizes within a population in terms of the estimated parameters of the logistic flowering function (equation (15.3.7)). There are problems applying this approach to real systems, as there is growth between the time flowering decisions are made and flowering occurs. However, the approach is still useful for interpreting model parameters.

LIBRARY, UNIVERSITY COLLEGE CHESTER

References

Aanes, S., Engen, S., Sæther, B.-E., Willebrand, T. & Marcstrom, V. (2002). Sustainable harvesting strategies of willow ptarmigan in a fluctuating environment. *Ecological Applications*, **12**, 281–290.

Abrams, P. A. & Ginzburg, L. R. (2000). The nature of predation: prey dependent, ratio dependent or neither? *Trends in Ecology and Evolution*, **15**, 337–341.

Adams, E. S. & Tschinkel, W. R. (2001). Mechanisms of population regulation in the fire ant *Solenopsis invicta*: an experimental study. *Journal of Animal Ecology*, **70**, 355–369.

Adelman, I. (1963). An econometric analysis of population growth. *American Economic Research*, **53**, 314–339.

Albon, S. D. & Clutton-Brock, T. H. (1988). Climate and the population dynamics of red deer in Scotland. In *Ecological Change in the Uplands*, ed. M. B. Usher and D. B. A. Thompson, pp. 93–117. Oxford: Blackwells.

Albon, S. D., Clutton-Brock, T. H. & Guinness, F. E. (1987). Early development and population dynamics in red deer. II. Density-independent effects and cohort variation. *Journal of Animal Ecology*, **56**, 69–81.

Albon, S. D., Coulson, T. N., Brown, D., Guinness, F. E., Pemberton, J. M. & Clutton-Brock, T. H. (2000). Temporal changes in key factors and key age groups influencing the population dynamics of female red deer. *Journal of Animal Ecology*, **69**, 1099–1110.

Albon, S. D., Guinness, F. E. & Clutton-Brock, T. H. (1983a). The influence of climatic variation on the birth weights of red deer, *Cervus elaphus*. *Journal of Zoology, London*, **200**, 295–298.

Albon, S. D., Mitchell, B., Huby, B. J. & Brown, D. (1986). Fertility in female red deer (*Cervus elaphus*): the effects of body composition, age and reproductive status. *Journal of Zoology, London*, **209**, 447–460.

Albon, S. D., Mitchell, B. & Staines, B. W. (1983b). Fertility and body weight in female red deer: a density dependent relationship. *Journal of Animal Ecology*, **52**, 969–980.

Allden, W. G. (1962). Rate of herbage intake and grazing time in relation to herbage availability. *Proceedings of the Australian Society for Animal Production*, **4**, 163–166.

Allee, W. C. (1931). *Animal Aggregations*, Chicago: University of Chicago Press.

Allee, W. C. (1938). *The Social Life of Animals*, New York: W. W. Norton & Co.

Allee, W. C. (1941). Integration of problems concerning protozoan populations with those of general biology. *American Naturalist*, **75**, 473–487.

Anderson, R. M., Donnelly, C. A., Ferguson, N. M., Woolhouse, M. E. J., Watt, C. J., Udy, H. J., MaWhinney, S., Duncan, S. P., Southwood, T. R. E., Wilesmith, J. W., Ryan, J. B. M., Hoinville, L. J., Hillerton, J. E., Austin, A. R. & Wells, G. A. H. (1996). Transmission dynamics and epidemiology of BSE in British cattle. *Nature*, **382**, 779–788.

Anderson, R. M. & May, R. M. (1978). Regulation and stability of host-parasite interactions. I. Regulatory processes. *Journal of Animal Ecology*, **47**, 219–249.

Anderson, R. M. & May, R. M. (1979). Population biology of infectious diseases. Part 1. *Nature*, **280**, 361–367.

Anderson, R. M. & May, R. M. (1991). *Infectious Disease of Humans: Dynamics and Control*, Oxford: Oxford University Press.

Andrewartha, H. G. & Birch, L. C. (1954). *The Distribution and Abundance of Animals*, Chicago: University of Chicago Press.

Arcese, P. & Smith, J. N. M. (1988). Effects of population density and supplemental food on reproduction in song sparrows. *Journal of Animal Ecology*, **57**, 119–136.

Arnold, W. (1990a). The evolution of marmot sociality. I Why disperse late? *Behavioural Ecology and Sociobiology*, **27**, 229–237.

Arnold, W. (1990b). The evolution of marmot sociality. II Costs and benefits of joint hibernation. *Behavioural Ecology and Sociobiology*, **27**, 239–246.

Arnqvist, G. & Wooster, D. (1995). Meta-analysis: synthesizing research findings in ecology and evolution. *Trends in Ecology and Evolution*, **10**, 236–240.

Baber, D. W. & Coblentz, B. E. (1986). Density, home range, habitat use and reproduction in feral pigs on Santa Catalina Island. *Journal of Mammalogy,* **67**, 512–525.

Bacon, P. J. & Andersen-Harild, P. (1989). Mute Swan. In *Lifetime Reproduction in Birds*, ed. I. Newton, pp. 363–386. New York: Academic Press.

Bacon, P. J. & Perrins, C. M. (1991). Long term population studies: the mute swan. *Acta XX Congressus Internationalis Ornithologici*, **20**, 1500–1513.

Bairlein, F. & Zink, G. (1979). Der bestand des weissstorchs *Ciconia ciconia* in Sudwestdeutschland: eine analyse der bestandsentwicklung. *Journal of Ornithology*, **120**, 1–11.

Barlow, J. (1992). Nonlinear and logistic growth in experimental populations of guppies. *Ecology*, **73**, 941–950.

Barlow, N. D. (1985). The interferential model re-examined. *Oecologia (Berlin)*, **66**, 307–308.

Barlow, N. D. (1991). Control of endemic bovine TB in New Zealand possum populations: results from a simple model. *Journal of Applied Ecology*, **28**, 794–809.

Barlow, N. D. & Clout, M. N. (1983). A comparison of 3-parameter, single-species population models, in relation to the management of brushtail possums in New Zealand. *Oecologia (Berlin)*, **60**, 250–258.

Barlow, N. D. & Kean, J. M. (1998). Simple models for the impact of rabbit calicivirus disease (RCD) on Australasian rabbits. *Ecological Modelling*, **109**, 225–241.

Barlow, N. D., Kean, J. M. & Briggs, C. J. (1997). Modelling the relative efficacy of culling and sterilisation for controlling populations. *Wildlife Research*, **24**, 129–141.

Barlow, N. D. & Norbury, G. L. (2001). A simple model for ferret population dynamics and control in semi-arid New Zealand habitats. *Wildlife Research*, **28**, 87–94.

Bartlett, M. S. (1957). Measles periodicity and community size. *Journal of the Royal Statistical Society A*, **123**, 48–70.

Bayliss, P. (1980). Kangaroos, plants and weather in the semi-arid. M.Sc. thesis, University of Sydney.

Bayliss, P. (1985a). The population dynamics of red and western grey kangaroos in arid New South Wales, Australia. II. The numerical response function. *Journal of Animal Ecology*, **54**, 127–135.

Bayliss, P. (1985b). The population dynamics of red and western grey kangaroos in arid New South Wales, Australia. I. Population trends and rainfall. *Journal of Animal Ecology*, **54**, 111–125.

Bayliss, P. (1987). Kangaroo dynamics. In *Kangaroos. Their Ecology and Management in the Sheep Rangelands of Australia*, ed. G. Caughley, N. Shepherd & J. Short, pp. 119–134. Cambridge: Cambridge University Press.

Bayliss, P. (1989). Population dynamics of magpie geese in relation to rainfall and density: implications for harvest models in a fluctuating environment. *Journal of Applied Ecology*, **26**, 913–924.

Bayliss, P. & Choquenot, D. (1998). Contribution of modelling to definition and implementation of management goals for overabundant marsupials. Proceedings of a Symposium held at the Society for Conservation Biology Conference, Sydney, Australia, July 1998. *Issues in Marsupial Conservation and Management Occasional Papers of the Marsupial CRC* No. 1, 69–75.

Bayliss, P. & Choquenot, D. (2002). The numerical response: rate of increase and food limitation in herbivores and predators. *Philosophical Transactions of the Royal Society London B*, **357**, 1233–1248.

Beaver, S. (1975). *Demographic Transition Theory Revisited*, Lexington: Lexington Books.

Beddington, J. R. & May, R. M. (1977). Harvesting natural populations in a randomly fluctuating environment. *Science*, **197**, 463–465.

Begon, M., Bowers, R. G., Sait, S. M. & Thompson, D. J. (1996a). Population dynamics beyond two species: hosts, parasitoids and pathogens. In *Frontiers of Population Ecology*, ed. R. B. Floyd, A. W. Sheppard & P. J. DeBarro, pp. 115–126. CSIRO: Melbourne.

Begon, M., Harper, J. H. & Townsend, C. R. (1996c). *Ecology. Individuals, Populations & Communities*, 3rd edn. Malden: Blackwell Science.

Begon, M., Sait, S. M. & Thompson, D. J. (1995). Persistence of a parasitoid-host system: refuges and generation cycles? *Proceedings of the Royal Society London B*, **260**, 131–137.

Begon, M., Sait, S. M. & Thompson, D. J. (1996b). Predator–prey cycles with period shifts between two- and three-species systems. *Nature*, **381**, 311–315.

Beissinger, S. R. (1995). Modelling extinction in periodic environments: Everglades water levels and snail kite population viability. *Ecological Applications*, **5**, 618–631.

Beissinger, S. R. & Westphal, M. I. (1998). On the use of demographic models of population viability in endangered species management. *Journal of Wildlife Management*, **62**, 821–841.

Bell, B. D. (1981). Breeding and condition of possums *Trichosurus vulpecula* in the Orongorongo Valley, near Wellington, New Zealand 1966–1975. In *Proceedings of the First Symposium on Marsupials in New Zealand*, No. 74, ed. B. D. Bell, pp. 87–139. Wellington: Victoria University of Wellington, New Zealand.

Belovsky, G. E. (1981). Food plant selection by a generalist herbivore: the moose. *Ecology*, **62**, 1020–1030.

Belovsky, G. E. (1984). Herbivore optimal foraging: a comparative test of three models. *American Naturalist*, **124**, 97–115.

Bennett, P. M. & Owens, I. P. F. (2002). *Evolutionary Ecology of Birds*. Oxford: Oxford University Press.

Benton, T. G., Grant, A. & Clutton-Brock, T. H. (1995). Does environmental stochasticity matter? Analysis of red deer life-histories on Rum. *Evolutionary Ecology*, **9**, 559–574.

Bergerud, A. T. (1980). Review of the population dynamics of caribou in North America. In *Second International Reindeer/Caribou Symposium, Norway, 1979*. ed. E. Reimers, E. Garre & S. Skjenneberg, pp. 556–581. Trondheim: Direktoratet for vilt og ferskvannsfisk.

Berryman, A. A. (1999). *Principles of Population Dynamics and their Application*, Cheltenham: Stanley Thornes.

Berryman, A. A., Michalski, J., Gutierrez, A. P. & Arditi, R. (1995). Logistic theory of food web dynamics. *Ecology*, **76**, 336–343.

Berryman, A. A. & Turchin, P. (1997). Detection of delayed density dependence: comment. *Ecology*, **78**, 318–320.

Birch, L. C. (1953). Experimental background to the study of the distribution and abundance of insects. I. The influence of temperature, moisture, and food on the innate capacity for increase of three grain beetles. *Ecology*, **34**, 698–711.

Birkhead, M. & Perrins, C. (1986). *The Mute Swan*, London: Croom Helm.

Bjørnstad, O. N., Begon, M., Stenseth, N. C., Falck, W., Sait, S. M. & Thompson, D. J. (1998). Population dynamics of the Indian meal moth: demographic stochasticity and delayed regulatory mechanisms. *Journal of Animal Ecology*, **67**, 110–126.

Bjørnstad, O. N. & Grenfell, B. T. (2001). Noisy clockwork: time series analysis of population fluctuations in animals. *Science*, **293**, 638–643.

Bjørnstad, O. N., Sait, S. M., Stenseth, N. C., Thompson, D. J. & Begon, M. (2001). The impact of specialised enemies on the dimensionality of host dynamics. *Nature*, **409**, 1001–1006.

Blueweiss, L., Fox, H., Kudzma, V., Nakashima, D., Peters, R. & Sams, S. (1978). Relationships between body size and some life history parameters. *Oecologia*, **37**, 257–272.

Blums, P., Bauga, I., Leja, P. & Mednis, A. (1993). Breeding populations of ducks at Engure Lake, Latvia. *Ring*, **15**, 165–169.

Blums, P., Mednis, A., Bauga, I., Nichols, J. D. & Hines, J. E. (1996). Age-specific survival and philopatry in three species of European ducks: a long-term study. *Condor*, **98**, 61–74.

Bongaarts, J. & Potter, R. (1983). *Fertility, Biology, and Behaviour*, New York: Academic Press.

Bonsall, M. B. & Hassell, M. P. (1997). Apparent competition structures ecological assemblages. *Nature*, **338**, 371–373.

Botero, G. (1588). *The Magnificence and Greatness of Cities*, Translated by R. Peterson, 1606. Reprinted 1979. Amsterdam: Theatrum Orbis Terrarum and Norwood, N.J.: Walter J. Johnson.

Both, C. (1998). Experimental evidence for density dependence of reproduction in great tits. *Journal of Animal Ecology*, **67**, 667–674.

Both, C. (2000). Density dependence of avian clutch size in resident and migrant species: is there a constraint on the predictability of competitor density? *Journal of Animal Ecology*, **31**, 412–417.

Boussès, P., Barbanson, B. & Chapuis, J. L. (1991). The Corsican mouflon (*Ovis ammon musimon*) on Kerguelen archipelago: structure and dynamics of the population. In *Proceedings of the International Symposium, Ongules/Ungulates 9*, ed. F. Spitz, G. Janeau, G. Gonzalez & S. Avlagnier, pp. 317–320. Toulouse, France.

Boutin, S. (1992). Predation and moose population dynamics: a critique. *Journal of Wildlife Management*, **56**, 116–127.

Box, G. E. P., Jenkins, G. M. & Reinsel, G. C. (1994). *Time Series Analysis, Forecasting and Control*, 3rd edn. Upper Saddle River, New Jersey: Prentice Hall.

Boyd, I. L. (1981a). Population changes and the distribution of a herd of feral goats (*Capra* sp.) on Rhum, Inner Hebrides, 1960–78. *Journal of Zoology, London*, **193**, 287–304.

Boyd, I. L. (1981b). The Boreray sheep of St. Kilda, Outer Hebrides, Scotland: the natural history of a feral population. *Biological Conservation*, **20**, 215–227.

Boyd, J. M. & Jewell, P. A. (1974). The Soay sheep and their environment: a synthesis. In *Island Survivors: the Ecology of the Soay Sheep of St. Kilda*, ed. P. A. Jewell, C. Milner & J. Morton Boyd, pp. 360–373. London: Athalone Press.

Brand, G. W., Fabris, G. J. & Arnott, G. H. (1985). Reduction of population growth in *Tisbe holothuriae* Humes (Copepoda: Harpacticoida) exposed to low cadmium concentrations. *Australian Journal of Marine & Freshwater Research*, **37**, 475–479.

Briggs, C. J., Sait, S. M., Begon, M., Thompson, D. J. & Godfray, H. C. J. (2000). What causes generation cycles in populations of stored-product moths? *Journal of Animal Ecology*, **69**, 352–366.

Briggs, S. V. & Holmes, J. E. (1988). Bag sizes of waterfowl in New South Wales and their relation to antecedent rainfall. *Australian Wildlife Research*, **15**, 459–468.

Brommer, J. E., Pietiainen, H. & Kolunen, H. (1998). The effect of age at first breeding on Ural owl lifetime reproductive success and fitness under cyclic food conditions. *Journal of Animal Ecology*, **67**, 359–369.

Brook, B. W., Lim, L., Harden, R. & Frankham, R. (1997). Does population viability analysis software predict the behaviour of real populations? A retrospective study of the Lord Howe island woodhen *Tricholimnas sylvestris* (Sclater). *Biological Conservation*, **82**, 119–128.

Broom, D. & Johnson, K. (1993). *Stress and Animal Welfare*, London: Chapman and Hall.

Brown, D. & Alexander, N. (1991). The analysis of the variance and covariance of products. *Biometrics*, **47**, 429–444.

Brown, P. R. & Singleton, G. R. (1999). Rate of increase as a function of rainfall for house mouse *Mus domesticus* populations in a cereal-growing region in southern Australia. *Journal of Applied Ecology*, **36**, 484–493.

Bryant, A. A. & Janz, D. W. (1996). Distribution and abundance of Vancouver Island marmots. *Canadian Journal of Zoology*, **74**, 667–677.

Bulmer, M. G. (1975). The statistical analysis of density dependence. *Biometrics*, **31**, 901–911.

Burgman, M., Ferson, S. & Akcakaya, H. (1992). *Risk Assessment in Conservation Biology*, London: Chapman & Hall.

Burnham, K. P. & Anderson, D. R. (1998). *Model Selection and Inference. A Practical Information-Theoretic Approach*, New York: Springer-Verlag.

Burnham, K. P., Anderson, D. R. & White, G. C. (1996). Meta-analysis of vital rates of the northern spotted owl. *Studies in Avian Biology*, **17**, 92–101.

Cain, M. L., Pacala, S. W., Silander, J. A. & Fortin, M. J. (1995). Neighbourhood models of clonal growth in the white clover *Trifolium repens*. *American Naturalist*, **145**, 888–917.

Cairns, S. & Grigg, G. (1993). Population dynamics of red kangaroo (*Macropus rufus*) in relation to rainfall in the South Australian pastoral zone. *Journal of Applied Ecology*, **30**, 444–458.

Cairns, S. C., Grigg, G. C., Beard, L. A., Pople, A. R. & Alexander, P. (2000). Western grey kangaroos, *Macropus fuliginosus*, in the South Australian pastoral zone: populations at the edge of their range. *Wildlife Research*, **27**, 309–318.

Caley, P. (1993). Population dynamics of feral pigs (*Sus scrofa*) in a tropical riverine habitat complex. *Wildlife Research*, **20**, 625–636.

Calow, P., Sibly, R. M. & Forbes, V. (1997). Risk assessment on the basis of simplified life-history scenarios. *Environmental Toxicology & Chemistry*, **16**, 1983–1989.

Campbell, N. A. & Reece, J. B. (2002). *Biology*, 6th edn. San Francisco: Benjamin Cummings.

Cannan, E. (1895). The probability of a cessation of the growth of population in England and Wales during the next century. *Economics Journal*, **5**, 505–515.

Case, T. (1999). *An Illustrated Guide to Theoretical Ecology*, Oxford: Oxford University Press.

Caswell, H. (1989). *Matrix Population Models*, Sunderland, Mass.: Sinauer Associates.

Caswell, H. (2000). Prospective and retrospective perturbation analyses: their roles in conservation biology. *Ecology*, **81**, 619–627.

Caswell, H. (2001). *Matrix Population Models*, 2nd edn. Sunderland, Mass.: Sinauer Associates.

Cattadori, I. M., Haydon, D. T., Thirgood, S. J. & Hudson, P. J. (2003). Are indirect measures of abundance a useful index of population density? The case of red grouse harvesting. *Oikas*, **100**, 439–446.

Caughley, G. (1970a). Population statistics of chamois. *Mammalia*, **34**, 194–199.

Caughley, G. (1970b). Eruption of ungulate populations, with emphasis on Himalayan thar. *Ecology*, **51**, 53–72.

Caughley, G. (1976). Wildlife management and the dynamics of ungulate populations. In *Applied Biology*, vol. 1, ed. T. H. Coaker, pp. 183–246. London: Academic Press.

Caughley, G. (1977). *Analysis of Vertebrate Populations*, New York: Wiley.

Caughley, G. (1980). *Analysis of Vertebrate Populations*, Reprinted with corrections. New York: Wiley.

Caughley, G. (1981). Overpopulation. In *Problems in Management of Locally Abundant Wild Mammals*, ed. P. A. Jewell, S. Holt & D. Hart, pp. 7–19. New York: Academic Press.

Caughley, G. (1987a). Introduction to the sheep rangelands. In *Kangaroos. Their Ecology and Management in the Sheep Rangelands of Australia*, ed. G. Caughley, N. Shepherd & J. Short, pp. 1–7. Cambridge: Cambridge University Press.

Caughley, G. (1987b). Ecological relationships. In *Kangaroos. Their Ecology and Management in the Sheep Rangelands of Australia*, ed. G. Caughley, N. Shepherd & J. Short, pp. 159–187. Cambridge: Cambridge University Press.

Caughley, G. (1994). Directions in conservation biology. *Journal of Animal Ecology*, **63**, 215–244.

Caughley, G., Grice, D., Barker, R. & Brown, B. (1988). The edge of the range. *Journal of Animal Ecology*, **57**, 771–785.

Caughley, G. & Gunn, A. (1993). Dynamics of large herbivores in deserts. *Oikos*, **67**, 47–55.

Caughley, G. & Gunn, A. (1995). *Conservation Biology in Theory and Practice*, Oxford: Blackwell Science.

Caughley, G. & Krebs, C. J. (1983). Are big mammals simply little mammals writ large? *Oecologia*, **59**, 7–17.

Caughley, G. & Lawton, J. (1981). Plant-herbivore systems. In *Theoretical Ecology. Principles and Applications*, 2nd edn. ed. R. M. May, pp. 132–166. Oxford: Blackwell Science.

Caughley, G., Shepherd, N. & Short, J. (1987). *Kangaroos: Their Ecology and Management in the Sheep Rangelands of Australia*, Cambridge: Cambridge University Press.

Caughley, G. & Sinclair, A. R. E. (1994). *Wildlife Ecology and Management*. Oxford: Blackwell Scientific Publications.

Caughley, J., Bayliss, P. & Giles, J. (1984). Trends in kangaroo numbers in western New South Wales and their relation to rainfall. *Australian Wildlife Research*, **11**, 415–422.

Chandini, T. (1988). Effects of different food (*Chlorella*) concentrations on the chronic toxicity of cadmium to survivorship, growth and reproduction of *Echiniska triserialis* (Crustacea: Cladocera). *Environmental Pollution*, **54**, 139–154.

Charnov, E. L. (1993). *Life History Invariants*, Oxford: Oxford University Press.

Cheville, N. F., McCullough, D. R. & Paulson, L. R. (1998). *Brucellosis in the Greater Yellowstone Area*, Washington DC: National Academy of Sciences, National Research Council.

Chitty, D. (1996). *Do Lemmings Commit Suicide? Beautiful Hypotheses and Ugly Facts*. New York: Oxford University Press.

Choquenot, D. (1991). Density-dependent growth, body condition and demography in feral donkeys: testing the food hypothesis. *Ecology*, **72**, 805–813.

Choquenot, D. (1998). Testing the relative influence of intrinsic and extrinsic variation in food availability on feral pig populations in Australia's rangelands. *Journal of Animal Ecology*, **67**, 887–907.

Choquenot, D. & Dexter, N. (1996). Spatial variation in food limitation: the effects of foraging constraints on the distribution and abundance of feral pigs in the rangelands. In *Frontiers of Population Ecology*, ed. R. B. Floyd, A. W. Sheppard & P. J. De Barro, pp. 531–546. Melbourne: CSIRO Publishing.

Choquenot, D. & Parkes, J. (2001). Setting thresholds for pest control: how does pest density affect resource viability? *Biological Conservation*, **99**, 29–46.

Claessen, D., de Roos, A. M. & Persson, L. (2000). Dwarf and giants: cannibalism and competition in size-structured populations. *American Naturalist*, **155**, 219–237.

Clark, C. W. (1973). The economics of overexploitation. *Science*, **181**, 630–634.

Clark, C. W. (1985). *Bioeconomic Modeling and Fisheries Management*, New York: Wiley.

Cleland, J. (1996). A regional review of fertility trends in developing countries: 1960 to 1995. In *The Future Population of the World: What Can we Assume Today?* Revised edn, ed. W. Lutz, pp. 47–72. London: Earthscan.

Clobert, J., Perrins, C. M., McCleery, R. H. & Gosler, A. G. (1988). Survival rate in the great tit, *Parus major* in relation to sex, age, and immigration status. *Journal of Animal Ecology*, **57**, 287–306.

Clout, M. N. & Merton, D. V. (1998). Saving the kakapo: the conservation of the world's most peculiar parrot. *Bird Conservation International*, **8**, 281–296.

Clutton-Brock, J. (1981). *Domesticated Animals from Early Times*, London: William Heinemann & British Museum of Natural History.

Clutton-Brock, T. H. & Albon, S. D. (1989). *Red Deer in the Highlands*, Oxford: Blackwell Scientific Publications.

Clutton-Brock, T. H., Albon, S. D. & Guinness, F. E. (1982a). Competition between female relatives in a matrilocal mammal. *Nature*, **300**, 178–180.

Clutton-Brock, T. H., Albon, S. D. & Guinness, F. E. (1985a). Parental investment and sex differences in juvenile mortality in birds and mammals. *Nature*, **313**, 131–133.

Clutton-Brock, T. H., Albon, S. D. & Guinness, F. E. (1988). Reproductive success in male and female red deer. In *Reproductive Success*, ed. T. H. Clutton-Brock, pp. 325–343. Chicago: University of Chicago Press.

Clutton-Brock, T. H., Coulson, T. N., Milner-Gulland, E. J., Thomson, D. & Armstrong, H. M. (2002). Sex differences in emigration and mortality affect optimal management of deer populations. *Nature*, **415**, 633–637.

Clutton-Brock, T. H., Guinness, F. E. & Albon, S. D. (1982b). *Red Deer: Behaviour and Ecology of Two Sexes*, Edinburgh: Edinburgh University Press.

Clutton-Brock, T. H., Illius, A. W., Wilson, K., Grenfell, B. T., MacColl, A. D. C. & Albon, S. D. (1997a). Stability and instability in ungulate populations: an empirical analysis. *American Naturalist*, **149**, 195–219.

Clutton-Brock, T. H., Major, M. & Guinness, F. E. (1985b). Population regulation in male and female red deer. *Journal of Animal Ecology*, **54**, 831–846.

Clutton-Brock, T. H., Price, O. F., Albon, S. D. & Jewell, P. A. (1991). Persistent instability and population regulation in Soay sheep. *Journal of Animal Ecology*, **60**, 593–608.

Clutton-Brock, T. H., Price, O. F., Albon, S. D. & Jewell, P. A. (1992). Early development and population fluctuation in Soay sheep. *Journal of Animal Ecology*, **61**, 381–396.

Clutton-Brock, T. H., Rose, K. E. & Guinness, F. E. (1997b). Density-related changes in sexual selection in red deer. *Proceedings of the Royal Society London B*, **264**, 1509–1516.

Coale, A. J. (1973). The demographic transition. In *Proceedings of the International Population Conference*, Vol. 1, Liege, pp. 53–71. Belgium: International Union for the Scientific Study of Population.

Coale, A. & Treadway, R. (1979). A summary of changing fertility in the provinces of Europe. Paper presented to the Summary Conference on European Fertility, Princeton, New Jersey, USA.

Coale, A. & Watkins, S. (1986). *The Decline of Fertility in Europe*, Princeton: Princeton University Press.

Cobbold, T. S. (1873). The grouse disease. *The Field*, London. p15.

Cole L. C. (1958). Sketches of general and comparative demography. *Cold Spring Harbor Symposium in Quantitative Biology*, **22**, 1–15.

Cooch, E. G. & Cooke, F. (1991). Demographic changes in a snow goose population – biological and management implications. In *Bird Population Studies: Their Relevance to Conservation and Management*, ed. C. M. Perrins, J.-D. Lebreton & G. J. M. Hirons, pp. 168–189. Oxford: Oxford University Press.

Collver, A., Speare, A. & Liu, P. K. C. (1967). Local variations of fertility in Taiwan. *Population Studies*, **20**, 329–342.

Coltman, D. W., Bancroft, D. R., Robertson, A., Smith, J. A., Clutton-Brock, T. H. & Pemberton, J. M. (1999a). Male reproductive success in a promiscuous mammal; behavioural estimates compared with genetic paternity. *Molecular Ecology*, **8**, 1199–1209.

Coltman, D. W., Pilkington, J. G., Smith, J. A. & Pemberton, J. M. (1999b). Parasite-mediated selection against inbred Soay sheep in a free-living island population. *Evolution*, **53**, 1259–1267.

Coltman, D. W., Smith, J. A., Bancroft, D. R., Pilkington, J. G., MacColl, A. D. C., Clutton-Brock, T. H. & Pemberton, J. M. (1999c). Density-dependent variation in lifetime breeding success and in natural and sexual selection in Soay rams. *American Naturalist*, **154**, 730–746.

Condit, R., Ashton, P. S., Baker, P., Bunyavejchewin, S., Gunatilleke, S., Gunatilleke, N., Hubbell, S. P., Foster, R. B., Itoh, A., LaFrankie, J. V., Lee, H. S., Losos, E., Manokaran, N., Sukumar, R. & Yamakura, T. (2000). Spatial patterns in the distribution of tropical tree species. *Science*, **288**, 1414–1418.

Costantino, R. F., Desharnais, R. A., Cushing, J. M. & Dennis, B. (1997). Chaotic dynamics in an insect population. *Science*, **275**, 389–391.

Cook, R. C. (1962). How many people have ever lived on earth? *Population Bulletin*, **18**, 1–19.

Coughenour, M. B. & Singer, F. J. (1996). Elk population processes in Yellowstone National Park under the policy of natural regulation. *Ecological Applications*, **6**, 573–593.

Coulson, T., Albon, S. D., Guinness, F., Pemberton, J. M. & Clutton-Brock, T. H. (1997). Population substructure, local density and calf winter survival in red deer, *Cervus elaphus*. *Ecology*, **78**, 852–863.

Coulson, T., Albon, S. D., Pilkington, G. & Clutton-Brock, T. H. (1999). Small scale spatial dynamics in a fluctuating ungulate population. *Journal of Animal Ecology*, **68**, 658–671.

Coulson, T., Catchpole, E. A., Albon, S. D., Morgan, B. J. T., Pemberton, J. M., Clutton-Brock, T. H., Crawley, M. J. & Grenfell, B. T. (2001). Age, sex, density, winter weather, and population crashes in Soay sheep. *Science*, **292**, 1528–1531.

Coulson, T., Milner-Gulland, E. J. & Clutton-Brock, T. (2000). The relative roles of density and climatic variation on population dynamics and fecundity rates in three contrasting ungulate species. *Proceedings of Royal Society London B*, **267**, 1771–1779.

Courchamp, F., Clutton-Brock, T. & Grenfell, B. (1999). Inverse density-dependence and the Allee effect. *Trends in Ecology and Evolution*, **14**, 405–410.

Cowan, P. E. & Waddington, D. C. (1990). Suppression of fruit production of the endemic forest tree, *Elaeocarpus dentatus*, by introduced marsupial brushtail possums, *Trichosurus vulpecula*. *New Zealand Journal of Botany*, **28**, 217–224.

Cowan, P. E. & Waddington, D. C. (1991). Litterfall under hinau, *Elaeocarpus dentatus*, in lowland podocarp/mixed hardwood forest, and the impact of brushtail possums, *Trichosurus vulpecula*. *New Zealand Journal of Botany*, **29**, 385–394.

Cramp, S. (1972). One hundred and fifty years of mute swans on the Thames. *Wildfowl*, **23**, 119–124.

Crawley, M. J. (1983). *Herbivory: The Dynamics of Animal–Plant Interactions*, Oxford: Blackwell Scientific Publications.

Crawley, M. J. (1990). The population dynamics of plants. *Philosophical Transactions of the Royal Society London B*, **330**, 125–140.

Croxall, J. P., Prince, P. A., Rothery, P. & Wood, A. G. (1997). Population changes in albatrosses at South Georgia. In *Albatross Biology and Conservation*, ed. G. Robertson & R. Gales, pp. 69–85. Chipping Norton: Surrey Beatty & Sons.

Cutright, P. & Kelly, W. R. (1978). Modernization and other determinants of national birth, death, and growth rates: 1958–1972. *Comparative Studies in Sociology*, **1**, 17–46.

Daniels, R. E. & Allan, J. D. (1981). Life table evaluation of chronic exposure to pesticide. *Canadian Journal of Fisheries & Aquatic Sciences*, **38**, 485–494.

Davis, K. (1954). The world demographic transition. *Annals of the American Academy of Political and Social Sciences*, **237**, 1–11.

Davis, K. (1991). Population and resources: fact and interpretation. In *Resources, Environment and Population: Present Knowledge*, ed. K. Davis & M. S. Bernstam, pp. 1–21. Oxford: Oxford University Press.

Davis, P. E. & Newton, I. (1981). Population and breeding of red kites in Wales over a 30-year period. *Journal of Animal Ecology*, **50**, 759–772.

Davis, S. A., Pech, R. P. & Catchpole, E. A. (2002). Populations in variable environments: the effect of variability in a species' primary resource. *Philosophical Transactions of the Royal Society London B*, **357**, 1249–1257.

Deevey, E. S. Jr. (1947). Life tables for natural populations of animals. *Quarterly Review of Biology*, **22**, 283–314.

De Kroon, H., Plaisier, A., Van Groenendael, J. & Caswell, H. (1986). Elasticity: the relative contribution of demographic parameters to population growth rate. *Ecology*, **67**, 1427–1431.

De Kroon, H., Van Groenendael, J. & Ehrlen, J. (2000). Elasticities: a review of methods and model limitations. *Ecology*, **81**, 607–618.

den Boer, P. J. & Reddingius, J. (1989). On the stabilization of animal numbers. Problems of testing. 2. Confrontation with data from the field. *Oecologia*, **79**, 143–149.

den Boer, P. J. & Reddingius, J. (1996). *Regulation and Stabilization Paradigms in Population Ecology*, London: Chapman & Hall.

Dennis, B. (1989). Allee effects: population growth, critical density and the chance of extinction. *Natural Resources Modeling*, **3**, 481–537.

Dennis, B., Mulholland, P. & Scott, J. M. (1991). Estimation of growth and extinction parameters for endangered species. *Ecological Monographs*, **61**, 115–143.

Dennis, B. & Otten, M. R. M. (2000). Joint effects of density dependence and rainfall on abundance of San Joaquin kit fox. *Journal of Wildlife Management*, **64**, 388–400.

Dennis, B. & Tapper, M. L. (1994). Density dependence in time series observations of natural populations: estimation and testing. *Ecological Monographs*, **64**, 205–224.

Dennis, R. (1995). Ospreys *Pandion haliaetus* in Scotland – a study of recolonization. *Vogelwelt*, **116**, 193–196.

Dhondt, A., Matthysen, E., Andriaensen, F. & Lambrechts, M. M. (1990). Population dynamics and regulation of a high density blue tit population. In *Population Biology of Passerine Birds*, ed. J. Blondel *et al.*, pp. 39–53. Berlin: Springer-Verlag.

Dieckmann, U., Law, R. & Metz, J. A. J. (2000). *The Geometry of Ecological Complexity*, Cambridge: Cambridge University Press.

Diserud, O. & Engen, S. (2000). A general and dynamic species abundance model, embracing the lognormal and the gamma models. *American Naturalist*, **155**, 497–511.

Dobson, A. P. & Hudson, P. J. (1992). Regulation and stability of a free-living host-parasite system, *Trichostrongylus tenuis* in red grouse. II: Population models. *Journal of Animal Ecology*, **61**, 487–498.

Dobson, A. P. & May, R. M. (1986). Patterns of invasion by pathogens and parasites. In *Ecology of Invasions of North America and Hawaii*, ed. H. A. Mooney & J. A Drake, pp. 58–76. New York: Springer-Verlag.

Dublin, H. (1995). Vegetation dynamics in the Serengeti-Mara ecosystem: the role of elephants, fire and other factors. In *Serengeti II. Dynamics, Management and*

Conservation of an Ecosystem, ed. A. R. E. Sinclair & P. Arcese, pp. 71–90. Chicago: University of Chicago Press.

Dublin, H., Sinclair, A. R. E. & McGlade, J. (1990). Elephants and fire as causes of multiple stable states for Serengeti–Mara woodlands. *Journal of Animal Ecology*, **59**, 1157–1164.

Durell, S., Goss-Custard, J. D. & Clarke, R. T. (2001). Modelling the population consequences of age- and sex-related differences in winter mortality in the oystercatcher, *Haematopus ostralegus*. *Oikos*, **95**, 69–77.

Easterling, M. R., Ellner, S. P. & Dixon, P. M. (2000). Size-specific sensitivity: applying a new structured population model. *Ecology*, **81**, 694–708.

Eberhardt, L. E., Eberhardt, L. L., Tiller, B. L. & Cadwell, L. L. (1996). Growth of an isolated elk population. *Journal of Wildlife Management*, **60**, 369–373.

Eberhardt, L. L. (1987). Population projections from simple models. *Journal of Applied Ecology*, **24**, 103–118.

Eberhardt, L. L. (1988). Testing hypotheses about populations. *Journal of Wildlife Management*, **52**, 50–56.

Eberhardt, L. L. (1998). Applying difference equations to wolf predation. *Canadian Journal of Zoology*, **76**, 380–386.

Eberhardt, L. L. & Peterson, R. O. (1999). Predicting the wolf–prey equilibrium point. *Canadian Journal of Zoology*, **77**, 494–498.

Edgerly, J. S. & Livdahl, T. P. (1992). Density-dependent interactions within a complex life cycle: the roles of cohort structure and mode of recruitment. *Journal of Animal Ecology*, **61**, 139–150.

Efford, M. (1998). Demographic consequences of sex-biased dispersal in a population of brushtail possums. *Journal of Animal Ecology,* **67**, 503–517.

Efford, M. (2000). Possum density, population structure, and dynamics. In *The Brushtail Possum. Biology, Impact and Management of an Introduced Marsupial*, ed. T. L. Montague, pp. 47–61. Lincoln, New Zealand: Manaaki Whenua Press.

Efford, M. (2001). Environmental stochasticity cannot save declining populations. *Trends in Ecology and Evolution*, **16**, 177.

Ellenberg, H. (1988). *Vegetation Ecology of Central Europe*, 4th edn (translated into English by G. K. Strutt). Cambridge: Cambridge University Press.

Ellner, S. (1985). ESS germination strategies in randomly varying environments. I. Logistic-type models. *Theoretical Population Biology*, **28**, 50–79.

Ellner, S. P., Bailey, B. A., Bobashev, G. V., Gallant, A. R., Grenfell, B. T. & Nychka, D. W. (1998). Noise and nonlinearity in measles epidemics: combining mechanistic and statistical approaches to population modeling. *American Naturalist*, **151**, 425–440.

Ellner, S. P., Kendall, B. E., Wood, S. N., McCauley, E. & Briggs, C. J. (1997). Inferring mechanism from time-series data: delay-differential equations. *Physica D*, **110**, 182–194.

Ellner, S. P. & Turchin, P. (1995). Chaos in a noisy world: new methods and evidence from time-series analysis. *American Naturalist*, **145**, 343–375.

Engen, S., Bakke, O. & Islam, A. (1998). Demographic and environmental stochasticity – concepts and definitions. *Biometrics*, **54**, 840–846.

Engen, S. & Sæther, B.-E. (2000). Predicting the time to quasi-extinction for populations far below their carrying capacity. *Journal of Theoretical Biology*, **205**, 649–658.

Engen, S., Sæther, B.-E. & Moller, A. P. (2001). Stochastic population dynamics and time to extinction of a declining population of barn swallows. *Journal of Animal Ecology*, **70**, 789–797.

Enright, N. J., Franco, M. & Silvertown, J. (1995). Comparing plant life-histories using elasticity analysis – the importance of life-span and the number of life-cycle stages. *Oecologia (Berlin)*, **104**, 79–84.

Ens, B. J., Kersten, M., Brenninkmeijer, A. & Hulscher, J. B. (1992). Territory quality, parental effort and reproductive success of oystercatchers (*Haematopus ostralegus*). *Journal of Animal Ecology*, **61**, 703–715.

Ens, B. J., Weising, F. J. & Drent, R. H. (1995). The despotic distribution and deferred maturity: two sides of the same coin. *American Naturalist*, **146**, 625–650.

Erb, J., Boyce, M. S. & Stenseth, N. C. (2001). Population dynamics of large and small mammals. *Oikos*, **92**, 3–12.

Euler, L. (1760). A general investigation into the mortality and multiplication of the human species. Translated by N. and B. Keyfitz and republished 1970 in *Theoretical Population Biology*, **1**, 307–314.

Fava, G. & Crotti, E. (1979). Effect of crowding on nauplii production during mating time in *Tisbe clodiensis* and *T. holothuriae* (Copepoda: Harpacticoida). *Helgolander wissenschaftliche Meeresun*, **32**, 466–475.

Festa-Bianchet, M. (1998). Condition-dependent reproductive success in bighorn ewes. *Ecology Letters*, **1**, 91–94.

Firebaugh, G. (1982). Population density and fertility in 22 Indian villages. *Demography*, **19**, 481–494.

Fischer, G., Van Velthuizen, H. T. & Nachtergaele, F. O. (1998). *Global Agro-Ecological Zones Assessment: Methodology & Results*. Interim Report, IR-98–110, Laxenburg, Austria: International Institute for Applied Systems Analysis.

Fisher, R. A. (1930). *The Genetical Theory of Natural Selection*, Oxford: Oxford University Press.

Forbes, V. & Calow, P. (1999). Is the per capita rate of increase a good measure of population-level effects in ecotoxicology? *Environmental Toxicology and Chemistry*, **18**, 1544–1556.

Forbes, V. E., Calow, P. & Sibly, R. M. (2001a). Are current species extrapolation models a good basis for ecological risk assessment? *Environmental Toxicology and Chemistry*, **20**, 442–447.

Forbes, V. E., Sibly, R. M. & Calow, P. (2001b). Determining toxicant impacts on density-limited populations: a critical review of theory, practice and results. *Ecological Applications*, **11**, 1249–1257.

Forbes, V. E., Sibly, R. M. & Linke-Gamenick, I. (2003). Joint effects of a toxicant and population density on population dynamics: an experimental study using *Capitella* sp. I (Polychaeta). *Ecological Applications* (in Press).

Forchhammer, M. C., Clutton-Brock, T. H., Lindstrom, J. & Albon, S. D. (2001). Climate and population density induce long-term cohort variation in a northern ungulate. *Journal of Animal Ecology*, **70**, 721–729.

Fowler, C. W. (1981). Density dependence as related to life history strategy. *Ecology*, **62**, 602–610.

Fowler, C. W. (1987a). Population dynamics as related to rate of increase per generation. *Evolutionary Ecology*, **2**, 197–204.

Fowler, C. W. (1987b). A review of density dependence in populations of large mammals. In *Current Mammology*, ed. H. H. Genoways, pp. 401–441. New York: Plenum Press.

Fowler, C. W. (1988). Population dynamics as related to rate of increase per generation. *Evolutionary Ecology*, **2**, 197–204.

Fowler, C. W. & Baker, J. D. (1991). A review of animal population dynamics at extremely reduced population levels. *Report of the International Whaling Commission*, **41**, 545–554.

Fox, A. & Hudson, P. J. (2001). Parasites reduce territorial behaviour in red grouse (*Lagopus lagopus scoticus*). *Ecology Letters*, **4**, 139–143.

Fox, D. R. & Ridsdill-Smith, T. J. (1996). Detecting density dependence. In *Frontiers of Population Ecology*, ed. R. B. Floyd, A. W. Sheppard & P. J. De Barro, pp. 45–51. Melbourne: CSIRO Publishing.

Frank, P. W., Boll, C. D. & Kelly, R. W. (1957). Vital statistics of laboratory cultures of *Daphnia pulex* DeGeer as related to density. *Physiological Zoology*, **30**, 287–305.

Franklin, A. B., Anderson, D. R., Gutierrez, R. J. & Burnham, K. P. (2000). Climate, habitat quality, and fitness in northern spotted owl populations in northwestern California. *Ecological Monographs*, **70**, 539–590.

Freeland, W. J. & Boulton, W. J. (1990). Feral water buffalo (*Bubalus bubalis*) in the major floodplains of the 'Top End', Northern Territory, Australia: population growth and the brucellosis and tuberculosis eradication campaign. *Australian Wildlife Research*, **17**, 411–420.

Frejka, T. (1994). Long-range global population projections: lessons learned. In *The Future Population of the World: What can we Assume Today?* ed. W. Lutz, pp. 3–15. London: Earthscan.

Fretwell, S. D. & Lucas, J. H. J. (1970). On territorial behaviour and other factors influencing habitat distribution in birds. *Acta Biotheoretica*, **19**, 16–36.

Friend, J. A. & Thomas, N. D. (1994). Reintroduction and the numbat recovery programme. In *Reintroduction Biology of Australian and New Zealand Fauna*, ed. M. Serena, pp. 180–198. Chipping Norton: Surrey Beatty & Sons.

Fromentin, J. M., Myers, R. A., Bjornstad, O. N., Stenseth, N. C., Gjosaeter, J. & Christie, H. (2001). Effects of density-dependent and stochastic processes on the regulation of cod populations. *Ecology*, **82**, 567–579.

Fryxell, J. M. (1987). Food limitation and demography of a migratory antelope, the white-eared kob. *Oecologia (Berlin)*, **72**, 83–91.

Fryxell, J. M., Hussell, D. J. T., Lambert, A. B. & Smith, P. C. (1991). Time lags and population fluctuations in white-tailed deer. *Journal of Wildlife Management*, **55**, 377–385.

Fujiwara, M. & Caswell, H. (2001). Demography of the endangered North Atlantic right whale. *Nature*, **414**, 537–541.

Gaillard, J.-M., Festa-Bianchet, M. & Yoccoz, N. G. (1998). Population dynamics of large herbivores: variable recruitment with constant adult survival. *Trends in Ecology and Evolution*, **13**, 58–63.

Gaillard, J.-M., Festa-Bianchet, M., Yoccoz, N. G., Loison, A. & Toigo, C. (2000). Temporal variation in fitness components and population dynamics of large herbivores. *Annual Review of Ecology and Systematics*, **31**, 367–393.

Gasaway, W. C., Boertje, R. D., Grangaard, D. V., Kelleyhouse, D. G., Stephenson, R. O. & Larsen, D. G. (1992). The role of predation in limiting moose at low densities in

Alaska and Yukon and implications for conservation. *Wildlife Monographs*, **120**, 1–59.

Gasaway, W. C., Stephenson, R. O., Davis, J. L., Shepherd, P. E. K. & Burris, O. E. (1983). Interrelationships of wolves, prey, and man in interior Alaska. *Wildlife Monographs*, No. 4, p. 50.

Gaudy, R. & Guerin, J. P. (1982). Population dynamics of *Tisbe holothuriae* (Copepoda: Harpacticoida) in exploited mass cultures. *Netherlands Journal of Sea Research*, **16**, 208–216.

Gill, J. A., Norris, K., Potts, P., Gunnarsson, T., Atkinson, P. W. & Sutherland, W. J. (2001b). The Buffer effect and large-scale population regulation in migratory birds. *Nature*, **412**, 436–438.

Gill, J. A., Sutherland, W. J. & Norris, K. (2001a). Depletion models can predict shorebird distribution at different spatial scales. *Proceedings of the Royal Society London B*, **246**, 369–376.

Gill, J. A., Sutherland, W. J. & Watkinson, A. R. (1996). A method to quantify the effects of human disturbance on animal populations. *Journal of Applied Ecology*, **33**, 786–792.

Gilpin, M. E. & Ayala, F. J. (1973). Global models of growth and competition. *Proceedings of the National Academy of Sciences USA*, **70**, 3590–3593.

Gilpin, M. E., Case, T. J. & Ayala, F. J. (1976). θ-selection. *Mathematical Biosciences*, **32**, 131–139.

Ginzburg, L. R. (1998). Assuming reproduction to be a function of consumption raises doubts about some popular predator–prey models. *Journal of Animal Ecology*, **67**, 325–327.

Godfray, H. C. J. (1994). *Parasitoids. Behavioural and Evolutionary Ecology*, Princeton: Princeton University Press.

Godfray, H. C. J. & Hassell, M. P. (1989). Discrete and continuous insect populations in tropical environments. *Journal of Animal Ecology*, **58**, 153–174.

Godfray, H. C. J. & Waage, J. K. (1991). Predictive modelling in biological control: the mango mealy bug (*Rastrococcus invadens*) and its parasitoids. *Journal of Applied Ecology*, **28**, 434–453.

Goss-Custard, J. D. (1996). *The Oystercatcher*, Oxford: Oxford University Press.

Goss-Custard, J. D. & Sutherland, W. J. (1997). Individual behaviour, populations and conservation. In *Behavioural Ecology*, 4th edn, ed. J. R. Krebs & N. B. Davies, pp. 373–395. Oxford: Blackwells.

Gotelli, M. J. (1991). Metapopulation models: the rescue effect, the propagule rain, and the core-satellite hypothesis. *American Naturalist*, **138**, 768–776.

Grant, A. (1998). Population consequences of chronic toxicity: incorporating density dependence into the analysis of life table response experiments. *Ecological Modelling*, **105**, 325–335.

Grant, A. & Benton, T. G. (1996). The impact of environmental variation on demographic convergence of Leslie matrix models: an assessment using Lyapunov characteristic exponents. *Theoretical Population Biology*, **50**, 18–30.

Grant, A. & Benton, T. G. (2000). Elasticity analysis for density-dependent populations in stochastic environments. *Ecology*, **81**, 680–693.

Green, R. E. & Hirons, G. J. M (1991). The relevance of population studies to the conservation of threatened birds. In *Bird Population Studies*, ed. C. M. Perrins, J.-D. Lebreton & G. J. M. Hirons, pp. 594–633. Oxford: Oxford University Press.

Green, R. E., Pienkowski, M. W. & Love, J. A. (1996). Long-term viability of the re-introduced population of the white-tailed sea eagle *Haliaeetus albicilla* in Scotland. *Journal of Applied Ecology*, **33**, 357–368.

Greenwood, J. J. D. & Baillie, S. R. (1991). Effects of density-dependence and weather on population changes of English passerines using a non-experimental paradigm. *Ibis*, **133** suppl. 1, 121–133.

Grenfell, B. T., Price, O. F., Albon, S. D. & Clutton-Brock, T. H. (1992). Overcompensation and population cycles in an ungulate. *Nature*, **355**, 823–826.

Grenfell, B. T., Wilson, K., Finkenstadt, B. F., Coulson, T. N., Murray, S., Albon, S. D., Pemberton, J. M., Clutton-Brock, T. H. & Crawley, M. J. (1998). Noise and determinism in synchronized sheep dynamics. *Nature*, **394**, 674–677.

Griffiths, D. (1998). Sampling effort, regression method, and the shape and slope of size-abundance relations. *Journal of Animal Ecology*, **67**, 795–804.

Groom, M. & Pascual, M. A. (1998). The analysis of population persistence: an outlook on the practice of viability analysis. In *Conservation Biology*, 2nd edn, ed. P. L. Fiedler & P. Kareiva, pp. 4–27. New York: Chapman & Hall.

Groombridge, J. J., Jones, C. G., Bruford, M. W. & Nichols, R. A. (2001). Evaluating the severity of the population bottleneck in the Mauritius kestrel *Falco punctatus* from ringing records using MCMC estimation. *Journal of Animal Ecology*, **70**, 401–409.

Grubb, P. (1974a). Population dynamics of the Soay sheep. In *Island Survivors: The Ecology of the Soay Sheep of St. Kilda*, ed. P. A. Jewell, C. Milner & J. M. Boyd, pp. 242–272. London: Athlone Press.

Grubb, P. (1974b). Mating activity and the social significance of rams in a feral sheep community. In *The Behaviour of Ungulates and its Relation to Management*, Vol. 1, ed. V. Geist & F. Walther, pp. 457–476. Morges, Switzerland: IUCN.

Grubb, P. (1974c). The rut and behaviour of Soay rams. In *Island Survivor: the Ecology of the Soay Sheep of St. Kilda*, ed. P. A. Jewell, C. Milner & J. M. Boyd, pp. 195–223. London: Athlone Press.

Grubb, P. & Jewell, P. A. (1974). Movement, daily activity and home range of Soay sheep. In *Island Survivor: the Ecology of the Soay Sheep of St. Kilda*, ed. P. A. Jewell, C. Milner & J. M. Boyd, pp. 160–194. London: Athlone Press.

Grubb, P. J. (1977). The maintenance of species-richness in plant communities: the importance of the regeneration niche. *Biological Reviews*, **52**, 107–145.

Gundersen, G., Aars, J., Andreassen, H. P. & Ims, R. A. (2001). Inbreeding in the field: an experiment on root vole populations. *Canadian Journal of Zoology*, **79**, 1901–1905.

Gurney, W. S. C., McCauley, E., Nisbet, R. M. & Murdoch, W. W. (1990). The physiological ecology of *Daphnia*: a dynamic model of growth and reproduction. *Ecology*, **71**, 716–732.

Gutierrez, A. P. & Baumgartner, J. U. (1984). Multitrophic level models of predator–prey energetics. II. A realistic model of plant–herbivore–parasitoid–predator interactions. *Canadian Entomologist*, **116**, 933–949.

Gutierrez, A. P., Neuenschwander, P. & van Alphen, J. J. M. (1993). Factors affecting biological control of cassava mealybug by exotic parasitoids – a ratio-dependent supply–demand driven model. *Journal of Applied Ecology*, **30**, 706–721.

Hackländer, K. & Arnold, W. (1999). Male-caused failure of female reproduction and its adaptive value in alpine marmots (*Marmota marmota*). *Behavioural Ecology*, **10**, 592–597.

Hakkarainen, H., Korpimaki, E., Koivunen, V. & Ydenberg, R. (2002). Survival of male Tengmalm's owls under temporally varying food conditions. *Oecologia (Berlin)*, **131**, 83–88.

Halterlein, B. & Sudbeck, P. (1996). Brutbestands-monitoring von kustenvogeln an der Deutschen nordseekuste. *Vogelwelt*, **117**, 277–285.

Hansen, F. T., Forbes, V. E. & Forbes, T. L. (1999a). The effects of chronic exposure to 4-n-nonylphenol on life-history traits and population dynamics of the polychaete *Capitella* sp. I. *Ecological Applications*, **9**, 482–495.

Hansen, F., Forbes, V. E. & Forbes, T. L. (1999b). Using elasticity analysis of demographic models to link toxicant effects on individuals to the population level: an example. *Functional Ecology*, **13**, 157–162.

Hansen, T. F., Stenseth, N. C., Henttonen, H. & Tast, J. (1999). Interspecific and intraspecific competition as causes of direct and delayed density dependence in a fluctuating vole population. *Proceedings of the National Academy of Sciences USA*, **96**, 986–991.

Hanski, I. (1998). Metapopulation dynamics. *Nature*, **396**, 41–49.

Hanski, I. (1999). *Metapopulation Biology*, Oxford: Oxford University Press.

Hanski, I., Woiwod, I. & Perry, J. (1993). Density dependence, population persistence, and largely futile arguments. *Oecologia (Berlin)*, **95**, 595–598.

Harris, M. P., Buckland, S. T., Russell, S. M. & Wanless, S. (1994). Year- and age-related variation in the survival of adult European shag over a 24-year period. *Condor*, **96**, 600–605.

Harrison, S. (1991). Local extinction in a metapopulation context: an empirical evaluation. *Biological Journal of the Linnean Society*, **42**, 73–88.

Harvey, P. H. & Pagel, M. (1991). *The Comparative Method in Evolutionary Biology*, Oxford: Oxford University Press.

Hassell, M. P. (2000). *The Spatial and Temporal Dynamics of Host–Parasitoid Interactions*, Oxford: Oxford University Press.

Hassell, M. P., Latto, J. & May, R. M. (1989). Seeing the wood for the trees: detecting density dependence from existing life-table studies. *Journal of Animal Ecology*, **58**, 883–892.

Hassell, M. P., Lawton, J. H. & May, R. M. (1976). Patterns of dynamical behaviour in single species populations. *Journal of Animal Ecology*, **45**, 471–486.

Hastings, A. (1996). *Population Biology. Concepts and Models*, Springer Verlag: New York.

Hastings, A. & Costantino, R. F. (1987). Cannibalistic egg–larva interactions in *Tribolium*: an explanation for the oscillations in population numbers. *American Naturalist*, **130**, 36–52.

Hastings, A., Hom, C. L., Ellner, S., Turchin, P. & Godfray, H. C. J. (1993). Chaos in ecology – is mother nature a strange attractor? *Annual Review of Ecology and Systematics*, **24**, 1–33.

Hauff, P. (1998). Bestandsentwicklung des seeadlers *Haliaeetus albicilla* in Deutschland seit 1980 mit einem ruckblick auf die vergangenen 100 jahre. *Vogelwelt*, **119**, 47–63.

Haydon, D. T., Shaw, D. J., Cattadori, I. M., Hudson, P. J. & Thirgood, S. J. (2002). Analysing noisy time series: describing regional variation in the cyclic dynamics of red grouse. *Proceedings of the Royal Society London B*, **269**, 1609–1617.

Heath, P. L. (1994). The development of *Tisbe holothuriae* as a live diet for larval flatfish rearing. PhD thesis, Heriot-Watt University.

Heer, D. M. (1966). Economic development and fertility. *Demography*, **3**, 423–444.

Hengge-Aronis, R. (1999). Interplay of global regulators and cell physiology in the general stress response of *Escherichia coli*. *Current Opinion in Microbiology*, **2**, 148–152.

Heppell, S. S., Caswell, H. & Crowder, L. B. (2000). Life histories and elasticity patterns: perturbation analysis for species with minimal demographic data. *Ecology*, **81**, 654–665.

Hermalin, W. I. & Lavely, W. R. (1979). Agricultural development and fertility change in Taiwan. Paper presented at the Annual Meeting of the Population Association of America. Philadelphia, PA, USA, April 1979.

Hestbeck, J. B. (1982). Population regulation of cyclic mammals: the social fence hypothesis. *Oikos*, **39**, 157–163.

Hestbeck, J. B. (1987). Multiple regulation states in populations of small mammals: a state transition model. *American Naturalist*, **129**, 520–532.

Heyde, C. C. & Cohen, J. E. (1985). Confidence intervals for demographic projections based on products of random matrices. *Theoretical Population Biology*, **27**, 120–153.

Hickling, G. J. & Pekelharing, C. J. (1989). Intrinsic rate of increase for a brushtail possum population in rata/kamahi forest, Westland. *New Zealand Journal of Ecology*, **12**, 117–120.

Hicks, W. (1974). Economic development and fertility change in Mexico, 1950–1970. *Demography*, **11**, 407–421.

Higgins, S. I., Bond, W. J. & Pickett, S. T. A. (2000). Predicting extinction risks for plants: environmental stochasticity can save declining populations. *Trends in Ecology and Evolution* **15**, 516–520.

Higgins, S. I., Bond, W. J. & Pickett, S. T. A. (2001). Reply to Efford. *Trends in Ecology and Evolution*, **16**, 177.

Hik, D. S. (1995). Does risk of predation influence population dynamics? Evidence from the cyclic decline of snowshoe hares. *Wildlife Research*, **22**, 115–129.

Hik, D. S., Jefferies, R. L. & Sinclair, A. R. E. (1992). Foraging by geese, isostatic uplift and asymmetry in the development of salt-marsh plant communities. *Journal of Ecology*, **80**, 395–406.

Hill, D. (1988). Population dynamics of the avocet (*Recurvirostra avosetta*) breeding in Britain. *Journal of Animal Ecology*, **57**, 669–683.

Hiraldo, F., Negro, J. J., Donazar, J. A. & Gaona, P. (1996). A demographic model for a population of the endangered lesser kestrel in southern Spain. *Journal of Applied Ecology*, **33**, 1085–1093.

Hladik, B. (1986). Bestandsanderungen des weissstorchs im nordosten des bohmisch-mahrischen hugellandes. In *White Stork. Status and Conservation*, ed. G. Rheinwald, J. Ogden & H. Sculz, pp. 77–80. Braunschweig: Dachverband Deutscher Avifaunisten.

Hobbs, N. T., Bowden, D. C. & Baker, D. L. (2000). Effects of fertility control on populations of ungulates: general, stage-structured models. *Journal of Wildlife Management*, **64**, 473–491.

Hoffmann, A. A. & Parsons, P. A. (1991). *Evolutionary Genetics and Environmental Stress*, Oxford: Oxford Science Publications.

Holling, C. S. (1959). The components of predation as revealed by a study of small mammal predation of the European pine sawfly. *Canadian Entomologist*, **91**, 293–320.

Holling, C. S. (1965). The functional response of predators to prey density and its role in mimicry and population regulation. *Memoirs of the Entomological Society of Canada*, **45**, 1–60.

Holling, C. S. (1966). The functional response of invertebrate predators to prey density. *Memoirs of the Entomological Society of Canada*, **48**, 3–86.

Holling, C. S. (1973). Resilience and stability of ecological systems. *Annual Review of Ecology and Systematics*, **4**, 1–23.

Holmes, J. C. (1995). Population regulation: a dynamic complex of interactions. *Wildlife Research*, **22**, 11–19.

Holyoak, M. & Lawler, S. P. (1996). Persistence of an extinction-prone predator–prey interaction through metapopulation dynamics. *Ecology*, **77**, 1867–1879.

Hone, J. (1994). *Analysis of Vertebrate Pest Control*, Cambridge: Cambridge University Press.

Hone, J. (1999). On rate of increase (r): patterns of variation in Australian mammals and the implications for wildlife management. *Journal of Applied Ecology*, **36**, 709–718.

Hoppenheit, M. (1976). Zur dynamik exploitierter populationen von *Tisbe holothuriae* (Copepoda: Harpacticoida). III. Reproduktion, geschlechtsverhaltnis, entwicklungsdauer und uberlebenszeit. *Helgolander wissenschaftliche Meeresuntersuchungen*, 28, 109–137.

Houston, D. B. (1982). *The Northern Yellowstone Elk: Ecology and Management*, New York: Macmillan.

Howells, O. & Edwards-Jones, G. (1997). A feasibility study of reintroducing wild boar *Sus scrofa* to Scotland: are existing woodlands large enough to support minimum viable populations? *Biological Conservation*, **81**, 77–89.

Hubbell, S. P. (2001). *The Unified Neutral Theory of Biodiversity and Biogeography*, Princeton: Princeton University Press.

Hudson, P. J. (1985). Harvesting red grouse in the north of England. In *Game Harvest Management*, ed. S. Beasom & S. F. Roberson, pp. 319–332. Kingsville, U.S.A: Texas A & I.

Hudson, P. J. (1986a). The effect of a parasitic nematode on the breeding production of red grouse. *Journal of Animal Ecology*, **55**, 85–94.

Hudson, P. J. (1986b). *Red Grouse. The Biology of a Managed Game Bird*. Fordingbridge: Game Conservancy Trust.

Hudson, P. J. (1986c). Bracken and ticks on grouse moors in the north of England. In *Bracken: Ecology, Land Use and Control Technology*, ed. R. T. Smith, pp. 161–170. Carnforth: Parthenon Press.

Hudson, P. J. (1992). *Grouse in Space and Time*, Fordingbridge: Game Conservancy Trust.

Hudson, P. J. & Dobson, A. P. (1996). Transmission dynamics and host–parasite interactions of *Trichostrongylus tenuis* in red grouse. *Journal of Parasitology*, **83**, 194–202.

Hudson, P. J. & Dobson, A. P. (2001). Harvesting unstable populations: Red grouse *Lagopus lagopus scoticus* (Lath.) in the United Kingdom. *Wildlife Biology*, **7**, 189–196.

Hudson, P. J., Dobson, A. P. & Newborn, D. (1985). Cyclic and non-cyclic populations of red grouse: a role for parasitism? In *Ecology and Genetics of Host–Parasite Interactions*, ed. D. Rollinson & R. M. Anderson, pp. 79–89. Academic Press: London.

Hudson, P. J., Dobson, A. P. & Newborn, D. (1992a). Do parasites make prey vulnerable to predation? Red grouse and parasites. *Journal of Animal Ecology*, **61**, 681–692.

Hudson, P. J., Dobson, A. P. & Newborn, D. (1998). Prevention of population cycles by parasite removal. *Science*, **228**, 2256–2258.

Hudson, P. J., Dobson, A. P. & Newborn, D. (1999). Population cycles and parasitism. *Science*, **286**, 2425.

Hudson, P. J., Dobson, A. P. & Newborn, D. (2002). Parasitic worms and the population cycles of red grouse (*Lagopus lagopus scoticus*). In *Population Cycles*, ed. A. Berryman. Oxford University Press: New York (in press).

Hudson, P. J. & Newborn, D. (1995). *A Handbook of Grouse and Moorland Management*, Fordingbridge: The Game Conservancy Trust.

Hudson, P. J., Newborn, D. & Dobson, A. P. (1992b). Regulation and stability of a free-living host-parasite system, *Trichostrongylus tenuis* in red grouse. I: Monitoring and parasite reduction experiments. *Journal of Animal Ecology*, **61**, 477–486.

Hudson, P. J. Newborn, D. N. & Robertson, P. J. (1997). Seasonal and geographical patterns of mortality in red grouse populations. *Wildlife Biology*, **2**, 79–88.

Hudson, P. J., Norman, R., Laurenson, M. K., Newborn, D., Gaunt, M., Reid, H. Gould, E., Bowers, R. & Dobson, A. P. (1995). Persistence and transmission of tick-borne viruses: *Ixodes ricinus* and louping-ill virus in red grouse populations. *Parasitology*, **111**, S49–S58.

Huff, D. E. & Varley, J. D. (1999). Natural regulation in Yellowstone National Park's northern range. *Ecological Applications*, **9**, 17–29.

Huggard, D. J. (1993). The effect of snow depth on predation and scavenging by wolves. *Journal of Wildlife Management*, **57**, 382–388.

Hunter, J. E. & Schmidt, F. L. (1990). *Methods of Meta-Analysis*, Newbury Park, California: Sage Publications.

Hutchinson, G. E. (1957). Concluding remarks. *Cold Spring Harbor Symposium on Quantitative Biology*, **22**, 415–427.

Hutchinson, G. E. (1978). *An Introduction to Population Ecology*, New Haven and London: Yale University Press.

Illius, A. W. & Gordon, I. J. (2000). Scaling up from functional response to numerical response in vertebrate herbivores. In *Herbivores; Between Plants and Predators*, ed. H. Oliff, V. K. Brown & R. H. Drent, pp. 397–425. Oxford: Blackwell Science.

Ivlev, V. S. (1961). *Experimental Ecology of the Feeding of Fishes*, New Haven: Yale University Press.

Janowitz, B. S. (1971). An empirical study of the effects of socio-economic development on fertility rate. *Demography*, **8**, 319–330.

Jenkins, D., Watson, A. & Miller, G. R. (1963). Population studies on red grouse, *Lagopus lagopus scoticus* (Lath.) in north-east Scotland. *Journal of Animal Ecology*, **32**, 317–376.

Jensen, A. L. (1999). Using simulation to verify life history relations indicated by time series analysis. *Environmetrics*, **10**, 237–245.

Jewell, P. A. & Grubb, P. (1974). The breeding cycle, the onset of oestrus and conception in Soay sheep. In *Island Survivors: The Ecology of the Soay Sheep of St. Kilda*, ed. P. A. Jewell, C. Milner & J. M. Boyd, pp. 224–241. London: Athlone Press.

Jewell, P. A., Milner, C. & Boyd, J. M. (1974). *Island Survivors: the Ecology of the Soay Sheep of St. Kilda*, London: Athlone Press.

Johnson, A. R., Green, R. E. & Hirons, G. J. M. (1991). Survival rates of greater flamingos in the west Mediterranean region. In *Bird Population Studies*, ed. C. M. Perrins, J.-D. Lebreton & G. J. M. Hirons, pp. 249–271. Oxford: Oxford University Press.

Jones, C. G., Groombridge, J. J. & Nichols, M. (2000). The genetic and population history of the Mauritius kestrel *Falco punctatus*. In *Raptors 2000; Proceedings of The World Conference on Birds of Prey and Owls*. Eilat, Israel, 2–4 April 2000.

Jonkers, D. A. (1986). Status and conservation of the white stork (*Ciconia ciconia* L.) in the Netherlands. In *White Stork. Status and Conservation*, ed. G. Rheinwald, J. Ogden & H. Sculz, pp. 45–54. Braunschweig: Dachverband Deutscher Avifaunisten.

Jonzén, N., Lundberg, P., Ranta, E. & Kaitala, V. (2002). The irreducible uncertainty of the demography-environment interaction in ecology. *Proceedings of the Royal Society London B*, **269**, 221–226.

Jouventin, P. & Guillotin, M. (1979). Socioecologie du skua antarctique a Ponte geologie. *Terre et Vie*, **38**, 109–127.

Kachi, N. (1983). Population dynamics and life history strategy of *Oenothera erythrosepala* in a sand dune system. PhD thesis, University of Tokyo.

Kachi, N. & Hirose, T. (1985). Population dynamics of *Oenothera glazioviana* in a sand-dune system with special reference to the adaptive significance of size-dependent reproduction. *Journal of Ecology*, **73**, 887–901.

Kaitala, V. & Ranta, E. (2001). Is the impact of environmental noise visible in the dynamics of age-structured populations? *Proceedings of the Royal Society London B*, **268**, 1769–1774.

Kaitala, V., Ylikarjula, J., Ranta, E. & Lundberg, P. (1997). Population dynamics and the colour of environmental noise. *Proceedings of the Royal Society London B*, **264**, 943–948.

Kareiva, P. (1989). Renewing the dialogue between theory and experiments in population ecology. In *Perspectives in Ecological Theory*, ed. J. Roughgarden, R. M. May & S. A. Levin, pp. 68–88. Princeton: Princeton University Press.

Karels, T. J. & Boonstra, R. (2000). Concurrent density dependence and independence in populations of arctic ground squirrels. *Nature*, **408**, 460–461.

Karlin, S. & Taylor, H. M. (1975). *A First Course in Stochastic Processes*, New York: Academic Press.

Karlin, S. & Taylor, H. M. (1981). *A Second Course in Stochastic Processes*, New York: Academic Press.

Kaunzinger, C. M. K. & Morin, P. J. (1998). Productivity controls food-chain properties in microbial communities. *Nature*, **395**, 495–497.

Kay, C. E. (1998). Are ecosystems structured from the top–down or bottom–up: a new look at an old debate. *Wildlife Society Bulletin*, **26**, 484–498.

Kendall, B. E., Briggs, C. J., Murdoch, W. W., Turchin, P., Ellner, S. P., McCauley, E., Nisbet, R. M. & Wood, S. N. (1999). Why do populations cycle? A synthesis of statistical and mechanistic modeling approaches. *Ecology*, **80**, 1789–1805.

Kendall, M. G. & Stuart, A. (1977). *The Advanced Theory of Statistics. Vol. 1, Distribution Theory*, 4th edn. New York: Hafner Publishing.

Kendall, M. G., Stuart, A. & Ord, J. K. (1983). *The Advanced Theory of Statistics. Vol. 3. Design and Analysis, and Time Series*, 4th edn. London: Griffin.

Kermack, W. O. & McKendrick, A. G. (1927). Contributions to the mathematical theory of epidemics. *Proceedings of the Royal Society London A*, **115**, 700–721.

Kingsland, S. E. (1996). Evolutionary theory and the foundations of population ecology: the work of A. J. Nicholson (1895–1969). In *Frontiers of Population Ecology*, ed. R. B. Floyd, A. W. Sheppard & P. J. De Barro, pp. 13–25. Melbourne: CSIRO Publishing.

Kinnear, J. E., Onus, M. L. & Bromilow, R. N. (1988). Fox control and rock-wallaby population dynamics. *Australian Wildlife Research*, **15**, 435–450.

Kinnear, J. E., Onus, M. L. & Bromilow, R. N. (1998). Fox control and rock-wallaby population dynamics: II, an update. *Wildlife Research*, **25**, 81–88.

Klemola, T., Korpimaki, E. & Koivula, M. (2002). Rate of population change in voles from different phases of the population cycle. *Oikos*, **96**, 291–298.

Klüttgen, B. & Ratte, H. T. (1994). Effects of different food doses on cadmium toxicity to *Daphnia magna*. *Environmental Toxicology and Chemistry*, **13**, 1619–1627.

Kohler, H. -P. (1997). Learning in social networks and contraceptive choice. *Demography*, **34**, 369–383.

Kokko, H. & Sutherland, W. J. (1998). Optimal floating strategies: consequences for density dependence and habitat loss. *American Naturalist*, **152**, 354–366.

Kokko, H., Sutherland, W. J. & Johnstone, R. A. (2001). The logic of territory choice: implications for conservation and source–sink dynamics. *American Naturalist*, **157**, 459–463.

Kooijman, S. A. L. M. (1993). *Dynamic Energy Budgets in Biological Systems*, Cambridge: Cambridge University Press.

Korpimäki, E. (1992). Fluctuating food abundance determines the lifetime reproductive success of male Tengmalm's owls. *Journal of Animal Ecology*, **61**, 103–111.

Korpimäki, E. (1994). Rapid or delayed tracking of multi-annual vole cycles by avian predators? *Journal of Animal Ecology*, **63**, 619–628.

Korpimäki, E. & Norrdahl, K. (1989). Predation of Tengmalm's owls: numerical responses, functional responses and dampening impact on population fluctuations of microtines. *Oikos*, **54**, 154–164.

Krebs, C. J. (1985). *Ecology: The Experimental Analysis of Distribution and Abundance*, 3rd edn. New York: Harper & Row.

Krebs, C. J. (1995). Two paradigms of population regulation. *Wildlife Research*, **22**, 1–10.

Krebs, C. J. (2002). Two complementary paradigms for analyzing population dynamics. *Philosophical Transactions of the Royal Society London*, **357**, 1211–1219.

Krebs, C. J., Boonstra, R., Boutin, S. & Sinclair, A. R. E. (2001c). What drives the 10-year cycle of snowshoe hares? *Bioscience*, **51**, 25–35.

Krebs, C. J., Boutin, S. & Boonstra, R. (2001a). *Ecosystem Dynamics of the Boreal Forest: the Kluane Project*, New York: Oxford University Press.

Krebs, C. J., Boutin, S., Boonstra, R., Sinclair, A. R. E., Smith, J. N. M., Dale, M. R. T., Martin, K. & Turkington, R. (1995). Impact of food and predation on the snowshoe hare cycle. *Science*, **269**, 1112–1115.

Krebs, C. J., Dale, M. R. T., Nams, V., Sinclair, A. R. E. & O'Donoghue, M. (2001b). Shrubs. In *Ecosystem Dynamics of the Boreal Forest: the Kluane Project*, ed. C. J. Krebs, S. Boutin & R. Boonstra, pp. 92–115. New York: Oxford University Press.

Krebs, C. J., Sinclair, A. R. E., Boonstra, R., Boutin, S., Martin, K. & Smith, J. N. M. (1999). Community dynamics of vertebrate herbivores: how can we untangle the web? In *Herbivores: Between Plants and Predators*, ed. H. Olff, V. K. Brown & R. H. Drent, pp. 447–473. Oxford: Blackwell Science.

Krebs, J. R. & Davies, N. B. (1993). *An Introduction to Behavioural Ecology*, 3rd edn. Oxford: Blackwell Science.

Kruuk, L. E. B., Clutton-Brock, T. H., Albon, S. D., Pemberton, J. M. & Guinness, F. E. (1999). Population density affects sex ratio variation in red deer. *Nature*, **399**, 459–461.

Kuhn, T. (1970). *The Structure of Scientific Revolutions*, Chicago: University of Chicago Press.

Lacy, R. C. (1993). VORTEX: A computer simulation model for population viability analysis. *Wildlife Research*, **20**, 45–64.

Lahaye, W. S., Gutierrez, R. J. & Akcakaya, R. H. (1994). Spotted owl metapopulation dynamics in southern California. *Journal of Animal Ecology*, **63**, 775–785.

Lande, R. (1982). A quantitative genetic theory of life-history evolution. *Ecology*, **63**, 607–615.

Lande, R. (1988). Demographic models of the northern spotted owl (*Strix occidentalis caurina*). *Oecologia (Berlin)*, **75**, 601–607.

Lande, R. (1993). Risks of population extinction from demographic and environmental stochasticity and random catastrophes. *American Naturalist*, **142**, 911–927.

Lande, R. (1998). Demographic stochasticity and Allee effect on a scale with isotrophic noise. *Oikos*, **83**, 353–358.

Lande, R., Engen, S. & Sæther, B.-E. (1998). Extinction times in finite metapopulation models with stochastic local dynamics. *Oikos*, **83**, 383–389.

Lande, R., Engen, S. & Sæther, B.-E. (1999). Spatial scale of population synchrony: environmental correlation versus dispersal and density regulation. *American Naturalist*, **154**, 271–281.

Lande, R., Engen, S. & Sæther, B.-E. (2003). *Stochastic Population Models in Ecology and Conservation*, Oxford: Oxford University Press.

Lande, R., Engen, S., Sæther, B.-E., Filli, F., Matthysen, E. & Weimerskirch, H. (2002). Estimating density dependence from population time series using demographic theory and life history data. *American Naturalist*, **159**, 321–337.

Larter, N. C., Sinclair, A. R. E., Ellsworth, T., Nishi, J. & Gates, C. C. (2000). Dynamics of reintroduction in an indigenous large ungulate: the wood bison of northern Canada. *Animal Conservation*, **3**, 299–309.

Laurenson, M. K., Hudson, P. J., McGuire, K., Thirgood, S. J. & Reid, H.W. (1998). Efficacy of acaricidal tags and pour-on as prophylaxis against ticks and louping-ill in red grouse. *Medical and Veterinary Entomology*, **11**, 389–393.

Lawler, S. P. & Morin, P. J. (1993). Food web architecture and population dynamics in laboratory microcosms of protists. *American Naturalist*, **141**, 675–686.

Laws, R. M., Parker, I. S. C. & Johnstone, R. C. B. (1975). *Elephants and their Habitats: the Ecology of Elephants in North Bunyoro, Uganda*, Oxford: Clarendon Press.

Leader-Williams, N. & Albon, S. D. (1988). Allocation of resources for conservation. *Nature*, **336**, 533–535.

Lebreton, J.-D. & Clobert, J. (1991). Bird population dynamics, management, and conservation: the role of mathematical modelling. In *Bird Population Studies*, ed. C. M. Perrins, J.-D. Lebreton & G. J. M. Hirons, pp. 105–125. Oxford: Oxford University Press.

Lee, A. K. & Cockburn, A. (1985). *Evolutionary Ecology of Marsupials*, Cambridge: Cambridge University Press.

Leet, D. R. (1977). Interrelations of population density, urbanization, literacy, and fertility. *Exploration and Economic History*, **14**, 388–401.

Lcirs, H., Stenseth, N. C., Nichols, J. D., Hines, J. E., Verhagen, R. & Verheyen, W. (1997). Stochastic seasonality and nonlinear density-dependent factors regulate population size in an African rodent. *Nature*, **389**, 176–180.

Lemke, T. O., Mack, J. A. & Houston, D. B. (1998). Winter range expansion by the northern Yellowstone elk herd. *Intermountain Journal of Science*, **4**, 1–9.

Leslie, P. H. (1945). On the use of matrices in certain population mathematics. *Biometrika*, **33**, 183–212.

Leslie, P. H. (1948). Some further notes on the use of matrices in population mathematics. *Biometrika*, **35**, 213–245.

Levin, L. A., Caswell, H., De Patra, K. D. & Creed, E. L. (1987). The life table consequences of larval development mode: an intraspecific comparison for planktotrophy and lecithotrophy. *Ecology*, **68**, 1877–1886.

Levine, J. M. & Rees, M. (2002). Coexistence and relative abundance in annual plant assemblages: the role of competition and colonization. *American Naturalist*, **160**, 452–467.

Lewellen, R. H. & Vessey, S. H. (1998). The effect of density dependence and weather on population size of a polyvoltine species. *Ecological Monographs*, **68**, 571–594.

Lewis, M. A. & Murray, J. D. (1993). Modelling territoriality and wolf-deer interactions. *Nature*, **366**, 738–740.

Lewontin, R. C. & Cohen, D. (1969). On population growth in a randomly varying environment. *Proceedings of National Academy of Sciences USA*, **62**, 1056–1060.

Lidicker, W. Z. Jr. (1994). Population ecology. In *Seventy-five Years of Mammalogy*, Special Publication. ed. E. C. Birney & J. R. Choate, pp. 323–347. Lawrence, Kansas: American Society of Mammalogists.

Liess, M. (2002). Population response to toxicants is altered by intraspecific interaction. *Environmental Toxicology and Chemistry*, **21**, 138–142.

Liley, D. (1999). Predicting the consequences of human disturbance, predation and sea-level rise for ringed plover populations. Ph.D. thesis University of East Anglia, Norwich.

Lima, M. & Jaksic, F. M. (1999). Population rate of change in the leaf-eared mouse: the role of density-dependence, seasonality and rainfall. *Australian Journal of Ecology*, **24**, 110–116.

Lindström, J., Coulson, T. N., Kruuk, L. E. B., Forchhammer, M. C., Coltman, D. & Clutton-Brock, T. H. (2002). Sex-ratio variation in Soay sheep. *Behavioural Ecology and Sociobiology*, **53**, 25–30.

Lingjaerde, O. C., Stenseth, N. C., Kristoffersen, A. B., Smith, R. H., Moe, S. J., Read, J. M., Daniels, S. & Simkiss, K. (2001). Exploring the density-dependent structure of blowfly populations by nonparametric additive modeling. *Ecology*, **82**, 2645–2658.

Linke-Gamenick, I., Forbes, V. E. & Mèndez, M. N. (2000). Effects of chronic fluoranthene exposure on sibling species of *Capitella* with different development modes. *Marine Ecology Progress Series*, **203**, 191–203.

Linke-Gamenick, I., Forbes, V. E. & Sibly, R. M. (1999). Density-dependent effects of a toxicant on life-history traits and population dynamics of a capitellid polychaete. *Marine Ecology Progress Series*, **184**, 139–148.

Litzbarski, B. & Litzbarski, H. (1996). Zür situation der grosstrappe *Otis tarda* in Deutschland. *Vogelwelt*, **117**, 213–224.

Livdahl, T. P. (1982). Competition within and between hatching cohorts of a treehole mosquito. *Ecology*, **63**, 1751–1760.

Lockwood, M. (1997). Sons of the soil? Population growth, environmental change and men's reproductive intentions in northern Nigeria. *International Journal of Population and Geography*, **3**, 305–322.

Lotka, A. J. (1925). *Elements of Physical Biology*, Baltimore: Williams & Wilkins.

Lovat, L. (1911). *The Grouse in Health and Disease*, London: Smith, Elder and Co.

Luckinbill, L. S. (1973). Coexistence in laboratory populations of *Paramecium aurelia* and its predator *Didinium nasutum*. *Ecology*, **54**, 1320–1327.

Lundberg, P., Ranta, E., Ripa, J. & Kaitala, V. (2000). Population variability in space and time. *Trends in Ecology and Evolution*, **15**, 460–464.

Lutz, W. (1994). *Population–Development–Environment: Understanding their Interactions in Mauritius*, Berlin: Springer.

Lutz, W. (1996). *The Future Population of the World: What can we Assume Today?* revised edn London: Earthscan.

Lutz, W. (1997). Determinants and consequences of the world's most rapid fertility decline on the island of Mauritius. Paper presented at the Annual Meeting of the Population Association of America. Washington, DC, USA. 27–29 March 1997.

Lutz, W. & Qiang, R. (2002). Determinants of human population growth. *Philosophical Transactions of the Royal Society London B*, **357**, 1197–1210.

Lutz, W., Sanderson, W. & Scherbov, S. (1996). Probabilistic population projections based on expert opinion. In *The Future Population of the World: What can we Assume Today?* revised edn, ed. W. Lutz, pp. 397–428. London: Earthscan.

Lutz, W., Sanderson, W. & Scherbov, S. (1997). Doubling of world population unlikely. *Nature*, **387**, 803–805.

Lutz, W., Sanderson, W. & Scherbov, S. (2001). The end of world population growth. *Nature*, **412**, 543–545.

McCarthy, M. A. (1996). Red kangaroo (*Macropus rufus*) dynamics: effects of rainfall, density dependence, harvesting and environmental stochasticity. *Journal of Applied Ecology*, **33**, 45–53.

McEvoy, P. B. & Coombs, E. M. (1999). Biological control of plant invaders: regional patterns, field experiments, and structured population models. *Ecological Applications*, **9**, 387–401.

McLeod, S. (1997). Is the concept of carrying capacity useful in a variable environment? *Oikos*, **79**, 529–542.

McNaughton, S. J. (1979a). Grassland-herbivore dynamics. In *Serengeti: Dynamics of an Ecosystem*, ed. A. R. E. Sinclair & M. Norton-Griffiths, pp. 46–83. Chicago: University of Chicago Press.

McNaughton, S. J. (1979b). Grazing as an optimization process: grass-ungulate relationships in the Serengeti. *American Naturalist*, **113**, 691–703.

McNaughton, S. J. (1983). Compensatory plant growth as a response to herbivory. *Oikos*, **40**, 329–336.

McQueen, D. J. (1998). Freshwater food web biomanipulation: a powerful tool for water quality improvement, but maintenance is required. *Lakes and Reservoirs: Research and Management*, **3**, 83–94.

Maas, S. (1997). Population dynamics and control of feral goats in a semi-arid environment. M. Applied Science thesis, University of Canberra, Canberra.

MacDonald, N. (1978). *Time Lags in Biological Models*, Berlin: Springer-Verlag.

Mack, J. A. & Singer, F. J. (1993). Predicted effects of wolf predation on northern range elk using POP-II models. In *Ecological Issues on Reintroducing Wolves into Yellowstone National Park*, Science Monograph 22, ed. R. S. Cook, pp. 306–326. Washington DC: National Parks Service.

Maguire, B. (1973). Niche response structure and the analytical potentials of its relationship to the habitat. *American Naturalist*, **107**, 213–246.

Malthus, T. R. (1798). *An essay on the principle of population*. London: J. Johnson.

Mangel, M. & Tier, C. (1994). Four facts every conservation biologist should know about persistence. *Ecology*, **75**, 607–614.

Marshall, J. S. (1978). Population dynamics of *Daphnia galeata mendotae* as modified by chronic cadmium stress. *Journal of the Fisheries Research Board Canada*, **35**, 461–469.

Mason, K. O. (1997). Explaining fertility transitions. *Demography*, **34**, 443–454.

May, R. M. (1973). *Stability and Complexity in Model Ecosystems*, Princeton: Princeton University Press.

May, R. M. (1974). *Stability and Complexity in Model Ecosystems*, 2nd edn. Princeton: Princeton University Press.

May, R. M. (1976). Simple mathematical models with very complicated dynamics. *Nature*, **261**, 459–467.

May, R. M. (1977). Thresholds and breakpoints in ecosystems with a multiplicity of stable states. *Nature*, **269**, 471–477.

May, R. M. (1981a). Models for two interacting populations. In *Theoretical Ecology. Principles and Applications*, 2nd edn, ed. R. M. May, pp. 78–104. Oxford: Blackwell Science.

May, R. M. (1981b). Models for single populations. In *Theoretical Ecology. Principles and Applications*, 2nd edn, ed. R. M. May, pp. 5–29. Oxford: Blackwell Science.

May, R. M. (2000). Ornithology: British birds by number. *Nature*, **404**, 559–560.

May, R. M., Conway, G. R., Hassell, M. P. & Southwood, T. R. E. (1974). Time delays, density-dependence and single-species oscillations. *Journal of Animal Ecology*, **43**, 747–770.

Mduma, S. A. R., Sinclair, A. R. E. & Hilborn, R. (1999). Food regulates the Serengeti wildebeest: a 40-year record. *Journal of Animal Ecology*, **68**, 1101–1122.

Mendenhall, W., Scheaffer, R. L. & Wackerly, D. D. (1981). *Mathematical Statistics with Applications*, 2nd edn. Boston, Massachusetts: Duxbury Press.

Messier, F. (1991). The significance and limiting and regulating factors on the demography of moose and white-tailed deer. *Journal of Animal Ecology*, **60**, 377–393.

Messier, F. (1994). Ungulate population models with predation: a case study with the North American moose. *Ecology*, **75**, 478–488.

Messier, F. & Crête, M. (1984). Body condition and population regulation by food resources in moose. *Oecologia (Berlin)*, **65**, 44–50.

Metz, J. A. J., Nisbet, R. M. & Geritz, S. A. H. (1992). How should we define 'fitness' for general ecological scenarios? *Trends in Ecology and Evolution* **7**, 198–202.

Miller, G. R. (1979). Quantity and quality of the annual production of shrubs and flowers of *Calluna vulgaris* in North-east Scotland. *Journal of Ecology*, **67**, 109–129.

Mills, L. S., Hayes, S. G., Baldwin, C., Wisdom, M. J., Citta, J., Mattson, D. J. & Murphy, K. (1996). Factors leading to different viability predictions for a grizzly bear data set. *Conservation Biology*, **10**, 863–873.

Milner, J. M., Elston, D. A. & Albon, S. D. (1999). Estimating the contributions of population density and climatic fluctuations to interannual variation in survival of Soay sheep. *Journal of Animal Ecology*, **68**, 1235–1247.

Milner-Gulland, E. J. (1994). A population model for the management of the saiga antelope. *Journal of Applied Ecology*, **31**, 25–39.

Mitchell, B. & Brown, D. (1974). The effects of age and body size on fertility in female red deer (*Cervus elaphus* L.). *Proceedings of the International Congress on Game Biology*, **11**, 89–98.

Mitchell, B., McCowan, D. & Nicholson, I. A. (1976). Annual cycles of body weight and condition in Scottish red deer, *Cervus elaphus*. *Journal of Zoology, London*, **180**, 107–127.

Møller, A. P. (2001).The effects of dairy farming on barn swallow *Hirundo rustica* abundance, distribution and reproduction. *Journal of Applied Ecology*, **38**, 378–389.

Monod, J. (1950). La technique de culture continue: Theorie et applications. *Annals of Institute Pasteur*, **79**, 390–410.

Montgomery, W. I. (1989). Population regulation in the wood mouse *Apodemus sylvaticus* I. Density dependence in the annual cycle of abundance. *Journal of Animal Ecology*, **58**, 465–476.

Morris, K., Orel, P. & Brazell, R. (1995). The effect of fox control on native mammals in the jarrah forest, Western Australia. *Proceedings of the 10th Australian Vertebrate Pest Control Conference*, ed. M. Statham, pp. 177–181. Hobart: Department of Primary Industries & Fisheries.

Morris, R. & Smith, H. (1988). *Wild South. Saving New Zealand's Endangered Birds*, Auckland: TVNZ & Century Hutchinson NZ Ltd.

Morrison, M. L. (2001). A proposed research emphasis to overcome the limits of wildlife-habitat relationship studies. *Journal of Wildlife Management*, **65**, 613–623.

Moss, R., Parr, R. & Lambin, X. (1994). Effects of testosterone on breeding density, breeding success and survival of red grouse. *Proceedings of the Royal Society London B*, **258**, 175–180.

Moss, R. & Watson, A. (1991). Population cycles and kin selection in red grouse *Lagopus lagopus scoticus*. *Ibis*, **133** supplement 1, 113–120.

Moss, R. & Watson, A. (2000). Population cycles in birds of the grouse family (Tetraonidae). *Advances in Ecological Research*, **32**, 54–111.

Murdoch, W. W. (1994). Population regulation in theory and practice. *Ecology*, **75**, 271–287.

Murdoch, W. W., Nisbet, R. M., Blythe, S. P., Gurney, W. S. C. & Reeve, J. D. (1987). An invulnerable age class and stability in delay-differential parasitoid–host models. *American Naturalist*, **129**, 263–282.

Murray, B. G. J. (1999). Can the population regulation controversy be buried and forgotten? *Oikos*, **84**, 148–152.

Murray, B. G. J. (2000). Dynamics of an age-structured population drawn from a random numbers table. *Austral Ecology*, **25**, 297–304.

Myers, J. H. (1993). Population outbreaks in forest Lepidoptera. *American Scientist*, **81**, 240–250.

Myers, R. A., Bowen, K. G. & Barrowman, N. J. (1999). Maximum reproductive rate of fish at low densities. *Canadian Journal of Fish and Aquatic Sciences*, **56**, 2404–2419.

Myers, R. A., MacKenzie, B. R., Bowen, K. G. & Barrowman, N. J. (2001). What is the carrying capacity for fish in the ocean? A meta-analysis of population dynamics of North Atlantic cod. *Canadian Journal of Fish and Aquatic Sciences*, **58**, 1464–1476.

Mylius, S. D. & Diekmann, O. (1995). On evolutionarily stable life histories, optimization and the need to be specific about density dependence. *Oikos*, **74**, 218–224.

National Research Council (Panel on Population Projections). (2000). *Beyond Six Billion: Forecasting the World's Population*, ed. J. Bongaarts & R. A. Bulatao. Washington, DC: National Academy Press.

Nelson, B. (1978). *The Gannet*, Berkhamsted: T. & A. D. Poyser.

Newsome, A., Pech, R., Smyth, R., Banks, P. & Dickman, C. (1997). *Potential impacts on Australian native fauna of rabbit calicivirus disease*. Canberra: Environment Australia.

Newton, I. (1998). *Population Limitation in Birds*, London: Academic Press.

Nichols, J. D., Hines, J. E. & Blums, P. (1997). Tests for senescent decline in annual survival probabilities of common potchards, *Aythya ferina*. *Ecology*, **78**, 1009–1018.

Nichols, J. D., Hines, J. E., Lebreton, J-D. & Pradel, R. (2000). Estimation of contributions to population growth: a reverse-time capture–recapture approach. *Ecology*, **81**, 3362–3376.

Nicholson, A. J. (1933). The balance of animal populations. *Journal of Animal Ecology*, **2**, 132–178.

Nicholson, A. J. & Bailey, V. A. (1935). The balance of animal populations. Part 1. *Proceedings of the Zoological Society of London*, **3**, 551–598.

Nisbet, R. M. (1997). Delay-differential equations for structured populations. In *Structured Population Models in Marine, Terrestrial, and Freshwater Systems*, ed. S. Tuljapurkar & H. Caswell, pp. 89–118. New York: Chapman & Hall.

Nisbet, R. M. & Gurney, W. S. C. (1983). The systematic formulation of population models for insects with dynamically varying instar duration. *Theoretical Population Biology*, **23**, 114–135.

Nisbet, R. M., Muller, E. B., Lika, K. & Kooijman, S. A. L. M. (2000). From molecules to ecosystems through dynamic energy budget models. *Journal of Animal Ecology*, **69**, 913–926.

Norris, K. & Stillman, R. (2002). Predicting the impact of environmental change. In *Conserving Bird Biodiversity*, ed. K. Norris & D. J. Pain, pp. 180–201. Cambridge: Cambridge University Press.

North, P. M. & Morgan, B. J. T. (1979). Modelling heron survival using weather data. *Biometrics*, **35**, 667–681.

Norton-Griffiths, M. (1979). The influence of grazing, browsing and fire on the vegetation dynamics of the Serengeti. In *Serengeti: Dynamics of an Ecosystem*, ed. A. R. E. Sinclair & M. Norton-Griffiths, pp. 310–352. Chicago: University of Chicago Press.

Notestein, F. W. (1945). Population – the long view. In *Food for the World*, ed. T. W. Schultz, pp. 36–57. Chicago: University of Chicago Press.

Noy-Meir, I. (1975). Stability of grazing systems: an application of predator–prey graphs. *Journal of Ecology*, **63**, 459–481.

O'Donoghue, M., Boutin, S., Krebs, C. J. & Hofer, E. J. (1997). Numerical responses of coyotes and lynx to the snowshoe hare cycle. *Oikos*, **80**, 150–162.

O'Neill, B. C., MacKellar, F. L. & Lutz, W. (2001). *Population and Climate Change*, Cambridge: Cambridge University Press.

O'Neill, B. C., Scherbov, S. & Lutz, W. (1999). The long-term effect of the timing of fertility decline on population size. *Population Development Research*, **25**, 749–756.

O'Roke, E. C. & Hammerston, F. N. Jr. (1948). Productivity and yield of the George Reserve deer herd. *Journal of Wildlife Management*, **12**, 78–86.

Owen, D. F. (1960). The nesting success of the heron *Ardea cinerea* in relation to the availability of food. *Proceedings of the Zoological Society of London*, **133**, 597–617.

Owen-Smith, N. (1990). Demography of a large herbivore, the greater kudu *Tragelaphus strepsiceros*, in relation to rainfall. *Journal of Animal Ecology*, **59**, 893–913.

Owen-Smith, N. (2002). *Adaptive Herbivore Ecology from Resources to Populations in Variable Environments*, Cambridge: Cambridge University Press.

Pacala, S. W., Canham, C. D., Saponara, J., Silander, J. A., Kobe, R. K. & Ribbens, E. (1996). Forest models defined by field measurements: estimation, error analysis and dynamics. *Ecological Monographs*, **66**, 1–43.

Pacala, S. W. & Silander, J. A. (1990). Field-tests of neighborhood population-dynamic models of 2 annual weed species. *Ecological Monographs*, **60**, 113–134.

Pain, D. J. & Pienkowski, M. W. (1997). *Farming and Birds in Europe*, London: Academic Press.

Paradis, E., Baillie, S. R., Sutherland, W. J. & Gregory, R. D. (2002) Exploring density dependence in demographic parameters in populations of birds at a large spatial scale. *Oikos*, **97**, 293–307.

Parr, R. (1992). The decline to extinction of a population of Golden Plover in north-east Scotland. *Ornis Scandinavia*, **23**, 152–158.

Pearl, R. & Reed, L. J. (1920). On the rate of growth of the population of the United States since 1790 and its mathematical representation. *Proceedings of the National Academy of Sciences (USA)*, **6**, 275–280.

Pech, R. & Hood, G. (1998). Foxes, rabbits, alternative prey and rabbit calicivirus disease: consequences of a new biological control agent for an outbreaking species in Australia. *Journal of Applied Ecology*, **35**, 434–453.

Pech, R. P., Hood, G. M., McIlroy, J. & Saunders, G. (1997). Can foxes be controlled by reducing their fertility? *Reproduction, Fertility and Development*, **9**, 41–50.

Pech, R. P., Hood, G. M., Singleton, G. R., Salmon, E., Forrester, R. I. & Brown, P. R. (1999). Models for predicting plagues of house mice (*Mus domesticus*) in Australia. In *Ecologically-based Rodent Management*, ed. G. R. Singleton, L. Hinds, H. Leirs & Z. Zhang, pp. 81–112. Canberra: ACIAR.

Pech, R. P., Sinclair, A. R. E., Newsome, A. E. & Catling, P. C. (1992). Limits to predator regulation of rabbits in Australia: evidence from predator removal experiments. *Oecologia (Berlin)*, **89**, 102–112.

Pemberton, J. M., Smith, J. A., Coulson, T. N., Marshall, T. C., Slate, T., Paterson, S., Albon, S. D. & Clutton-Brock, T. H. (1996). The maintenance of genetic polymorphism in small island populations: large mammals in the Hebrides. *Philosophical Transactions of the Royal Society London B*, **351**, 745–752.

Pen, I. & Weising, F. J. (2000). Optimal floating and queuing strategies: the logic of territory choice. *American Naturalist*, **155**, 512–526.

Percival, S. M., Sutherland, W. J. & Evans, P. R. (1996). A spatial depletion model of the responses of grazing wildfowl to changes in availability of intertidal vegetation during the autumn and winter. *Journal of Applied Ecology*, **33**, 979–993.

Percival, S. M., Sutherland, W. J. & Evans, P. R. (1998). Intertidal habitat loss and wildfowl numbers: applications of a spatial depletion model. *Journal of Applied Ecology*, **35**, 57–63.

Peterson, R. O. (1999). Wolf–moose interaction on Isle Royale: the end of natural regulation? *Ecological Applications*, **9**, 10–16.

Peterson, R. O., Page, R. E. & Dodge, K. M. (1984). Wolves, moose and the allometry of population cycles. *Science*, **224**, 1350–1352.

Pettifor, R. A., Caldow, R. W. G., Rowcliffe, J. M., Goss-Custard, J. D., Black, J. M., Hodder, K. H., Houston, A. I., Lang, A. & Webb, J. (2000b). Spatially explicit, individual-based, behavioural models of the annual cycle of two migratory goose populations. *Journal of Applied Ecology*, **37**, 103–135.

Pettifor, R. A., Norris, K. & Rowcliffe, M. (2000a). Incorporating behaviour in predictive models for conservation. In *Behaviour and Conservation*, ed. L. M. Gosling & W. J. Sutherland, pp. 198–220. Cambridge: Cambridge University Press.

Picozzi, N. (1966). Grouse management in relation to the management and geology of heather moors. *Journal of Applied Ecology*, **5**, 483–488.

Pimm, S. L. (1991). *The Balance of Nature? Ecological Issues in the Conservation of Species and Communities*, Chicago: University of Chicago Press.

Pimm, S. L. & Redfearn, A. (1988). The variability of population densities. *Nature*, **334**, 613–614.

Pollard, E., Lakhani, K. H. & Rothery, P. (1987). The detection of density-dependence from a series of annual censuses. *Ecology*, **68**, 2046–2055.

Polunin, O. & Walters, M. (1985). *A Guide to the Vegetation of Britain and Europe*, Oxford: Oxford University Press.

Population Census Office under the State Council, Department of Population, Social, Science and Technology Statistics, & National Bureau of Statistics of the People's Republic of China. (2001). *Major Figures on 2000 Population Census of China*. Beijing: China Statistics Press.

Porter, J. M. & Coulson, J. C. (1987). Long-term changes in recruitment to the breeding group, and the quality of recruits at a kittiwake *Rissa tridactyla* colony. *Journal of Animal Ecology*, **56**, 675–689.

Power, M. E. (1992). Top–down and bottom–up forces in food webs: do plants have primacy? *Ecology*, **73**, 733–746.

Pratt, H. D., Bruner, P. L. & Berrett, D. G. (1987). *The Birds of Hawaii and the Tropical Pacific*, Princeton: Princeton University Press.

Pulliam, H. R. (1988). Sources, sinks and population regulation. *American Naturalist*, **132**, 652–661.

Putman, R. J., Langbein, J., Hewison, A. J. M. & Sharma, S. K. (1996). Relative roles of density-dependence and density-independent factors in population dynamics of British deer. *Mammalian Review*, **26**, 81–101.

Rand, D. A., Wilson, H. B. & McGlade, J. M. (1994). Dynamics and evolution – evolutionarily stable attractors, invasion exponents and phenotype dynamics. *Philosophical Transactions of the Royal Society London*, **343**, 261–283.

Ranta, E., Lundberg, P., Kaitala, V. & Laakso, J. (2000). Visibility of the environmental noise modulating population dynamics. *Proceedings of the Royal Society London B*, **267**, 1851–1856.

Rees, M. & Rose, K. E. (2002). Evolution of flowering strategies in Oenothera glazioviana: an integral projection model approach. *Proceedings of the Royal Society London B*, **269**, 1509–1515.

Rees, M., Bjornstad, O. N. & Kelly, D. (2002). Snow grass, chaos and the evolution of mast seeding. *American Naturalist*, **160**, 44–59.

Rees, M., Grubb, P. J. & Kelly, D. (1996). Quantifying the impact of competition and spatial heterogeneity on the structure and dynamics of a four-species guild of winter annuals. *American Naturalist*, **147**, 1–32.

Rees, M., Mangel, M., Turnbull, L. A., Sheppard, A. & Briese, D. (2000). The effects of heterogeneity on dispersal and colonisation in plants. In *Ecological Consequences of Environmental Heterogeneity*, ed. M. J. Hutchings, E. A. John & A. J. A. Stewart, pp. 237–265. Oxford: Blackwell.

Reid, H. W., Duncan, J. S., Phillips, J. D. P., Moss, R. & Watson, A. (1978). Studies of louping ill virus in wild red grouse (*Lagopus lagopus scoticus*). *Journal of Hygiene*, **81**, 321–329.

Renshaw, E. (1991). *Modelling Biological Populations in Space and Time*, Cambridge: Cambridge University Press.

Reynolds, P. E. (1998). Dynamics and range expansion of a reestablished muskox population. *Journal of Wildlife Management*, **62**, 734–744.

Reynolds, R. E. & Sauer, J. R. (1991). Changes in mallard breeding populations in relation to production and harvest rates. *Journal of Wildlife Management*, **55**, 483–487.

Ricker, W. E. (1982). Size and age of British Columbia sockeye salmon (*Oncorhynchus nerka*) in relation to environmental factors and the fishery. *Canadian Technical Reports in Fisheries and Aquatic Sciences*, **1115**, 1–117.

Ricklefs, R. E. (1979). *Ecology*, 2nd edn. New York: Chiron Press.

Ricklefs, R. E. & Miller, G. L. (1999). *Ecology*, 4th edn. New York: Freeman & Co.

Ripa, J., Lundberg, P. & Kaitala, V. (1998). A general theory of environmental noise in ecological food webs. *American Naturalist*, **151**, 256–263.

Robertson, G. (1987a). Plant dynamics. In *Kangaroos: Their Ecology and Management in the Sheep Rangelands of Australia*, ed. G. Caughley, N. Shepherd & J. Short, pp. 50–68. Cambridge: Cambridge University Press.

Robertson, G. (1987b). Effect of drought and high summer rainfall on biomass and composition of grazed pastures in western New South Wales. *Australian Rangeland Journal*, **9**, 79–85.

Robinson, P. A. & Redford, K. H. (1991). Sustainable harvest of neotropical forest mammals. In *Neotropical Wildlife Use and Conservation*, ed. J. G. Robinson & K. H. Redford, pp. 415–429. Chicago: University of Chicago Press.

Rochat, J. & Gutierrez, A. P. (2001). Weather-mediated regulation of olive scale by two parasitoids. *Journal of Animal Ecology*, **70**, 476–490.

Rogers, J. C. (1984). The association between the North Atlantic Oscillation and the Southern Oscillation in the northern hemisphere. *Monthly Weather Review*, **112**, 1999–2015.

Roland, J. (1994). After the decline: what maintains low winter moth density after successful biological control? *Journal of Animal Ecology*, **63**, 392–398.

Root, K. V. (1998). Evaluating the effects of habitat quality, connectivity and catastrophes on a threatened species. *Ecological Applications*, **8**, 854–865.

Rose, K. E., Clutton-Brock, T. H. & Guinness, F. E. (1998). Cohort variation in male survival and mating success in red deer, *Cervus elaphus*. *Journal of Animal Ecology*, **67**, 979–986.

Rose, K. E., Rees, M. & Grubb, P. J. (2002). Evolution in the real world: stochastic variation and the determinants of fitness in *Carlina vulgaris*. *Evolution*, **56**, 1416–1430.

Rosenzweig, M. L. (1971). Paradox of enrichment: destabilization of exploitation ecosystems in ecological time. *Science*, **171**, 385–387.

Rosenzweig, M. L. & MacArthur, R. H. (1963). Graphical representation and stability conditions of predator–prey interactions. *American Naturalist*, **47**, 209–223.

Royama, T. (1981). Fundamental concepts and methodology for the analysis of animal population dynamics, with particular reference to univoltine species. *Ecological Monographs*, **51**, 473–493.

Royama, T. (1992). *Analytical Population Dynamics*, London: Chapman & Hall.

Sæther, B.-E. (1988). Pattern of covariation between life-history traits of European birds. *Nature*, **331**, 616–617.

Sæther, B.-E. (1997). Environmental stochasticity and population dynamics of large herbivores: a search for mechanisms. *Trends in Ecology and Evolution*, **12**, 143–149.

Sæther, B.-E. & Bakke, O. (2000). Avian life history variation and contribution of demographic traits to the population growth rate. *Ecology*, **81**, 642–653.

Sæther, B.-E. & Engen, S. (2002a). Pattern of variation in avian population growth rates. *Philosophical Transactions of the Royal Society London*, **357**, 1185–1195.

Sæther, B.-E. & Engen, S. (2002b). Including uncertainties in population viability analysis. In *Population Viability Analysis*, ed. S. R. Beissinger & D. R. McCullough, pp. 191–212. Chicago: University of Chicago Press.

Sæther, B.-E., Engen, S., Islam, A., McCleery, R. H. & Perrins, C. (1998). Environmental stochasticity and extinction risk in a population of a small songbird, the great tit. *American Naturalist*, **151**, 441–450.

Sæther, B.-E., Engen, S. & Lande, R. (1996). Density-dependence and optimal harvesting of fluctuating populations. *Oikos*, **76**, 40–46.

Sæther, B.-E., Engen, S., Lande, R., Arcese, P. & Smith, J. N. M. (2000a). Estimating the time to extinction in an island population of song sparrows. *Proceedings of the Royal Society London B*, **267**, 621–626.

Sæther, B.-E., Engen, S. & Matthysen, E. (2002). Demographic characteristics and population dynamical patterns of solitary birds. *Science*, **295**, 2070–2073.

Sæther, B.-E., Ringsby, T. H., Bakke, O. & Solberg, E. J. (1999). Spatial and temporal variation in demography of a house sparrow population. *Journal of Animal Ecology*, **68**, 628–637.

Sæther, B.-E., Tufto, J., Engen, S., Jerstad, K., Rostad, O. W. & Skatan, J. E. (2000b). Population dynamical consequences of climate change for a small temperate songbird. *Science*, **287**, 854–856.

Sait, S. M., Begon, M. & Thompson, D. J. (1994a). The effects of a sublethal baculovirus infection in the Indian meal moth, *Plodia interpunctella*. *Journal of Animal Ecology*, **63**, 541–550.

Sait, S. M., Begon, M. & Thompson, D. J. (1994b). Long-term population dynamics of the Indian meal moth *Plodia interpunctella* and its granulosis virus. *Journal of Animal Ecology*, **63**, 861–870.

Sait, S. M., Begon, M., Thompson, D. J., Harvey, J. A. & Hails, R. S. (1997). Factors affecting host selection in an insect host–parasitoid interaction. *Ecological Entomology*, **22**, 225–230.

Sait, S. M., Liu, W. C., Thompson, D. J., Godfray, H. C. J. & Begon, M. (2000). Invasion sequence affects predator–prey dynamics in a multi-species interaction. *Nature*, **405**, 448–450.

Saitoh, T., Stenseth, N. C. & Bjornstad, O. N. (1997). Density dependence in fluctuating grey-sided vole populations. *Journal of Animal Ecology*, **66**, 14–24.

Sasaki, A. & Ellner, S. (1995). The evolutionarily stable phenotype distribution in a random environment. *Evolution*, **49**, 337–350.

Sauer, J. R. & Boyce, M. S. (1983). Density dependence and survival of elk in northwestern Wyoming. *Journal of Wildlife Management*, **47**, 31–37.

Schaffer, W. M. (1985). Order and chaos in ecological systems. *Ecology*, **66**, 93–106.

Scheffer, M., Carpenter, S., Foley, J. A., Folkes, C. & Walker, B. (2001). Catastrophic shifts in ecosystems. *Nature*, **413**, 591–596.

Schmitz, O. J. & Sinclair, A. R. E. (1997). Multiple ecosystem states: rethinking the role of deer in forest ecosystem dynamics. In *The Science of Overabundance*, ed. W. McShea, H. B. Underwood & J. H. Rappole, pp. 201–223. Washington DC: Smithsonian Institute Press.

Schnakenwinkel, G. (1970). Studien an der population des austernfischers (*Haematopus ostralegus*) auf mellum. *Vogelwarte*, **25**, 336–355.

Seamans, M. E., Gutierrez, R. J., May, C. A. & Peery, M. Z. (1999). Demography of two Mexican spotted owl populations. *Conservation Biology*, **13**, 744–754.

Seamans, M. E., Gutierrez, R. J., Moen, C. A. & Peery, M. Z. (2001). Spotted owl demography in the central Sierra Nevada. *Journal of Wildlife Management*, **65**, 425–431.

Shaw, D. J. & Dobson, A. P. (1995). Patterns of macroparasite abundance and aggregation in wildlife populations: a quantitative review. *Parasitology*, **111**, S111–S133.

Shaw, D. J., Grenfell, B. T. & Dobson, A. P. (1998). Patterns of macroparasite aggregation in wildlife host populations. *Parasitology*, **117**, 597–610.

Shea, K. & Kelly, D. (1998). Estimating biocontrol agent impact with matrix models: *Carduus nutans* in New Zealand. *Ecological Applications*, **8**, 824–832.

Shenk, T. M., White, G. C. & Burnham, K. P. (1998). Sampling variance effects in detecting density dependence from temporal trends in natural populations. *Ecological Monographs*, **68**, 445–463.

Short, J. (1985). The functional response of kangaroos, sheep and rabbits in an arid grazing system. *Journal of Applied Ecology*, **22**, 435–447.

Short, J. (1987). Factors affecting food intake of rangeland herbivores. In *Kangaroos: Their Ecology and Management in the Sheep Rangelands of Australia*, ed. G. Caughley, N. Shepherd & J. Short, pp. 84–99. Cambridge: Cambridge University Press.

Sibly, R. M. (1999). Efficient experimental designs for studying stress and population density in animal populations. *Ecological Applications*, **9**, 496–503.

Sibly, R. M. & Calow, P. (1989). A life-cycle theory of responses to stress. *Biological Journal of the Linnean Society*, **37**, 101–116.

Sibly, R. M., Hansen, F. T. & Forbes, V. E. (2000a). Confidence limits for population growth rate of organisms with two-stage life histories. *Oikos*, **88**, 335–340.

Sibly, R. M. & Hone, J. (2002). Population growth rate and its determinants: an overview. *Philosophical Transactions of the Royal Society London*, **357**, 1153–1170.

Sibly, R. M., Newton, I. & Walker, C. H. (2000c). Effects of dieldrin on population growth rates of sparrowhawks 1963–1986. *Journal of Applied Ecology*, **37**, 540–546.

Sibly, R. M. & Smith, R. H. (1998). Identifying key factors using λ-contribution analysis. *Journal of Animal Ecology*, **67**, 17–24.

Sibly, R. M., Williams, T. D. & Jones, M. B. (2000b). How environmental stress affects density dependence and carrying capacity in a marine copepod. *Journal of Applied Ecology*, **37**, 388–397.

Silvertown, J., Franco, M., Pisanty, I. & Mendoza, A. (1996). Interpretation of elasticity matrices as an aid to the management of life cycle components to the finite rate of increase in woody and herbaceous perennials. *Journal of Ecology*, **81**, 465–476.

Sinclair, A. R. E. (1977). *The African Buffalo*, Chicago: University of Chicago Press.

Sinclair, A. R. E. (1989). Population regulation in animals. In *Ecological Concepts*, ed. J. M. Cherrett, pp. 197–241. Oxford: Blackwell.

Sinclair, A. R. E. (1995). Equilibria in plant-herbivore interactions. In *Serengeti II: Dynamics, Management and Conservation of an Ecosystem*, ed. A. R. E. Sinclair & P. Arcese, pp. 91–113. Chicago: University of Chicago Press.

Sinclair, A. R. E. (1996). Mammal populations: fluctuation, regulation, life history theory and their implications for conservation. In *Frontiers of Population Ecology*, ed. R. B. Floyd, A. W. Sheppard & P. J. De Barro, pp. 127–154. Melbourne: CSIRO Publishing.

Sinclair, A. R. E., Dublin, H. & Borner, M. (1985). Population regulation of Serengeti wilderbeest: a test of the food hypothesis. *Oecologia (Berlin)*, **65**, 266–268.

Sinclair, A. R. E. & Krebs, C. J. (2002). Complex numerical responses to top–down and bottom–up processes in vertebrate populations. *Philosophical Transactions of the Royal Society London*, **357**, 1221–1231.

Sinclair, A. R. E., Olsen, P. D. & Redhead, T. D. (1990). Can predators regulate small mammal populations?: evidence from house mouse outbreaks in Australia. *Oikos*, **59**, 382–392.

Sinclair, A. R. E., Pech, R. P., Dickman, C. R., Hik, D., Mahon, P. & Newsome, A. E. (1998). Predicting effects of predation on conservation of endangered prey. *Conservation Biology*, **12**, 564–575.

Singer, F. J., Harting, A., Symonds, K. K. & Coughenour, M. B. (1997). Density dependence, compensation, and environmental effects on elk calf mortality in Yellowstone National Park. *Journal of Wildlife Management*, **61**, 12–25.

Singer, F. J. & Mack, J. A. (1999). Predicting the effects of wildfire and carnivore predation on ungulates. In *Carnivores in Ecosystems: The Yellowstone Experience*, ed. T. W. Clark, A. P. Curlee, S. C. Minta & P. M. Karieva, pp. 189–237. New Haven, Connecticut: Yale University Press.

Singer, F. J. & Norland, J. E. (1994). Niche relationships within a guild of ungulate species in Yellowstone National Park, Wyoming, following release from artificial controls. *Canadian Journal of Zoology*, **72**, 1383–1394.

Singer, F. J., Swift, D. M., Coughenour, M. B. & Varley, J. D. (1998). Thunder on the Yellowstone revisited: an assessment of management of native ungulates by natural regulation, 1968–1993. *Wildlife Society Bulletin*, **26**, 375–390.

Skeat, A. (1990). Feral buffalo in Kakadu National Park: survey methods, population dynamics and control. M. Applied Science thesis, University of Canberra, Canberra.

Skogland, T. (1983). The effects of density dependent resource limitation on size of wild reindeer. *Oecologia (Berlin)*, **60**, 156–168.

Skogland, T. (1985). The effects of density dependent resource limitation on the demography of wild reindeer. *Journal of Animal Ecology*, **54**, 359–374.

Skogland, T. (1991). Ungulate foraging strategies: optimization for avoiding predation or competition for limiting resources? *Transactions of the 18th International Union Game Biologists*, pp. 161–167. Krakow, Poland.

Skov, H. (1986). Der status des weisstorchs in Danemark. In *White Stork. Status and Conservation*, ed. G. Rheinwald, J. Ogden & H. Sculz, pp. 55–59. Braunschweig: Dachverband Deutscher Avifaunisten.

Smith, C., Reynolds, J. D. & Sutherland, W. J. (2000). Population consequences of reproductive decisions. *Proceedings of the Royal Society London B*, **267**, 1327–1334.

Smith, C. C. & Fretwell, S. D. (1974). The optimal balance between size and number of offspring. *American Naturalist*, **108**, 499–506.

Smith, F. E. (1963). Population dynamics in *Daphnia magna* and a new model for population growth. *Ecology*, **44**, 651–663.

Smith, J. N. M. (1988). Determinants of lifetime reproductive success in the song sparrow. In *Reproductive Success*, ed. T. H. Clutton-Brock, pp. 154–172. Chicago: University of Chicago Press.

Smith, J. N. M. & Arcese, P. (1986). How does territorial behaviour influence breeding bird numbers? In *Behavioural Ecology and Population Biology*, ed. L. C. Drickamer, pp. 89–94. Toulouse, France: International Ethology Conference.

Smith, J. N. M., Taitt, M. J., Rogers, C. M., Arcese, P., Keller, L. F., Cassidy, A. L. E. V. & Hochachka, W. M. (1996). A metapopulation approach to the population biology of the song sparrow *Melospiza melodia*. *Ibis*, **138**, 120–128.

Smith, V. H. (1983). Low nitrogen to phosphorus ratios favor dominance by blue-green algae in lake phytoplankton. *Science*, **221**, 669–671.

Solberg, E. J., Sæther, B.-E., Strand, O. & Loison, A. (1999). Dynamics of a harvested moose population in a variable environment. *Journal of Animal Ecology*, **68**, 186–204.

Solomon, M. E. (1949). The natural control of animal populations. *Journal of Animal Ecology*, **18**, 1–35.

Solow, A. R. (1998). On fitting a population model in the presence of observation error. *Ecology*, **79**, 1463–1466.

Stearns, S. C. (1992). *The Evolution of Life Histories*, Oxford: Oxford University Press.

Steen, H. & Haydon, D. (2000). Can population growth rates vary with the spatial scale at which they are measured? *Journal of Animal Ecology*, **69**, 659–671.

Stenseth, N. C., Falck, W., Chan, K. S., Bjornstad, O. N., O'Donoghue, M., Tong, H., Boonstra, R., Boutin, S., Krebs, C. J. & Yoccoz, N. G. (1998). From patterns to processes: phase and density dependencies in the Canadian lynx cycle. *Proceedings of the National Academy of Sciences of USA* **95**, 15430–15435.

Stephens, P. A., Frey-Roos, F., Arnold, W. & Sutherland, W. J. (2002). Model complexity and population predictions. The alpine marmot as a case study. *Journal of Animal Ecology*, **71**, 343–361.

Stephens, P. A. & Sutherland, W. J. (1999). Consequences of the Allee effect for behaviour, ecology and conservation. *Trends in Ecology and Evolution*, **14**, 401–405.

Stephens, P. W. & Krebs, J. R. (1986). *Foraging Behaviour*, Princeton: Princeton University Press.

Stillman, R. A., Goss-Custard, J. D., West, A. D., Durell, S. E. A. le V. D., Caldow, R. W. G., McGrorty, S. & Clarke, R. T. (2000). Predicting mortality in novel environments: tests and sensitivity of a behaviour-based model. *Journal of Applied Ecology*, **37**, 564–588.

Stillman, R. A., Goss-Custard, J. D., West, A. D., Durell, S. E. A. le V. D., McGrorty, S., Caldow, R. W. G., Norris, K. J., Johnstone, I. G., Ens, B. J., Van der Meer, J. & Triplet, P. (2001). Predicting shorebird mortality and population size under different regimes of shellfishery management. *Journal of Applied Ecology*, **38**, 857–868.

Strong, D. R. (1986). Density-vague population change. *Trends in Ecology and Evolution*, **1**, 39–42.

Sugihara, G. & May, R. M. (1990). Non-linear forecasting as a way of distinguishing chaos from measurement error in time series. *Nature*, **344**, 734–741.

Sutherland, W. J. (1996a). Predicting the consequences of habitat loss for migratory populations. *Proceedings of the Royal Society London B*, **263**, 1325–1327.

Sutherland, W. J. (1996b). *From Individual Behaviour to Population Ecology*, Oxford: Oxford University Press.

Sutherland, W. J. (1998). The effect of local change in habitat quality on populations of migratory species. *Journal of Applied Ecology*, **35**, 418–421.

Sutherland, W. J. & Allport, G. A. (1994). A spatial depletion model of the interaction between bean geese and wigeon with the consequences for habitat management. *Journal of Animal Ecology*, **63**, 51–59.

Sutherland, W. J. & Anderson, C. W. (1993). Predicting the distribution of individuals and the consequences of habitat loss: the role of prey depletion. *Journal of Theoretical Biology*, **160**, 223–230.

Sutherland, W. J. & Norris, K. (2002). Behavioural models of population growth rates: implications for conservation and prediction. *Philosophical Transactions of the Royal Society London*, **357**, 1273–1284.

Sutherland, W. J. & Watkinson, A. R. (2001). Policy making within ecological uncertainty: lessons from badgers and GM crops. *Trends in Ecology and Evolution*, **16**, 261–263.

Tabah, L. (1989). From one demographic transition to another. *Population Bulletin of the United Nations*, **28**, 1–24.

Tanner, J. T. (1975). The stability and intrinsic growth rates of prey and predator populations. *Ecology*, **56**, 855–867.

Taylor, B. L. & Gerrodette, T. (1993). The uses of statistical power in conservation biology: the vaquita and northern spotted owl. *Conservation Biology*, **7**, 489–500.

Taylor, I. (1994). *Barn Owls. Predator–Prey Relationships and Conservation*, Cambridge: Cambridge University Press.

Temple, S. A. (1977). The status and conservation of endemic kestrels on Indian Ocean islands. In *World Conference on Birds of Prey*, ed. R. D. Chancellor pp. 74–82. Report of Proceedings, Vienna 1975. Basingstoke: Taylor and Francis.

Thirgood, S. J., Redpath, S. M., Haydon, D., Rothery P., Newton, I. & Hudson, P. J. (2000). Habitat loss and raptor predation: disentangling long- and short-term causes of red grouse declines. *Proceedings of the Royal Society London B*, **267**, 651–656.

Thomas, C. D. & Kunin, W. E. (1999). The spatial structure of populations. *Journal of Animal Ecology*, **68**, 647–657.

Thompson, P. M. & Ollason, J. C. (2001). Lagged effects of ocean climate on fulmar population dynamics. *Nature*, **413**, 417–420.

Tilghman, N. G. (1989). Impacts of white-tailed deer on forest regeneration in northwestern Pennsylvania. *Journal of Wildlife Management*, **53**, 524–532.

Tilman, D. (1982). *Resource Competition and Community Structure*, Princeton: Princeton University Press.

Tilman, D. (1993). Species richness of experimental productivity gradients – how important is colonization limitation? *Ecology*, **74**, 2179–2191.

Tilman, D. (1994). Competition and biodiversity in spatially structured habitats. *Ecology*, **75**, 2–16.

Tilman, D. (1996). Biodiversity – population versus ecosystem stability. *Ecology*, **77**, 350–363.

Tilman, D. (1997). Community invasibility, recruitment limitation, and grassland biodiversity. *Ecology*, **78**, 81–92.

Tilman, D., Mattson, M. & Langer, S. (1981). Competition and nutrient kinetics along a temperature gradient: an experimental test of a mechanistic approach to niche theory. *Limnology and Oceanography*, **26**, 1020–1033.

Tschinkel, W. R. (1998). The reproductive biology of fire ant societies. *BioScience*, **48**, 593–605.

Tuljapurkar, S. (1990). *Population Dynamics in Variable Environments*, Berlin: Springer-Verlag.

Tuljapurkar, S. (1997). Stochastic matrix models. In *Structured Population Models in Marine, Terrestrial and Freshwater Systems*, ed. S. Tuljapurkar & H. Caswell, pp. 59–87. New York: Chapman & Hall.

Turchin, P. (1990). Rarity of density dependence or population regulation with lags? *Nature*, **344**, 660–663.

Turchin, P. (1991). Reconstructing endogenous dynamics of a laboratory *Drosophila* population. *Journal of Animal Ecology*, **60**, 1091–1098.

Turchin, P. W. (1995). Population regulation: old arguments and a new synthesis. In *Population Dynamics: New Approaches and Synthesis*, ed. N. Cappucino & P. W. Price, pp. 19–40. San Diego: Academic Press.

Turchin, P. (1999). Population regulation: a synthetic view. *Oikos*, **84**, 153–159.

Turchin, P. (2001). Does population ecology have general laws? *Oikos*, **94**, 17–26.

Turchin, P. & Hanski, I. (1997). An empirically based model for latitudinal gradient in vole population dynamics. *American Naturalist*, **149**, 842–874.

Turchin, P. & Ostfeld, R. S. (1997). Effects of density and season on the population rate of change in the meadow vole. *Oikos*, **78**, 355–361.

Turchin, P. & Taylor, A. D. (1992). Complex dynamics in ecological time series. *Ecology*, **73**, 289–305.

Turnbull, L. A., Rees, M. & Crawley, M. J. (1999). Seed mass and the competition/colonization trade-off: a sowing experiment. *Journal of Ecology*, **87**, 899–912.

Underhill-Day, J. C. (1984). Population and breeding biology of marsh harriers in Britain since 1900. *Journal of Applied Ecology*, **21**, 773–787.

United Nations (2001). *World Population Prospects. The 2000 Revision*, New York: United Nations.

US Bureau of the Census (2000). International database, 10 May 2000 edition. See http://www.census.gov/ipc/www/worldpop.html.

USEPA (1992). Framework for Ecological Risk Assessment. EPA/630-R-92/001. Washington, D.C.: US Environmental Protection Agency, Risk Assessment Forum. 41 pp.

Van de Walle, E. (1992). Fertility transition, conscious choice and numeracy. *Demography*, **29**, 487–502.

Van Eerden, M. R. & Gregersen, J. (1995). Long-term changes in the northwest European population of cormorants *Phalacrocorax carbo sinensis*. *Ardea*, **83**, 61–79.

Van Hyning, J. M. (1974). Stock-recruitment relationships for Columbia River chinook salmon. *Rapports et Proces-Verbaux, Reun. Cons. Int. Perm. Explor. Mer.*, **164**, 89–97.

Van Leeuwen, C. J. & Hermens, J. L. M. (1995). *Risk Assessment of Chemicals: An Introduction*, Dordrecht: Kluwer.

Van Tienderen, P. H. (2000). Elasticities and the link between demographic and evolutionary dynamics. *Ecology*, **81**, 666–679.

van Vuren, D. (1984). Diurnal activity and habitat use of feral pigs on Santa Cruz Island, California. *California Fish and Game*, **70**, 140–144.

Vandermeer, J. (1978). Choosing category size in a stage projection matrix. *Oecologia (Berlin)*, **32**, 79–84.

Varley, G. C., Gradwell, G. R. & Hassell, M. P. (1973). *Insect Population Ecology*, Oxford: Blackwells.

Verhulst, P. F. (1838). Notice sur la loi que la population suit dans son accroisissement. *Corresp. Math. Phys., A. Quetelet*, **10**, 113–121.

Veromann, H. (1986). Thirty-two year population trends of the white stork in Estonian S. S. R. In *White Stork. Status and Conservation*, ed. G. Rheinwald, J. Ogden & H. Sculz, pp. 153–158. Braunschweig: Dachverband Deutscher Avifaunisten.

Vessey-Fitzgerald, D. F. (1968). Grazing succession among East African game animals. *Journal of Mammalogy*, **41**, 161–172.

Volterra, V. (1926). Variations and fluctuations of the numbers of individuals in animal species living together. Reprinted 1931 in R. N. Chapman, *Animal Ecology*. New York: McGraw-Hill.

Walker, B. H., Ludwig, D., Holling, C. S. & Peterman, R. M. (1981). Stability of semi-arid savanna grazing systems. *Journal of Ecology*, **69**, 473–498.

Walker, C. H., Hopkin, S. P., Sibly, R. M. & Peakall, D. (2001). *Principles of Ecotoxicology*, 2nd edn. London: Taylor & Francis.

Walters, C. (1997). Challenges in adaptive management of riparian and coastal ecosystems. *Conservation Ecology*, **1**, 1–21.

Walters, C. J. (1987). Nonstationarity of production relationships in exploited populations. *Canadian Journal of Fisheries and Aquatic Sciences*, **44** (Supplement 2), 156–165.

Walthall, W. K. & Stark, J. D. (1997). Comparison of two population-level ecotoxicological endpoints: the intrinsic (r_m) and instantaneous (r_i) rates of increase. *Environmental Toxicology and Chemistry*, **16**, 1068–1083.

Watson, A., Moss, R. & Parr, R. (1984). Effects of food enrichment on numbers and spacing behaviour of red grouse. *Journal of Animal Ecology*, **53**, 663–678.

Watt, K. E. F. (1959). A mathematical model for the effect of densities of attacked and attacking species on the number attacked. *Canadian Entomology*, **91**, 129–144.

Weimerskirch, H., Brothers, N. & Jouventin, P. (1997). Population dynamics of wandering albatross *Diomedea exulans* and Amsterdam albatross *D. amsterdamensis* in the Indian Ocean and their relationships with long-line fisheries: conservation implications. *Biological Conservation*, **79**, 257–270.

Weiner, J., Martinez, S., MullerScharer, H., Stoll, P. & Schmid, B. (1997). How important are environmental maternal effects in plants? A study with *Centaurea maculosa*. *Journal of Ecology*, **85**, 133–142.

Wesselingh, R. A. & de Jong, T. J. (1995). Bidirectional selection on threshold size for flowering in *Cynoglossum officinale* (Hounds tongue). *Heredity*, **74**, 415–424.

Wesselingh, R. A., de Jong, T. J., Klinkhamer, P. G. L., Vandijk, M. J. & Schlatmann, E. G. M. (1993). Geographical variation in threshold size for flowering in *Cynoglossum officinale*. *Acta Botanica Neerlandica*, **42**, 81–91.

Westoff, C. F. (1996). Reproductive preferences and future fertility in developing countries. In *The Future Population of the World: What can we Assume Today?* revised edn, ed. W. Lutz, pp. 73–87. London: Earthscan.

Whelpton, P. K. (1936). An empirical model of calculating future population. *Journal of the American Statistical Association*, **31**, 457–473.

White, P. J. & Garrott, R. A. (1999). Population dynamics of kit foxes. *Canadian Journal of Zoology*, **77**, 486–493.

Williams, D. W. & Liebhold, A. M. (1995). Detection of delayed density dependence: effects of autocorrelation in an exogenous factor. *Ecology*, **76**, 1005–1008.

Williams, K., Parer, I., Coman, B., Burley, J. & Braysher, M. (1995). *Managing Vertebrate Pests: Rabbits.* Canberra: Australian Government Publishing Service.

Williams, T. D. (1997). Life-cycle parameters of *Tisbe battagliai* (Copepoda: Harpacticoida) as indicators of chronic toxicity. Ph.D. thesis, University of Plymouth.

Wilson, K. W., Bjørnstad, O. N., Dobson, A. P., Merler, S., Poglayen, G., Randolph, S. E., Read, A. F. & Skorping, A. (2001). Heterogeneities in macroparasite infections: patterns and processes. In *The Ecology of Wildlife Diseases*, ed. P. J. Hudson, A. Rizzoli, B. T. Grenfell, J. A. P. Heesterbeek & A. P. Dobson, pp. 6–44. Oxford: Oxford University Press.

Windberg, L. A., Ebbert, S. M. & Kelly, B. T. (1997). Population characteristics of coyotes (*Canis latrans*) in the Northern Chihuahuan desert of New Mexico. *The American Midland Naturalist*, **138**, 197–207.

Winkel, W. (1996). Der braunschweiger hohlenbruterprogramm des instituts fur vogelforschung 'Vogelwarte Helgoaland'. *Vogelwelt*, **117**, 269–275.

Winner, R. W., Keeling, T., Yeager, R. & Farrell, M. P. (1977). Effect of food type on the acute and chronic toxicity of copper to *Daphnia magna*. *Freshwater Biology*, **7**, 343–349.

Wisdom, M. J., Mills, L. S. & Doak, D. F. (2000). Life stage simulation analysis: estimating vital-rate effects on population growth for conservation. *Ecology*, **81**, 628–641.

Woiwod, I. P. & Hanski, I. (1992). Patterns of density dependence in moths and aphids. *Journal of Animal Ecology*, **61**, 619–629.

Wolff, J. (1997). Population regulation in mammals: an evolutionary perspective. *Journal of Animal Ecology*, **66**, 1–13.

Wood, S. N. (2001). Partially specified ecological models. *Ecological Monographs*, **71**, 1–25.

Woodgerd, W. (1963). Population dynamics of bighorn sheep on Wildhorse Island. *Journal of Wildlife Management*, **28**, 381–391.

World Bank (2000). *World Development Indicators*. (CD-ROM), Washington, D.C.: The World Bank.

World Bank (2001). *World Development Indicators*. (CD-ROM), Washington, D.C.: The World Bank.

Yao, Q., Tong, H., Finkenstadt, B. & Stenseth, N. C. (2000). Common structure in panels of short ecological time-series. *Proceedings of the Royal Society London B*, **267**, 2459–2467.

Yasuba, Y. (1962). *Birth Rates of the White Population in the United States, 1800–1860*, Baltimore: John Hopkins Press.

Yin Hua & Lin Xiaohong. (1996). *Population and Socio-economic Data of China by Province*. Data User Service, CPIRC/UNFPA. Data User Service Series No. 7. Beijing: China Population Publishing House.

Zeng, Z., Nowierski, R. M., Taper, M. L., Dennis, B. & Kemp, W. P. (1998). Complex population dynamics in the real world: modeling the influence of time-varying parameters and time lags. *Ecology*, **79**, 2193–2209.

Zhang, Q. & Uhlig, G. (1993). Dry-weight and chemical-composition in relation to population density of cultivated *Tisbe holothuriae* (Copepoda: Harpacticoida). *Helgolander Meeresun*, **47**, 221–227.

Glossary of abbreviations

ESS: evolutionarily stable strategy
GDP: gross domestic product
GM: genetically modified
GPDD: global population dynamics database
IBM: individual-based model
IFD: ideal free distribution
LDC: less developed country
MDC: more developed country
MSP: minimum safe population
NAO: North Atlantic Oscillation
TFR: total fertility rate

Author index

Subject index